Bacterial Interactions with Dental and Medical Materials

Bacterial Interactions with Dental and Medical Materials

Editor

Mary Anne Melo

MDPI • Basel • Beijing • Wuhan • Barcelona • Belgrade • Manchester • Tokyo • Cluj • Tianjin

Editor
Mary Anne Melo
Department of General Dentistry,
University of Maryland School of Dentistry
USA

Editorial Office
MDPI
St. Alban-Anlage 66
4052 Basel, Switzerland

This is a reprint of articles from the Special Issue published online in the open access journal *Journal of Functional Biomaterials* (ISSN 2079-4983) (available at: https://www.mdpi.com/journal/jfb/special_issues/Bacterial_Interactions_Dental_Medical_Materials).

For citation purposes, cite each article independently as indicated on the article page online and as indicated below:

LastName, A.A.; LastName, B.B.; LastName, C.C. Article Title. *Journal Name* **Year**, *Volume Number*, Page Range.

ISBN 978-3-0365-0020-1 (Hbk)
ISBN 978-3-0365-0021-8 (PDF)

Cover image courtesy of Mary Anne Melo.

© 2020 by the authors. Articles in this book are Open Access and distributed under the Creative Commons Attribution (CC BY) license, which allows users to download, copy and build upon published articles, as long as the author and publisher are properly credited, which ensures maximum dissemination and a wider impact of our publications.

The book as a whole is distributed by MDPI under the terms and conditions of the Creative Commons license CC BY-NC-ND.

Contents

About the Editor vii

Mary Anne Melo
Bacterial Interactions with Dental and Medical Materials
Reprinted from: *J. Funct. Biomater.* **2020**, *11*, 83, doi:10.3390/jfb11040083 1

Hui Lu and Xiaoming Jin
Novel Orthodontic Cement Comprising Unique Imidazolium-Based Polymerizable Antibacterial Monomers
Reprinted from: *J. Funct. Biomater.* **2020**, *11*, 75, doi:10.3390/jfb11040075 5

Thomas E. Paterson, Rui Shi, Jingjing Tian, Caroline J. Harrison, Mailys De Sousa Mendes, Paul V. Hatton, Zhou Li and Ilida Ortega
Electrospun Scaffolds Containing Silver-Doped Hydroxyapatite with Antimicrobial Properties for Applications in Orthopedic and Dental Bone Surgery
Reprinted from: *J. Funct. Biomater.* **2020**, *11*, 58, doi:10.3390/jfb11030058 21

Rayan B. Yaghmoor, Wendy Xia, Paul Ashley, Elaine Allan and Anne M. Young
Effect of Novel Antibacterial Composites on Bacterial Biofilms
Reprinted from: *J. Funct. Biomater.* **2020**, *11*, 55, doi:10.3390/jfb11030055 39

Leopoldo Torres Jr and Diane R. Bienek
Use of Protein Repellents to Enhance the Antimicrobial Functionality of Quaternary Ammonium Containing Dental Materials
Reprinted from: *J. Funct. Biomater.* **2020**, *11*, 54, doi:10.3390/jfb11030054 53

Paola Andrea Mena Silva, Isadora Martini Garcia, Julia Nunes, Fernanda Visioli, Vicente Castelo Branco Leitune, Mary Anne Melo and Fabrício Mezzomo Collares
Myristyltrimethylammonium Bromide (MYTAB) as a Cationic Surface Agent to Inhibit *Streptococcus mutans* Grown over Dental Resins: An In Vitro Study
Reprinted from: *J. Funct. Biomater.* **2020**, *11*, 9, doi:10.3390/jfb11010009 67

Heba Mitwalli, Abdulrahman A. Balhaddad, Rashed AlSahafi, Thomas W. Oates, Mary Anne S. Melo, Hockin H. K. Xu and Michael D. Weir
Novel CaF_2 Nanocomposites with Antibacterial Function and Fluoride and Calcium Ion Release to Inhibit Oral Biofilm and Protect Teeth
Reprinted from: *J. Funct. Biomater.* **2020**, *11*, 56, doi:10.3390/jfb11030056 81

Nikos N. Lygidakis, Elaine Allan, Wendy Xia, Paul F. Ashley and Anne M. Young
Early Polylysine Release from Dental Composites and Its Effects on Planktonic *Streptococcus mutans* Growth
Reprinted from: *J. Funct. Biomater.* **2020**, *11*, 53, doi:10.3390/jfb11030053 99

Masaya Shimabukuro, Akari Hiji, Tomoyo Manaka, Kosuke Nozaki, Peng Chen, Maki Ashida, Yusuke Tsutsumi, Akiko Nagai and Takao Hanawa
Time-Transient Effects of Silver and Copper in the Porous Titanium Dioxide Layer on Antibacterial Properties
Reprinted from: *J. Funct. Biomater.* **2020**, *11*, 44, doi:10.3390/jfb11020044 109

Evgeniy Papynov, Oleg Shichalin, Igor Buravlev, Anton Belov, Arseniy Portnyagin, Vitaliy Mayorov, Evgeniy Merkulov, Taisiya Kaidalova, Yulia Skurikhina, Vyacheslav Turkutyukov, Alexander Fedorets and Vladimir Apanasevich
CaSiO$_3$-HAp Structural Bioceramic by Sol-Gel and SPS-RS Techniques: Bacteria Test Assessment
Reprinted from: J. Funct. Biomater. **2020**, *11*, 41, doi:10.3390/jfb11020041 121

Minh N. Luong, Laurie Huang, Daniel C. N. Chan and Alireza Sadr
In Vitro Study on the Effect of a New Bioactive Desensitizer on Dentin Tubule Sealing and Bonding
Reprinted from: J. Funct. Biomater. **2020**, *11*, 38, doi:10.3390/jfb11020038 141

Elena García-Gareta, Justyna Binkowska, Nupur Kohli and Vaibhav Sharma
Towards the Development of a Novel Ex Ovo Model of Infection to Pre-Screen Biomaterials Intended for Treating Chronic Wounds
Reprinted from: J. Funct. Biomater. **2020**, *11*, 37, doi:10.3390/jfb11020037 153

Andrei C. Ionescu, Gloria Cazzaniga, Marco Ottobelli, Franklin Garcia-Godoy and Eugenio Brambilla
Substituted Nano-Hydroxyapatite Toothpastes Reduce Biofilm Formation on Enamel and Resin-Based Composite Surfaces
Reprinted from: J. Funct. Biomater. **2020**, *11*, 36, doi:10.3390/jfb11020036 163

Samira Esteves Afonso Camargo, Azeem S. Mohiuddeen, Chaker Fares, Jessica L. Partain, Patrick H. Carey IV, Fan Ren, Shu-Min Hsu, Arthur E. Clark and Josephine F. Esquivel-Upshaw
Anti-Bacterial Properties and Biocompatibility of Novel SiC Coating for Dental Ceramic
Reprinted from: J. Funct. Biomater. **2020**, *11*, 33, doi:10.3390/jfb11020033 181

Isadora Martini Garcia, Vicente Castelo Branco Leitune, Antonio Shigueaki Takimi, Carlos Pérez Bergmann, Susana Maria Werner Samuel, Mary Anne Melo and Fabrício Mezzomo Collares
Cerium Dioxide Particles to Tune Radiopacity of Dental Adhesives: Microstructural and Physico-Chemical Evaluation
Reprinted from: J. Funct. Biomater. **2020**, *11*, 7, doi:10.3390/jfb11010007 195

Diane R. Bienek, Anthony A. Giuseppetti, Stanislav A. Frukhtbeyn, Rochelle D. Hiers, Fernando L. Esteban Florez, Sharukh S. Khajotia and Drago Skrtic
Physicochemical, Mechanical, and Antimicrobial Properties of Novel Dental Polymers Containing Quaternary Ammonium and Trimethoxysilyl Functionalities
Reprinted from: J. Funct. Biomater. **2020**, *11*, 1, doi:10.3390/jfb11010001 205

Hamid Mortazavian, Guillaume A. Picquet, Jānis Lejnieks, Lynette A. Zaidel, Carl P. Myers and Kenichi Kuroda
Understanding the Role of Shape and Composition of Star-Shaped Polymers and their Ability to Both Bind and Prevent Bacteria Attachment on Oral Relevant Surfaces
Reprinted from: J. Funct. Biomater. **2019**, *10*, 56, doi:10.3390/jfb10040056 223

About the Editor

Mary Anne Melo is Clinical Associate Professor and Division Director of Operative Dentistry at the School of Dentistry, University of Maryland, Baltimore, Maryland. Dr. Melo attended dental school at the University of Fortaleza, Brazil; completed a residency in operative dentistry and a master's and Ph.D. at the Federal University of Ceara; and pursued fellowship training in dental materials at the University of Maryland before joining the operative faculty.

Dr. Melo is a clinician, researcher, and educator. In her teaching work, she directs and participates in restorative dentistry predoctoral and postgraduate courses and mentors students in master's and Ph.D. research projects. Her main areas of research focus primarily on (1) oral biofilm control with nanotechnologies by investigating anti-biofilm nanotechnologies for the oral healthcare field, (2) biofilm modeling with the development of new in vitro caries/biofilm models, (3) evaluation of the antimicrobial properties of biomaterials, and (4) translational research for the development of clinically relevant therapeutic strategies for the control of dental caries.

She has published over 120 peer-review scientific articles; serves as an ad hoc reviewer and editor in several scientific journals in dentistry, medicine, and biomaterials. Dr. Melo is a fellow of the Academy of Dental Materials and a member of the Academy of Operative Dentistry, the International Association for Dental Research, the Society for Color and Appearance in Dentistry, and the American Academy of Cosmetic Dentistry. Dr. Melo's unique qualifications as a researcher with clinical and materials science perspectives have successfully aligned her role with advances in the dental field.

Editorial

Bacterial Interactions with Dental and Medical Materials

Mary Anne Melo [1,2]

1. Division of Operative Dentistry, Department of General Dentistry, University of Maryland School of Dentistry, 650 W. Baltimore St, Baltimore, MD 21201, USA; MMelo@umaryland.edu
2. Dental Biomedical Science Ph.D. Program, University of Maryland School of Dentistry, 650 W. Baltimore St, Baltimore, MD 21201, USA

Received: 19 November 2020; Accepted: 19 November 2020; Published: 23 November 2020

Fundamental scientific understanding of oral diseases associated with tissue-contacting dental and medical devices is primordial to facilitate pathways for their translation to clinical use. The interaction of bacteria with biomaterials' surfaces has critical clinical implications due to biofilm formation and biofouling. Although biofilms play an important positive role in various ecosystems, they also have many adverse effects, including biofilm-related infections in medical and dental settings.

Biofilms account for up to 80% of the total number of bacteria-related infections, including endocarditis, cystic fibrosis, secondary dental caries, periodontitis, rhinosinusitis, osteomyelitis, non-healing chronic wounds, meningitis, kidney infections, and prosthesis- and implantable device-related infections.

Worldwide, populations of all ages suffer from oral biofilm-trigged diseases, such as dental caries and tooth surrounding tissue infections impacting the human body. New dental materials and biomaterials, currently being introduced, under development, or proposed, are expected to benefit the oral health status. These materials will provide a wide range of diverse functions, from promoting osteogenesis to bacterial biofilm formation inhibition.

In the Special Issue "Bacterial Interactions with Dental and Medical Materials," encouraging findings on tissue-contacting biomaterials to control biofilms and susceptibility to bacterial colonization and understand mechanisms, clinical perspectives beneficial to healthcare were discussed.

As cariogenic bacterial grows inside the mouth due to lack of caries disease management, the tooth–material interface continues to degrade by bacterial acids leading to increasing premature failure of tooth filling. Bienek et al. [1], Mena Silva et al. [2], Yaghmoor et al. [3], Lu and Jin [4], Lygidakis et al. [5], and Mitwalli et al. [6] investigated the application of polymerizable antibacterial monomers based on di-imidazolium or quaternary ammonium compounds as an antibacterial strategy for resin-based materials in efforts toward caries-related biofilm control.

Looking for the control of and reduction in oral biofilm formation, initiated by bacterial species living in polymicrobial, pathogenic colonies at or below the gingival margin, Mortazavian et al. [7], Afonso Camargo et al. [8], and Torres and Bienek [9] presented antibiofilm strategies targeting adhesion to a substrate via brush-like structures of highly packed polymer chains and silicon carbide coating that could physically repel bacteria to result in a significant reduction in attached bacteria.

The search for new bioactive compounds to prevent dental caries development and progression has led researchers to focus their attention on the use of nanotechnologies, especially hydroxyapatite (HAp), metals, and metal oxide nanoparticles. Garcia et al. [10] investigate the effects of different loadings of cerium dioxide on their radiopacity and degree of conversion of dental adhesives. Nanotechnology applications have offered the opportunity to modulate the formation of dental biofilms using nanoparticles with bioactive effects. Luong et al. [11] have assessed novel nanohydroxyapatite-based desensitizer and its effect on dental adhesives' bond strength.

Ionescu et al. [12] have looked to the synergistic antibacterial performance of toothpaste containing nano-hydroxyapatite substituted with metal ions. Shimabukuro et al. [13] have investigated both the durability of the antibacterial effect and the surface change of Ag- and Cu-incorporated porous titanium dioxide (TiO_2) layer; silver (Ag) and copper (Cu) have been incorporated into the titanium (Ti) surface to realize their antibacterial properties. Ag- and Cu-incorporated TiO_2 layers were formed by micro-arc oxidation (MAO) treatment using the electrolyte with Ag and Cu ions. Their collective findings indicated the importance of the time-transient effects of Ag and Cu. This knowledge will help design antibacterial implants based on Ag and Cu's surface changes. Paterson et al. [14] incorporate silver-doped nano-hydroxyapatite into electrospun scaffolds for applications in bone repair.

Papynov et al. [15] have developed a combination of calcium monosilicate, β-wollastonite ($CaSiO_3$), and hydroxyapatite (HAp), which is a complete analog of a living bone. This approach can pave the way to the fabrication of biocompatible ceramics for bone tissue engineering; thus, contributing another flexible strategy for the synthesis of biomaterials with broad intended applications in traumatology, orthopedics, dentistry, maxillofacial surgery, and other areas of medicine for the recovery, replacement, and reconstruction of the damaged tissue.

As we observe an increasing number of investigations addressing new antibacterial materials and devices, we also face a significant challenge in evaluating the new materials. There is a limited understanding of the mechanisms involved in bacteria–materials interactions in the oral environment. The gaps in the development and application of reliable methodological approaches for the characterization of this new generation of antibacterial materials require considering the complexity and heterogeneity of the disease process investigated.

Garcia-gareta et al. [16] developed a live ex vivo model of persistent infection that can be used for pre-screening biomaterials intended for treating chronic wounds for their antimicrobial and angiogenic potential. This model is relatively simple, quick, and low-cost and mimics the in vivo situation more closely than traditionally used antimicrobial tests using agar plates and dilution assays. Additionally, keeping under the principles of the National Centre for the Replacement Refinement and Reduction of Animals in Research, this model does not require administrative procedures for obtaining ethics committee approval for animal experimentation.

In this Special Issue, there are 16 papers by authors from 9 countries in Asia, Europe, and Latin and North America. Their investigations represent a wide range of aspects related to the current scenario on bacteria interactions with biomaterial surfaces and give timely examples of research activities that can be observed around the globe. We hope that this work will be inspirational for further research on understanding oral diseases associated with tissue-contacting dental and medical devices, new antibiofilm agents, and the relevance of assessing the material using clinically significant biofilm models.

Funding: This research received no external funding.

Conflicts of Interest: The authors declare no conflict of interest.

References

1. Bienek, D.R.; Giuseppetti, A.A.; Frukhtbeyn, S.A.; Hiers, R.D.; Florez, F.L.E.; Khajotia, S.S.; Skrtic, D. Physicochemical, Mechanical, and Antimicrobial Properties of Novel Dental Polymers Containing Quaternary Ammonium and Trimethoxysilyl Functionalities. *J. Funct. Biomater.* **2020**, *11*, 1. [CrossRef]
2. Silva, P.A.M.; Garcia, I.M.; Nunes, J.; Visioli, F.; Leitune, V.C.B.; Melo, M.A.; Collares, F.M. Myristyltrimethylammonium Bromide (MYTAB) as a Cationic Surface Agent to Inhibit Streptococcus mutans Grown over Dental Resins: An In Vitro Study. *J. Funct. Biomater.* **2020**, *11*, 9. [CrossRef]
3. Yaghmoor, R.B.; Xia, W.; Ashley, P.; Allan, E.; Young, A.M. Effect of Novel Antibacterial Composites on Bacterial Biofilms. *J. Funct. Biomater.* **2020**, *11*, 55. [CrossRef]
4. Lu, H.; Jin, X. Novel Orthodontic Cement Comprising Unique Imidazolium-Based Polymerizable Antibacterial Monomers. *J. Funct. Biomater.* **2020**, *11*, 75. [CrossRef]

5. Lygidakis, N.N.; Allan, E.; Xia, W.; Ashley, P.F.; Young, A.M. Early Polylysine Release from Dental Composites and Its Effects on Planktonic Streptococcus mutans Growth. *J. Funct. Biomater.* **2020**, *11*, 75. [CrossRef]
6. Mitwalli, H.; Balhaddad, A.A.; AlSahafi, R.; Oates, T.W.; Melo, M.A.S.; Xu, H.H.K.; Weir, M.D. Novel CaF2 Nanocomposites with Antibacterial Function and Fluoride and Calcium Ion Release to Inhibit Oral Biofilm and Protect Teeth. *J. Funct. Biomater.* **2020**, *11*, 56. [CrossRef]
7. Mortazavian, H.; Picquet, G.A.; Lejnieks, J.; Zaidel, L.A.; Myers, C.P.; Kuroda, K. Understanding the Role of Shape and Composition of Star-Shaped Polymers and their Ability to Both Bind and Prevent Bacteria Attachment on Oral Relevant Surfaces. *J. Funct. Biomater.* **2019**, *10*, 56. [CrossRef]
8. Camargo, S.E.A.; Mohiuddeen, A.S.; Fares, C.; Partain, J.L.; Carey, P.H.; Ren, F.; Hsu, S.-M.; Clark, A.E.; Esquivel-Upshaw, J.F. Anti-Bacterial Properties and Biocompatibility of Novel SiC Coating for Dental Ceramic. *J. Funct. Biomater.* **2020**, *11*, 33. [CrossRef]
9. Torres, L., Jr.; Bienek, D.R. Use of Protein Repellents to Enhance the Antimicrobial Functionality of Quaternary Ammonium Containing Dental Materials. *J. Funct. Biomater.* **2020**, *11*, 54. [CrossRef]
10. Garcia, I.M.; Leitune, V.C.B.; Takimi, A.S.; Bergmann, C.P.; Samuel, S.M.W.; Melo, M.A.; Collares, F.M. Cerium Dioxide Particles to Tune Radiopacity of Dental Adhesives: Microstructural and Physico-Chemical Evaluation. *J. Funct. Biomater.* **2020**, *11*, 7. [CrossRef]
11. Luong, M.N.; Huang, L.; Chan, D.C.N.; Sadr, A. In Vitro Study on the Effect of a New Bioactive Desensitizer on Dentin Tubule Sealing and Bonding. *J. Funct. Biomater.* **2020**, *11*, 38. [CrossRef]
12. Ionescu, A.C.; Cazzaniga, G.; Ottobelli, M.; Garcia-Godoy, F.; Brambilla, E. Substituted Nano-Hydroxyapatite Toothpastes Reduce Biofilm Formation on Enamel and Resin-Based Composite Surfaces. *J. Funct. Biomater.* **2020**, *11*, 36. [CrossRef]
13. Shimabukuro, M.; Hiji, A.; Manaka, T.; Nozaki, K.; Chen, P.; Ashida, M.; Tsutsumi, Y.; Nagai, A.; Hanawa, T. Time-Transient Effects of Silver and Copper in the Porous Titanium Dioxide Layer on Antibacterial Properties. *J. Funct. Biomater.* **2020**, *11*, 44. [CrossRef]
14. Paterson, T.E.; Shi, R.; Tian, J.; Harrison, C.J.; De Sousa Mendes, M.; Hatton, P.V.; Li, Z.; Ortega, I. Electrospun Scaffolds Containing Silver-Doped Hydroxyapatite with Antimicrobial Properties for Applications in Orthopedic and Dental Bone Surgery. *J. Funct. Biomater.* **2020**, *11*, 58. [CrossRef]
15. Papynov, E.; Shichalin, O.; Buravlev, I.; Belov, A.; Portnyagin, A.; Mayorov, V.; Merkulov, E.; Kaidalova, T.; Skurikhina, Y.; Turkutyukov, V.; et al. CaSiO3-HAp Structural Bioceramic by Sol-Gel and SPS-RS Techniques: Bacteria Test Assessment. *J. Funct. Bioceram.* **2020**, *11*, 41. [CrossRef]
16. García-Gareta, E.; Binkowska, J.; Kohli, N.; Sharma, V. Towards the Development of a Novel Ex Ovo Model of Infection to Pre-Screen Biomaterials Intended for Treating Chronic Wounds. *J. Funct. Biomater.* **2020**, *11*, 37. [CrossRef]

Publisher's Note: MDPI stays neutral with regard to jurisdictional claims in published maps and institutional affiliations.

© 2020 by the author. Licensee MDPI, Basel, Switzerland. This article is an open access article distributed under the terms and conditions of the Creative Commons Attribution (CC BY) license (http://creativecommons.org/licenses/by/4.0/).

Article

Novel Orthodontic Cement Comprising Unique Imidazolium-Based Polymerizable Antibacterial Monomers

Hui Lu * and Xiaoming Jin *

Dentsply Sirona, R&D—Consumables Product Group, 38 W Clarke Ave, Milford, DE 19963, USA
* Correspondence: Hui.Lu@dentsplysirona.com (H.L.); Xiaoming.Jin@dentsplysirona.com (X.J.)

Received: 12 June 2020; Accepted: 14 October 2020; Published: 17 October 2020

Abstract: White spot lesions (WSLs) can develop quickly and compromise the successful outcome of the orthodontic treatment. Orthodontic bonding cement with the capability to prevent or mitigate WSLs could be beneficial, especially for patients with high risk of caries. This study explored novel mono- and di-imidazolium-based polymerizable antibacterial monomers and evaluated orthodontic cement compositions comprising such novel monomers. Their antibacterial potentials, mechanical properties, and shear bond strength (SBS) to bovine enamel were investigated. Statistical tests were applied to SBS and mechanical tests (one-way ANOVA and Tukey's test). For antibacterial resins C (ABR-C) and E (ABR-E), their minimum inhibitory concentration (MIC) and minimum bactericidal concentration (MBC) against cariogenic *Streptococcus mutans* bacterial strain UA159 were found to be 4 µg/mL and 8 µg/mL, respectively. The loss of dry mass from completely demineralized dentin beams in buffer solutions pre-dipped into ABR-C and ABR-E resins is much less than that in control buffer (artificial saliva) only. For unfilled resins comprising up to 12 wt % ABR-C, no significant decreases in flexural strength or modulus were observed. For experimental cements incorporating 1–4 wt % ABR-C, there was no drastic compromise to the SBS to enamel except for 3 wt % ABR-C. Furthermore, their SBS was all comparable to the commercially available orthodontic cements. The ISO-22196 antimicrobial test against *S. aureus* showed significant levels of antibacterial effects—up to over 5 logs of microorganism reduction exhibited by ABR-C-containing experimental cements. The imidazolium-based polymerizable monomers could be utilized to functionalize orthodontic bonding cement with steady antibacterial activity and develop a potential strategy to counteract WSLs.

Keywords: white spot lesions; antibacterial; biofilms; dental; orthodontic

1. Introduction

Beyond just straightening teeth and achieving better esthetics, proper orthodontic intervention can result in greatly improved oral and psychological health [1,2]. A major undesirable outcome during orthodontic treatment is the development of white spot lesions (WSLs) around orthodontic brackets, especially near the gingival margin [3–6]. The boosted accumulation of biofilms around the brackets further lowers pH around these sites. Despite advancements in patient education, WSLs can still develop rapidly and could compromise the successful outcome of the treatment, even resulting in premature termination of treatment in severe cases [7–9]. It has been reported that between roughly 26% and 70% of patients using fixed orthodontic appliances exhibited various levels of WSLs during the orthodontic treatment [4,6,7]. In another study, the prevalence of WSLs was reported to be 38% in the 6-month group, whereas it was 46% in the 12-month group [9].

Due to the more retentive, complicated surfaces around brackets and higher occurrences for plaque or biofilm adherence, it has become quite challenging to maintain good oral hygiene, especially for less compliant patient groups [8,10–12]. It has been reported that fixed orthodontic appliances can affect the self-cleansing capabilities of the intraoral system, owing to the interactions of saliva and teeth surfaces. Even for patients with clear aligners, wearing aligners 20~22 h daily could limit the natural cleansing and neutralizing effects of saliva [13]. The fixed orthodontic appliances can even alter the oral microflora and increase the levels of acidogenic plaque bacteria, i.e., *Streptococcus mutans* (*S. mutans*), and *lactobacilli* in saliva and dental biofilm during active wearing of such appliances throughout lengthy orthodontic treatment procedures, with an average duration around two to three years [14–19].

The interface between tooth and orthodontic bonding adhesive or cement, many of them based on (meth)acrylate chemistry, could also become the victim of bacterial attack and subsequent degradation [20,21]. Finer et al.'s findings revealed increased degradation of resin-based composite and bonding agent by *S. mutans* UA159 vs. control. Esterase activities at levels that could degrade adhesives and resin composite have also been linked to *S. mutans* [20]. While bonding the irregularly shaped orthodontic appliances onto the tooth surface, the residual excess of orthodontic cements can also stimulate the extra accumulation of dental biofilm around brackets. The biofilm degradation of this important tooth–cement–bracket interface may contribute to the premature debonding of the brackets, especially given the compounding interactions with the microorganisms that could lead to biocorrosion on the surfaces of metal-based brackets and appliances [22].

Given the extensive prevalence of WSLs occurring during orthodontic treatments, various strategies have been utilized to prevent or mediate demineralization and occurrence of WSL formation. The use of fluoride via various forms, i.e., mouth rinse, gel, topical varnish, toothpaste, etc., has been the predominant remedy applied and studied [7,14,23,24]. Fluoride ions have been well documented to enhance remineralization of enamel and mitigate mineral loss during acid dissolution, by replacing the hydroxyl groups in hydroxyapatite and forming more acid-resistant fluorapatite [25]. The findings consolidated by Bergstrand and Twetman concluded that the use of topical fluoride varnish along with fluoride toothpaste has been the best evidence-based approach to prevent WSLs. The mean prevented portion based on six clinical trials was 42.5% with a range from 4% to 73% [23]. These studies provided convincing support for routine professional applications of fluoride varnish around the bracket base during orthodontic treatment [24–28].

The application of quaternary ammonium compound (QAC) has been actively investigated to introduce antibacterial activity into the orthodontic bonding system. A unique type of antimicrobial polymer is demonstrated by incorporating quaternary ammonium polyethyleneimine (QPEI) nanoparticles. QPEI nanoparticle-containing composites have been reported to exhibit antibacterial activity against salivary bacteria and in vivo antibiofilm activity [29]. Experimental orthodontic bonding cements with 1% incorporated insoluble QPEI and polycationic polyethyleneimine nanoparticles were reported to exhibit stable, long-lasting antibacterial properties against *S. mutans* [30,31]. A variety of QAC-containing functional monomers have been systematically studied, including the widely studied methacryloyloxydodecylpyridinium bromide (MDPB) [32,33]; dimethylaminohexadecyl methacrylate (DMAHDM), a mono-methacrylate QAC with an alkyl chain length of 16 [34,35]; dimethylaminododedecyl methacrylate (DMADDM), a mono-methacrylate QAC with an alkyl chain length of 12 [36–38]; and 2-methacryloxylethyl hexadecyl methyl ammonium bromide (MAE-HB), a crosslinkable, di-methacrylate QAC with an alkyl chain length of 16 [39,40]. Significant antibacterial activities and biofilm reductions have been demonstrated.

Recently, a series of polymerizable imidazolium-based antibacterial resins (ABRs) were successfully designed in our laboratory [41,42]. To the best of our knowledge, no such novel crosslinkable imidazolium-based monomers have been investigated for their application towards an orthodontic bonding system with the potential to prevent and mitigate WSLs. The aims of this study were to investigate imidazolium-based ABR's antibacterial activities and impacts on the adhesive and mechanical properties of experimental orthodontic bonding cement formulations.

2. Results

By coupling with different mono-ols, diols, polyols (such as triols), monoamines, diamines, and/or polyamines, a range of methacrylate resins with (poly)imidazole moieties were prepared. During the processes of synthesizing imidazolium-based polymerizable resins from imidazole-containing monomers, a facile process based on imidazole and (meth)acrylated resins was developed. As demonstrated in Schemes 1 and 2, a variety of imidazole-containing polymerizable monomers were able to be prepared accordingly.

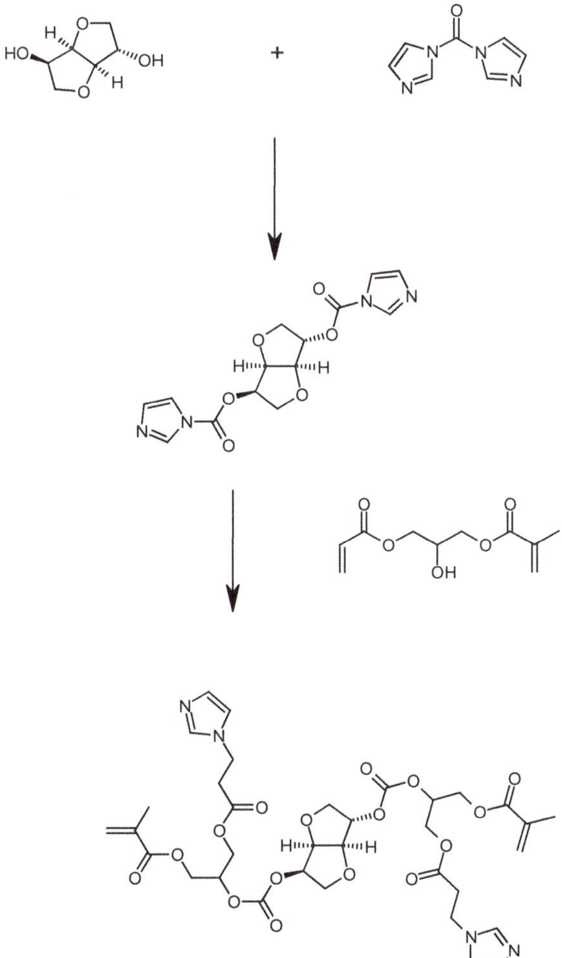

Scheme 1. Isosorbide-based bisimidazole–dimethacrylate resins.

The polymerizable imidazole-based resins were further converted into polymerizable imidazolium-based monomers by reacting them with a variety of halogenated alkyls. A variety of polymerizable imidazolium-based antibacterial resins (ABRs) were successfully synthesized as exhibited in Schemes 3 and 4. The primary evaluations of these novel ABRs' antibacterial activity include the determination of minimum inhibitory concentration (MIC) and minimum bactericidal concentration (MBC) against cariogenic *S. mutans* bacterial strain UA159.

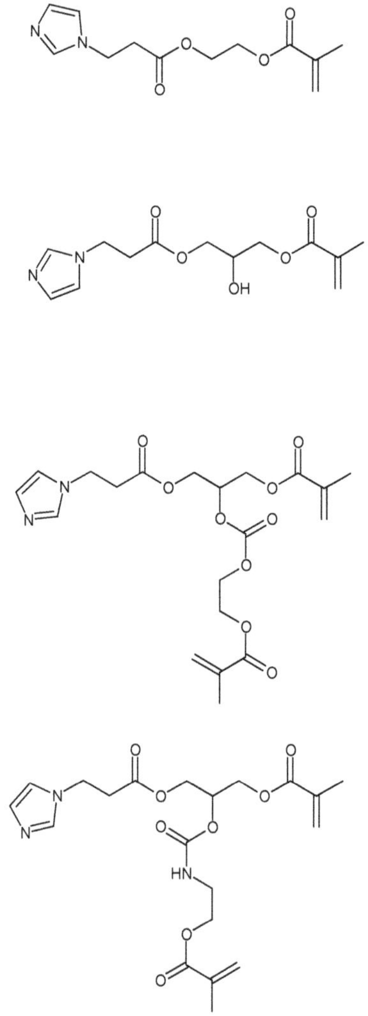

Scheme 2. Various polymerizable mono-imidazole monomers synthesized.

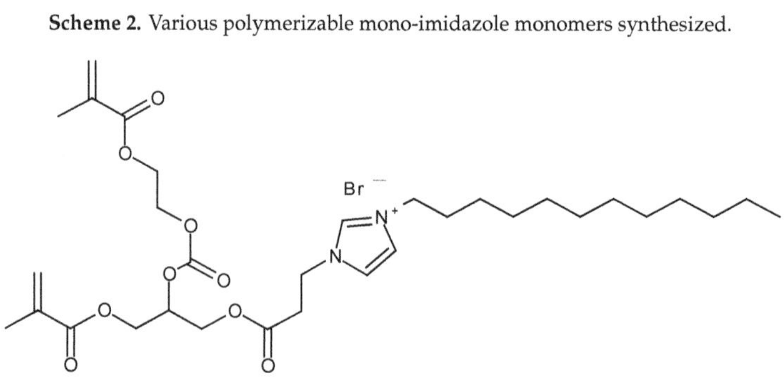

Scheme 3. Polymerizable alkylimidazoliumbromide–dimethacrylate monomer, antibacterial resin C (ABR-C).

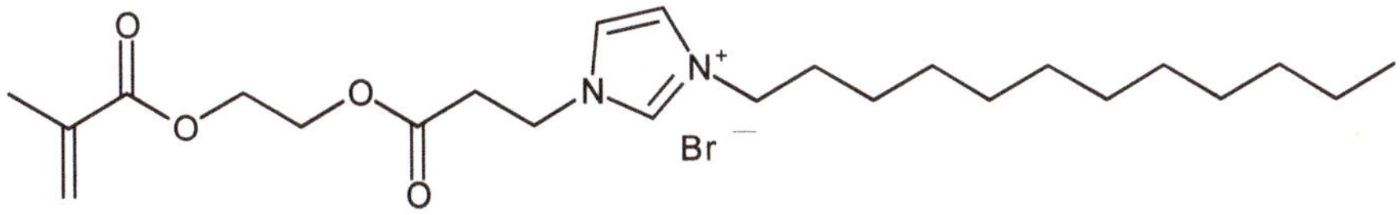

Scheme 4. Polymerizable alkylimidazoliumbromide–methacrylate monomer ABR-E.

Table 1 presents the results of MIC and MBC for representative ABR resins and SDR Resin (a proprietary dimethacrylate monomer, without imidazole moiety, synthesized by Dentsply Sirona) in comparison to frequently used controls, namely chlorhexidine (CHX) as wide-spectrum antimicrobial agent and triethylene glycol dimethacrylate (TEGDMA). TEGDMA was chosen as control monomer because it is widely used in the composition of resins for dental-related applications, including but not limited to restorative, adhesive, and orthodontic compositions. In addition, it was not known to have antibacterial potencies against common cariogenic oral bacteria strains. Furthermore, by using a modified JIS Z 2801 test method, a preliminary antibacterial test against *S. mutans* strain UA159 was used to evaluate the antibacterial effectiveness for such formulated compositions. After 24 h contact, a bacterial reduction of 99.88% was exhibited by test sample containing ABR-E (Scheme 4), as compared to control. This indicated that imidazolium-based polymerizable resin could be highly effective in suppressing the growth of bacteria such as *S. mutans*.

Table 1. Minimum inhibitory concentration (MIC) and minimum bactericidal concentration (MBC) of novel antibacterial resins ABR-E and ABR-C against *Streptococcus mutans* strain UA159, as compared to SDR Resin, triethylene glycol dimethacrylate (TEGDMA), and chlorhexidine.

Test Compound	Stock Solution	Tested Concentrations	Bacterial Inoculum Size (CFU/mL)	MIC	MBC
ABR-E	2.5%	0–0.1%	9×10^7	4 µg/mL (0.0004%)	8 µg/mL (0.0008%)
ABR-C	1%	0–0.1%	9×10^7	4 µg/mL (0.0004%)	8 µg/mL (0.0008%)
SDR Resin	0.25%	0–0.1%	9×10^7	No bacterial growth inhibition	
TEGDMA	100%	0–0.1%	9×10^7	No bacterial growth inhibition	
Chlorhexidine	2.5 mg/mL	0–125 mg/mL	9×10^7	2 mg/mL	4 mg/mL

The loss of dry mass from completely demineralized dentin beams (2 mm × 1 mm × 6 mm) in buffer solutions pre-dipped into ABR-C and ABR-E resins, versus in control buffer (artificial saliva) without pre-dip, was studied. The amount of soluble collagen in these demineralized dentin beams was measured after 7 days of incubation at 37 °C. As Figure 1 indicates, dentin beams without pre-dip lost about 15.5% of their dry mass due to degradations of endogenous matrix metalloproteinases (MMPs) and cathepsins. When the dentin beams were pre-dipped in 1–4% of imidazolium-based ABR resins (ABR-C, ABR-E) for 30 s and then dropped in incubation medium, the dentin beams only lost about 3.5–5% of their dry mass. Furthermore, the hydroxyproline (HYP), an amino acid unique to collagen, content in the medium was also measured. The collagen peptide fragments that were solubilized by the endogenous proteases of the dentin matrix were hydrolyzed to amino acids in 6N HCl and then analyzed for hydroxyproline. In the control group, the proteases released about 10.5 µg HYP/mg dry dentin after 7 days; beams pre-dipped in 1% or 4% of ABR-C released less than half as much hydroxyproline and the beam treated with ABR-E yielded a similar result.

As the designed imidazolium-based resin still comprises a methacrylate functional group, no compatibility or copolymerization issues were found when mixed with typical dimethacrylate-type dental resins, such as 1,6-bis[methacryloyloxyethoxycarbonylamino]-2,4,4-trimethylhexane (UDMA), ethoxylated bisphenol A dimethacrylate (EBPADMA), and TEGDMA. Even with 16 wt % loading of ABR-C into the control orthodontic resin, no mixability issue was observed and the resulting resin mixture was found uniform. The 3-point bending flexural strength and modulus of unfilled orthodontic resins containing 4~16 wt% ABR-C, along with control, were measured following ISO-4049 method and the results are presented in Table 2.

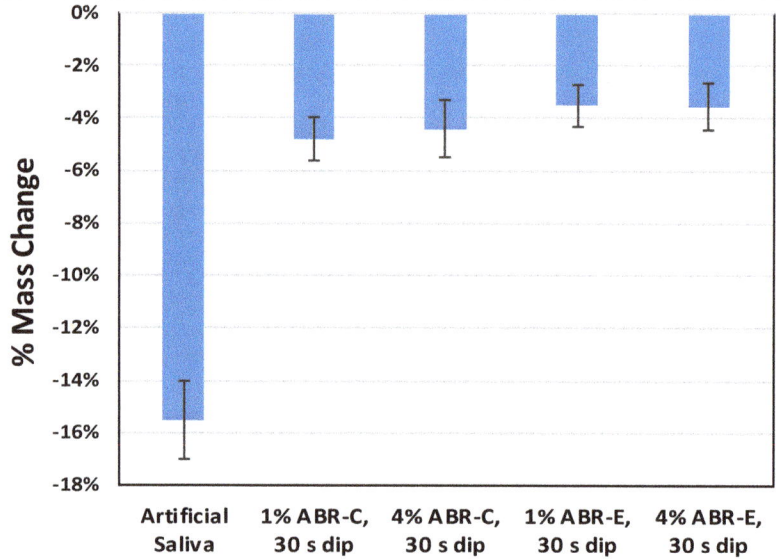

Figure 1. Effect of pre-dipping in antibacterial resins ABR-C and ABR-E on loss of dry mass of demineralized dentin beams after incubation for 7 days at 37 °C.

Table 2. Flexural strength and flexural modulus of exp. orthodontic resins (unfilled) that contain various concentrations of imidazolium-based polymerizable antibacterial monomer ABR-C.

Resin Sample	Control	4%_ABR-C	8%_ABR-C	12%_ABR-C	16%_ABR-C
Antibacterial Monomer in Resin	0	4 wt%	8 wt%	12 wt%	16 wt%
Flexural Strength, MPa (s.d.)	97 (8) [A]	88 (5) [A,B]	91 (4) [A]	95 (3) [A]	81 (4) [B]
Flexural Modulus, MPa (s.d.)	2443 (87) [a]	2388 (183) [a]	2394 (139) [a]	2293 (82) [a,b]	2125 (125) [b]

Within the same row, means that do not share the same superscript(s) are significantly different ($\alpha = 0.05$).

The 75 wt % filled experimental orthodontic cements incorporating various concentrations of antibacterial monomer ABR-C were also studied. The current investigation examined flexural and compressive strengths, flexural modulus, ambient light sensitivity, and notched-edge shear bond strength (NE-SBS) to bovine enamel. Results along with statistical analysis are presented in Table 3. For reference, the shear bond strength to bovine enamel, using the same NE-SBS method, and ambient light sensitivity of five commercially available orthodontic bonding cements were also evaluated, as presented in Figures 2 and 3, respectively.

Antibacterial testing was also conducted at an independent and good laboratory practice (GLP) complied testing institution. As shown in Table 4, for orthodontic cement paste formulations that incorporated imidazolium-based dimethacrylate antibacterial monomer (ABR-C), the ISO-22196 antimicrobial testing results against ATCC 6538 showed significant levels of antibacterial effects—up to over 5 logs of microorganism reduction—when compared with control. Such highly effective bactericidal effects for the imidazolium-based polymerizable resins were promising due to a relatively low-level loading and significantly reduced probability of leaching out owing to the dimethacrylate functional groups.

Table 3. Adhesive and physical properties of orthodontic cements (filled with 75% inorganic filler) that contain various concentrations of imidazolium-based polymerizable antibacterial monomer ABR-C.

Orthodontic Cement	Control	1%_ABR-C	2%_ABR-C	3%_ABR-C	4%_ABR-C
Antibacterial Monomer in Cement	0	1 wt%	2 wt%	3 wt%	4 wt%
Resin Conc.	25.0%	25.0%	25.0%	25.0%	25.0%
Ambient Light Sensitivity	2:35"	2:00"	2:05"	2:00"	2:15"
Compressive Strength, MPa (s.d.)	373 (25) [A]	361 (12) [A,B]	346 (10) [B]	352 (5) [A,B]	307 (15) [C]
Flexural Strength, MPa (s.d.)	150 (14) [A]	139 (6) [A]	114 (10) [B]	114 (10) [B]	88 (7) [C]
Flexural Modulus, MPa (s.d.)	11165 (531) [a]	10848 (383) [a,b]	10319 (549) [a,b]	10607 (629) [a,b]	10066 (445) [b]
NE-SBS to Etched Enamel, MPa (s.d.)	32.8 (2.7) [A,B]	36.7 (4.5) [A]	28.8 (4.3) [B,C]	25.4 (3.5) [C]	31.0 (5.8) [A,B,C]
SBS to Enamel Range, MPa	29.5 ~ 37.2	31.9 ~ 43.4	21.9 ~ 32.7	21.5 ~ 29.5	23.1 ~ 38.0
SBS to Enamel C.V.	8.1%	12.1%	14.8%	14.0%	18.7%

Within the same row, means that do not share the same superscript(s) are significantly different ($\alpha = 0.05$).

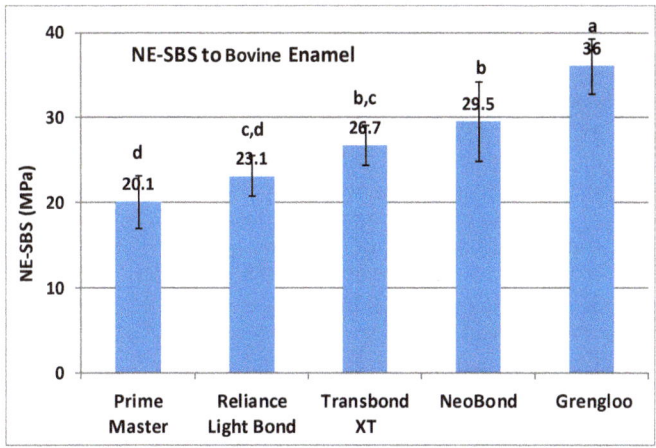

Figure 2. Notched-edge shear bond strength to bovine enamel of commercial orthodontic cements. Values that do not share the same superscript(s) are significantly different ($\alpha = 0.05$).

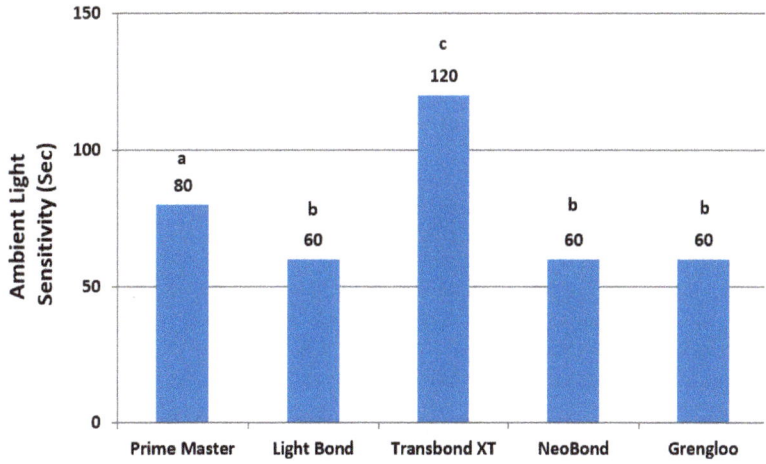

Figure 3. Ambient light sensitivity of commercial orthodontic cements. Values that do not share the same superscript are significantly different ($\alpha = 0.05$).

Table 4. ISO-22196 antimicrobial test of orthodontic cements (filled with 75 wt % inorganic filler) that contain various concentrations of imidazolium-based polymerizable antibacterial monomer using S. aureus 6538. The limit of detection for this assay is 5 CFU/carrier.

Test Micro-Organism	Contact Time	Carrier Type	Anti-Bacterial Monomer in Cement	CFU/Carrier	Percent Reduction Compared to Control at Contact Time	Log_{10} Reduction Compared to Control at Contact Time
	Time Zero	Control	0	1.00×10^6	N/A	
S. aureus 6538	24 h	ATL Control	0	8.50×10^5		
		2%_ABR-C	2 wt%	1.00×10	99.9988%	4.93
		4%_ABR-C	4 wt%	1.50×10	99.998%	4.75
		1%_ABR-C	1 wt%	4.72×10^3	99.44%	2.26
		3%_ABR-C	3 wt%	<5.00	> 99.9994%	> 5.23

3. Discussion

As demonstrated by MIC and MBC test results reported in Table 1, ABR-C and ABR-E displayed distinct antibacterial activities against cariogenic S. mutans cells. Not surprisingly, the MIC values of ABR-C and ABR-E were higher than that of benchmark wide-spectrum bactericidal chlorhexidine, which is also highly leachable and does not offer sustained antibacterial activity once leached out. Also as expected, the 8 µg/mL MBC of ABR-C and ABR-E is higher than CHX's MBC of 4 µg/mL, whereas no bactericidal activities were observed at all for SDR Resin and TEGDMA. It has been shown that, for an organic QAS compound, 10~14 carbon atoms for the N-alkyl group are associated with higher antimicrobial activity [43,44]. Another study also showed that imidazolium's N-alkyl's chain length can play a critical role in antitumor activity and cytotoxicity; C-12 was found exhibiting high antitumor activity against ~60 tumor cell lines as well as low cytotoxicity in most cases. Longer chain length could improve antitumor potency but also increase imidazolium's cytotoxicity [45]. As demonstrated in their chemical structures in Schemes 3 and 4, respectively, both ABR-C and ABR-E have a 12-carbon atom chain length on their N-alkyl group, which helps in explaining their identical MIC and MBC values.

The dentin mass loss results (Figure 1) indicate that the ABR resins, even at low concentrations, are effective against matrix metalloproteinases (MMPs) for minimizing the collagen degradation. MMPs belong to a larger group of proteases known as the metzincin superfamily. These collagenases are capable of degrading triple-helical fibrillary collagens, such as those in dentin, into distinctive fragments. More specifically, certain MMPs are considered to contribute to the gradual degradation of collagen fibrils in hybrid layers established during dentin bonding. A wide range of MMPs, including MMP-2, MMP-8, MMP-9, and MMP-20, have been detected in the human dentin matrix [46]. It is well reported that CHX has broad anti-MMP activity in addition to antimicrobial capability. However, the long-term in vivo anti-MMP activity of CHX may not be as effective, which has been attributed to possible leaching out of the CHX [47,48]. It has also been reported that cationic quaternary ammonium methacrylates may exhibit dentin MMP inhibition comparable with that of CHX but that higher concentrations were required [49].

The synthetic strategy in this investigation focused on covalently linking antibacterial compounds that contain functional groups with strong antibacterial activities to a variety of backbones and polymerizable groups compatible with conventional dental resin systems. The antibacterial capability of the active compound could be either enhanced or reduced by polymerization. This depends on how the compound eradicates bacteria, either by disrupting bacterial membrane or via depleting the bacterial food supply. In order for an antibacterial resin to be a viable option for industrial-scale production and application, there are several essential requirements that need to be fulfilled: (1) synthesis of ABR should be straightforward and not cost-prohibitive; ideally, well-established synthetic routes are preferred; (2) ABR should have good storage stability (sufficient shelf life); (3) end products that contain ABR should be able to maintain their antibacterial activities through recurrent bacterial challenges; (4) ideally, the antibacterial effect can be achieved with relatively small dosing and minimum bacterial resistance in the target pathogenic microorganisms; (5) no drastic compromises to the other vital properties should be introduced by ABR. As shown in Table 2, even at up to 12 wt % loading level, there are no significant decreases in flexural strength or flexural modulus with the incorporation of the imidazole-based polymerizable antibacterial monomer ABR-C, as compared to control. The flexural strength of the resin mixture with 16 wt % antibacterial monomer showed lower flexural strength, but flexural modulus still retained 84% value as compared to control.

During this investigation, it was found that a group of polymethacrylate resins containing at least one imidazole moiety could be readily prepared via appropriate hybrid methacrylate–acrylate resins or polyacrylate resins with proper control of the conversion of the imidazole addition. This is an effective approach to incorporating an imidazole moiety into a polymerizable resin as novel acid-free functional resins. Moreover, such imidazole-containing polymerizable resins may be further chemically modified by reacting with a variety of halogenated alkyls to form polymerizable resins with ionic moiety of imidazolium, which results in a unique class of polymerizable ionic liquid resins.

Sufficient mechanical properties of the orthodontic bonding cement are important as the bending force and torque during teeth straightening and aligning could be quite significant. There is a slight reduction of flexural strength with a higher loading amount of ABR-C; however, flexural modulus can be maintained with up to 3 wt % of ABR-C loading (Table 3). A similar trend was also observed for compressive strength. Ambient light sensitivity (ALS) is another important property for light curable orthodontic cement or adhesive, as adequate ALS is needed for the bracket to be carefully placed and aligned at the correct position and angle, before it is light cured. As also exhibited in Table 3, the ALS values of ABR-C-containing orthodontic cements are all at 2 min or higher, which compared well with commercially available orthodontic cements, as shown in Figure 3.

Orthodontic cement with the capability to prevent the development of WSLs without compromising enamel bond strength would be desirable. Orthodontic treatments last an average of two to three years, during which time orthodontic cement must be capable of bonding to the enamel with high bond strength in order to resist masticatory loads. As demonstrated in Table 3, for experimental orthodontic cements incorporating 1–4 wt % ABR-C, there is no drastic compromise to the SBS to enamel except for 3 wt % ABR-C. More importantly, their SBS values are all comparable to those of the commercially available orthodontic cements products (Figure 2), with mean SBS ranging from 20.1 MPa to 36.0 MPa. However, this bond strength to enamel should not be too high either, to prevent damages to the enamel while removing the bracket.

The strong effectiveness in eradicating bacteria for the imidazolium-based polymerizable resins was further demonstrated by the formulated experimental orthodontic cements. As exhibited in Table 4, even at a low-level loading (1 wt %) of such imidazolium-based polymerizable monomer, a reduction of over 2 logs of microorganisms can be achieved by ABR-C. When the loading level of ABR-C was increased from 2 wt % to 4 wt %, reductions of around 5 logs or higher were observed. Moreover, with optimized compositions, not only can a highly effective antibacterial capability be achieved, but balanced mechanical properties can also be maintained (Table 3). The potent antibacterial property of this novel polymerizable imidazolium resin offers another crucial benefit—no severe cytotoxicity was introduced [50]. Conventional QAS-based polymerizable resins, on the other hand, could be less effective and a high dose loading (up to 30 wt %) is needed, which frequently leads to significant drops in mechanical properties and elevated cytotoxicity [51,52].

Effective antimicrobial agents are expected to regulate the oral biofilm at levels compatible with good oral health while not sacrificing the beneficial properties of the resident oral microflora. Compounds comprising imidazolium moieties have exhibited various antibacterial, antioxidant, and antifungal properties; imidazolium salts' overall mechanism of antibacterial function is similar to that of QAS—mainly by disturbing the planktonic cell membrane [52]. Nonetheless, comprehensive studies by Koo et al. have shown that ABR-MC (ABR-modified composite) impaired biofilm initiation by disrupting bacterial accumulation, cell colonization, and subsequent biofilm development, if left unintervened [50]. The anti-biofilm properties of ABR-MC were assessed using an EPS-matrix producing *S. mutans* in an experimental biofilm model. Using high-resolution confocal fluorescence imaging and biophysical methods, they observed severely reduced biomass, remarkable disruption of bacterial accumulation, and defective 3D matrix structure on the surface of ABR-MC. Mechanism-wise, it was speculated that the distinct difference in the basicity between N-substitute imidazole (pK_{aH} of 7.20 for 1-methylimidazole) and aliphatic tertiary amine (pK_{aH} of 10.7 for triethylamine) might be the major contributor to the increased polarity and higher potential in its charged conjugated moiety (imidazolium). However, the exact mechanism contributing to the much more pronounced bactericidal effects as compared to conventional QAS compounds will require further investigation. Other areas of relevant investigation include the further study of the hydrolytical stability of these antibacterial monomers.

4. Materials and Methods

4.1. Materials

Antibacterial resins (ABR-C and ABR-E) were prepared in house; the detailed preparation route can be found in a previous paper [50]. SDR Resin and adhesion-promoting resin (APR, Dentsply proprietary resin) were supplied by Dentsply Sirona's internal production (Milford, DE, USA). Other conventional dental resins and ingredients used in formulations were purchased from commercial sources, including 1,6-bis [methacryloyloxyethoxycarbonylamino]-2,4,4-trimethylhexane (UDMA), ethoxylated bisphenol A dimethacrylate (EBPADMA), trimethylolpropane trimethacrylate (TMPTMA), camphorquinone (CQ), dimethylaminobenzonitrile (DMABN), diphenyl (2,4,6-trimethylbenzoyl) phosphine oxide (TPO), and butylated hydroxytoluene (BHT). Inorganic fillers were milled and silanated in house, including barium fluoroalumino borosilicate glass (BFBG), bariumalumino borosilicate glass (BABG), and AEROSIL fumed silica OX-50 (silanated OX-50). All fillers were surface treated by γ-methacryloxypropyl-trimethoxysilane.

Commercially available orthodontic cements studied include the following: Prime Master (Sun Medical Co., Shiga, Japan), Light Bond (Reliance Orthodontic Products, Alsip, IL, USA), Transbond XT (3M Unitek, Monrovia, CA, USA), NeoBond (Dentsply Sirona, Milford, DE, USA), and Grengloo (Ormco, Glendora, CA, USA).

Preparation of experimental orthodontic resin and cement. Experimental dental resin mixture with various concentrations of polymerizable antibacterial resin (0 (control), 4, 8, 12, 16 wt %) was mixed with other resins, namely UDMA (28~34 wt %), EBPADMA (20~26 wt %), APR (28~32 wt %), TMPTMA (6~8 wt %), CQ (~0.1 wt %), DMABN (~0.2 wt %), and BHT (~0.07 wt %), to a uniform resin blend at 55 °C using an overhead mechanical stirrer (Fisher Scientific, Pittsburgh, PA, USA). Experimental orthodontic cements were compounded with 75 wt % of inorganic filler mixture using a Ross double planetary mixer at 55 °C to a uniform paste (Ross Mixer, Charles Ross & Son Company, Hauppauge, NY, USA). The inorganic filler used in this study consists of three types of fillers in the filler blend: ~33 wt % BABG, ~65 wt % BFBG-2, and ~2 wt % silanated OX-50.

4.2. Evaluation Methods

MIC and MBC Testing. Three monomers were tested against controls, i.e., ABR-E, ABR-C, and SDR Resin (a proprietary dimethacrylate monomer, without imidazole moiety, synthesized by Dentsply Sirona). The monomers were diluted in UltraPure distilled water (Gibco, Gaithersburg, MD, USA) before use. Different concentrations of the three tested monomers, along with two controls, i.e., wide-spectrum antimicrobial agent chlorhexidine (CHX) and conventional monomer TEGDMA, were incubated statically with the cariogenic organism *S. mutans* strain UA159 using mid-exponential cells for 48 h at 37 °C in the presence of 5% CO_2. A fixed initial inoculum size of 9×10^7 CFU/mL of bacterial cells was inoculated per well and 2-fold dilutions of the experimental monomers, CHX, and TEGDMA were added. The following concentrations were tested: 0 to 125 μg/mL for CHX; 0 to 0.1% for the experimental monomers ABR-E, ABR-C, SDR Resin, and the control monomer TEGDMA. The MIC was determined as the lowest test concentration needed to ensure that culture did not grow over 10% of the relative cell, as determined by visual inspection of the growth inhibition of each well compared to that of the control well (without test compounds). For the MBC test, cells were serially diluted and spot-plated on agar plates; MBC values were determined by using the actual reduction of the bacterial viable cell counts. The MBC was determined as the lowest concentration that allows less than 0.1% of the original inoculum to survive.

Dentin Mass Loss. Completely demineralized dentin beams (2 mm × 1 mm × 6 mm) were prepared. Dentin beams were obtained from extracted molars after removing the enamel and superficial dentin. The beams were then submerged in 10% phosphoric acid for 18 h at 25 °C to completely demineralize the dentin. They were incubated in a control buffer (artificial saliva) to see how much of the insoluble collagen in demineralized dentin could be solubilized in 7 days of incubation at 37 °C. The loss of dry mass from completely demineralized dentin beams (2 mm × 1 mm × 6 mm) in buffer solutions pre-dipped into ABR-C and ABR-E resins for 30 s, versus in control buffer (artificial saliva) without pre-dip, was studied. The amount of soluble collagen in these demineralized dentin beams was measured after 7 days of incubation at 37 °C.

Flexural Strength. Specimens were prepared based on ISO 4049:2019. Samples were carefully injected into 25 mm × 2 mm × 2 mm stainless steel molds, then covered by Mylar film and cured with a Spectrum 800® (Dentsply Sirona, Milford, DE, USA) halogen lamp uniformly across the entire length of the specimen, using an irradiance of 850 mW/cm^2 for 5 × 20 s. After curing, specimens were immersed in DI water for 24 h at 37 °C prior to the 3-pt bending flexural test. Flexural test was performed on an Instron Universal Tester (Model 4400R, Norwood, MA, USA countr using a crosshead speed of 0.75 mm/min. Six specimens per sample set were prepared and tested.

Compressive Strength. Samples were carefully injected into ⌀4 mm × 7 mm stainless steel molds, sandwiched between two Mylar films. Light curing of the specimen was carried out with a Spectrum 800® (Dentsply Sirona) halogen lamp at an irradiance of 850 mW/cm^2 for 20 s on both ends. After curing, specimens were immersed in DI water for 24 h at 37 °C, before being polished to ⌀4 mm × 6 mm specimen size with 600 grit sand paper. The compression test was performed on an Instron Universal Tester (Model 4400R) using a crosshead speed of 5 mm/min. Six specimens per sample set were prepared and tested.

Ambient Light Sensitivity (ALS). ALS was tested on a Suntest unit (model CPS 56007014, Heraeus Noblelight, Buford, GA, USA) with a UV filter that was calibrated to provide illuminance of 8000 ± 1000 lux tested by a Lux Meter HD400 (Extech, Nashua, NH, USA). A spheroidal mass of approximately 0.03 gm of the material was weighed to be tested on a glass microscope slide, and the slide was positioned with a matte black cover on top of the cell with a UV filter in place. The Suntest light was turned on, and one should wait at least 10 s until the light output has stabilized. The matte black cover was removed and the timer was started simultaneously. After exposure, the slide with the sample was removed and the second microscope slide was immediately pressed against the material with a slight shearing action to produce a thin layer. The material was visually inspected to see whether it was physically homogeneous. If the material has begun to set, clefts and voids appear in the specimen when the thin layer is being produced. The time was recorded when the sample started to set and appeared physically non-homogeneous. In all, 3 runs were conducted for each sample.

Notched-Edge Shear Bond Strength. Caries-free and freshly extracted bovine incisors were carefully sectioned longitudinally through the distal, occlusal, and mesial surfaces using a precision low speed saw (TechCut 4™, Allied High Tech Products, Rancho Dominguez, CA, USA). The sectioned bovine incisors, with buccal surface exposed, were then mounted in a cylindrical block using cold-cure acrylics. The exposed surface was then coarse ground on a model trimmer until a flat enamel surface was exposed, followed by wet ground on a grinding wheel under running water using 120-grit and 320-grit SiC sanding papers until the surface was even and smooth when visually inspected. Enamel was acid etched using 34% phosphoric etching gel for 15 s, then thoroughly rinsed for 20 s. The notched-edge bonding jig contained a cylindrical plastic mold, resulting in samples with a defined bonding area (diameter of 2.38 mm). The antibacterial orthodontic cement restorative composite was then carefully placed into the center of the mold and good contact with the substrate was ensured. After light curing with a Spectrum 800® (Dentsply Sirona) halogen lamp at 850 mW/cm^2 for 20 s, the specimen was then carefully removed from the mold. Specimens were stored in DI water for 24 h at 37 °C prior to SBS testing. The SBS test was performed on an Instron Universal Tester (Model 4400R) using a crosshead speed of 1 mm/min. Seven specimens per sample set were prepared and tested.

JIS Z 2801. This test was conducted at Antimicrobial Test Laboratories (ATL, Round Rock, TX, USA), an independent and GLP complied testing institution. JIS Z 2801 was adopted as ISO standard (ISO-22196). This test is designed to quantitatively test the ability of hard surfaces to inhibit the growth of microorganisms over a 24-h period of contact. The test microorganism selected for this test was *Streptococcus mutans* (*S mutans*, ATCC® 25175). The following protocol was used by ATL: (1) test microorganism *S mutans*, ATCC® 25175 was prepared by growth in tryptic soy broth for 24 h; (2) the suspension of the test microorganism was standardized by dilution in a Dey-Engley neutralizing broth (D/E Broth) at 0.9 mL blanks; (3) control and test surfaces were inoculated with microorganisms, and then the microbial inoculum was covered with a thin sterile film; (4) microbial concentrations were determined at "time zero" by elution followed by dilution and plating to tryptic soy agar (Difco); (5) a control was run to verify that the neutralization/elution method effectively neutralized the antimicrobial agent in the antimicrobial surface being tested; (6) the inoculated covered control and antimicrobial test surfaces were allowed to incubate undisturbed in a humid environment for 24 h at 36 ± 1 °C; and (7) after incubation, microbial concentrations were determined and the reduction of microorganisms relative to the control surface was calculated as follows:

$$R \text{ (Average Log Reduction)} = \text{Log} (B/C), \quad (1)$$

where

B = Average number of viable cells on the control pieces after 24 h.
C = Average number of viable cells on the test pieces after 24 h.

ISO-22196 Antimicrobial Test. This test was conducted at Antimicrobial Test Laboratories (ATL, Round Rock, TX, USA), an independent and GLP complied testing institution. ISO method 22196 is a quantitative test designed to assess the performance of materials' antimicrobial capabilities on hard non-porous surfaces. The method can be conducted using contact times ranging from 10 minutes up to 24 h. For the ISO 22196 test, non-antimicrobial control surfaces are used as the baseline for calculations of microbial reduction. The test microorganism selected for this test was *Staphylococcus aureus* 6538 (*S. aureus* 6538). This bacterium is a Gram-positive, spherical-shaped, facultative anaerobe. *Staphylococcus* species are known to demonstrate resistance to antibiotics such as methicillin and is commonly used in standard test methods as a model for Gram-positive bacteria. *S. aureus* pathogenicity can range from commensal skin colonization to more severe diseases such as pneumonia and toxic shock syndrome (TSS). The following protocol was used by ATL: (1) *S. aureus* 6538 prepared by growth in tryptic soy broth for 18 h and the suspension of the test microorganism was standardized by dilution in broth at 1:500 by volume; (2) control and test surfaces were inoculated with microorganisms, and then the microbial inoculum was covered with a sterile thin film; (3) microbial concentrations were determined at "time zero" by elution, followed by dilution and plating to tryptic soy agar; (4) a control was run to verify that the neutralization/elution method effectively neutralized the antimicrobial agent in the antimicrobial surface being tested; (5) the inoculated covered control and test surfaces were allowed to incubate undisturbed in a humid environment at 36 ± 1 °C for 24 h; and (6) after incubation, microbial concentrations were determined and the reduction of microorganisms relative to the control surface was calculated as follows:

$$\text{Percent Reduction} = \left(\frac{B-A}{B}\right) \times 100$$

$$Log_{10} \text{ Reduction} = Log\left(\frac{B}{A}\right)$$

where

B = Number of viable test microorganisms on the control carriers after the contact time.

A = Number of viable test microorganisms on the test carriers after the contact time.

Statistical Analysis. Statistical analysis was conducted with Minitab 17 Statistical Software (Minitab®, State College, PA, USA). Normality of data was checked using the Anderson–Darling test. Mechanical properties and shear bond strength data were analyzed with one-way analysis of variance (ANOVA) and Tukey's test. Significance levels of 0.05 were used for ANOVA and Tukey's tests.

5. Conclusions

This study explored the potential of incorporating unique imidazolium-based polymerizable antibacterial monomers into orthodontic bonding cements. For unfilled resins comprising up to 12 wt % ABR-C, no significant decreases in flexural strength or modulus were observed. For experimental cements incorporating 1–4 wt % ABR-C, there is no drastic compromise to the SBS to enamel except for 3 wt % ABR-C; moreover, their SBS values are all comparable to those of the commercially available orthodontic cements. The ISO-22196 antimicrobial test against *S. aureus* showed significant levels of antibacterial effects—up to over 5 logs of microorganism reduction exhibited by ABR-C-containing experimental cements. Although MIC and MBC were identified, and a range of antibacterial activities of imidazolium-based orthodontic cements were demonstrated, clinical and long-term studies are still important to validate the efficacy of this unique series of imidazolium resins towards preventing and mitigating WSLs for patients.

Author Contributions: H.L. and X.J. contributed equally to conceptualization; methodology; validation; formal analysis; investigation; resources; data curation; writing—original draft preparation; writing—review and editing; visualization; supervision; and project administration. All authors have read and agreed to the published version of the manuscript.

Funding: This research received no external funding.

Acknowledgments: We would like to thank Anuradha Prakki and her team at University of Toronto for assistance with the MIC and MAC tests. We would also like to thank David Pashley and his team at Augusta University for the dentin mass loss and HYP content study. We also greatly appreciate the laboratory assistance provided by Donna Reid and Brandon Zhang.

Conflicts of Interest: The authors declare no conflict of interest.

References

1. Nascimento, V.; Conti, A.; Cardoso, M.; Valarelli, D.; Almeida-Pedrin, R. Impact of orthodontic treatment on self-esteem and quality of life of adult patients requiring oral rehabilitation. *Angle Orthod.* **2016**, *86*, 839–845. [CrossRef] [PubMed]
2. Grewal, H.; Sapawat, P.; Modi, P.; Aggarwal, S. Psychological impact of orthodontic treatment on quality of life—A longitudinal study. *Int. Orthod.* **2019**, *17*, 269–276. [CrossRef]
3. Wishney, M. Potential risks of orthodontic therapy: A critical review and conceptual framework. *Aust. Dent. J.* **2017**, *62*, 86–96. [CrossRef] [PubMed]
4. Heymann, G.C.; Grauer, D. A contemporary review of white spot lesions in orthodontics. *J. Esthet. Restor. Dent.* **2013**, *25*, 85–95. [CrossRef] [PubMed]
5. Chapman, J.A.; Roberts, W.E.; Eckert, G.J.; Kula, K.S.; González-Cabezas, C. Risk factors for incidence and severity of white spot lesions during treatment with fixed orthodontic appliances. *Am. J. Orthod. Dentofac. Orthop.* **2010**, *138*, 188–194. [CrossRef]
6. Buschang, P.H.; Chastain, D.; Keylor, C.L.; Crosby, D.; Julien, K. Incidence of white spot lesions among patients treated with clear aligners and traditional braces. *Angle Orthod.* **2019**, *89*, 359–364. [CrossRef]
7. Tasios, T.; Papageorgiou, S.N.; Papadopoulos, M.A.; Tsapas, A.; Haidich, A.B. Prevention of orthodontic enamel demineralization: A systematic review with meta-analyses. *Orthod. Craniofac. Res.* **2019**, *22*, 225–235. [CrossRef]
8. Leeper, D.K.; Noureldin, A.; Julien, K.; Campbell, P.M.; Buschang, P.H. Risk assessments in orthodontic patients developing white spot lesions. *J. Investig. Clin. Dent.* **2019**, *10*, e12470. [CrossRef]

9. Tufekci, E.; Dixon, J.S.; Gunsolley, J.C.; Lindauer, S.J. Prevalence of white spot lesions during orthodontic treatment with fixed appliances. *Angle Orthod.* **2011**, *81*, 206–210. [CrossRef]
10. Khalaf, K. Factors affecting the formation, severity and location of white spot lesions during orthodontic treatment with fixed appliances. *J. Oral Maxillofac. Res.* **2014**, *5*. [CrossRef]
11. Joen, D.; An, J.; Lim, B.; Ahn, S. Orthodontic bonding procedures significantly influence biofilm composition. *Prog. Orthod.* **2020**, *21*, 1–9. [CrossRef] [PubMed]
12. Baeshen, H.A.; Rangmar, S.; Kjellberg, H.; Birkhed, D. Dental Caries and Risk Factors in Swedish Adolescents about to Start Orthodontic Treatment with Fixed Appliances. *J. Contemp. Dent. Pract.* **2019**, *20*, 538–562. [CrossRef]
13. Srivastava, K.; Tikku, T.; Khanna, R.; Sachan, K. Risk factors and management of white spot lesions in orthodontics. *J. Orthod. Sci.* **2013**, *2*, 43–49. [CrossRef]
14. Sonesson, M.; Svensäter, G.; Wickström, C. Glucosidase activity in dental biofilms in adolescent patients with fixed orthodontic appliances—A putative marker for white spot lesions—A clinical exploratory trial. *Arch. Oral Biol.* **2019**, *102*, 122–127. [CrossRef] [PubMed]
15. Sinclair, P.M.; Berry, C.W.; Bennett, C.L.; Israelson, H. Changes in gingival and gingival flora with bonding and banding. *Angle Orthod.* **1987**, *57*, 271–278. [PubMed]
16. Ristic, M.; Svabic, M.V.; Sasic, M.; Zelic, O. Clinical and microbiological effects of fixed orthodontic appliances on periodontal tissues in adolescents. *Orthod. Craniofac. Res.* **2007**, *10*, 187–195. [CrossRef]
17. Paolantonio, M.; Festa, F.; Di Placido, G.; D'Attilio, M.; Catamo, G.; Piccolomini, R. Site-specific subgingival colonization by Actinobacillus actinomycetencomitans in orthodontic patients. *Am. J. Orthod. Dentofac. Orthop.* **1999**, *115*, 423–428. [CrossRef]
18. Lucchese, A.; Bondemark, L.; Marcolina, M.; Manuelli, M. Changes in oral microbiota due to orthodontic appliances: A systematic review. *J. Oral Microbiol.* **2018**, *10*, 1476645. [CrossRef]
19. Freitas, A.; Marquezan, M.; Nojima, M.; Alviano, D.; Maia, L. The influence of orthodontic fixed appliances on the oral microbiota: A systematic review. *Dent. Press J. Orthod.* **2014**, *19*, 46–55. [CrossRef]
20. Bourbia, M.; Ma, D.; Cvitkovitch, D.; Santerre, J.P.; Finer, Y. Cariogenic Bacteria Degrade Dental Resin Composites and Adhesives. *J. Dent. Res.* **2013**, *92*, 989–994. [CrossRef] [PubMed]
21. Stewart, C.A.; Finer, Y. Biostable, antidegradative and antimicrobial restorative systems based on host-biomaterials and microbial interactions. *Dent. Mater.* **2019**, *35*, 36–52. [CrossRef] [PubMed]
22. Mystkowska, J.; Niemirowicz-Laskowska, K.; Łysik, D.; Tokajuk, G.; Dąbrowski, J.R.; Bucki, R. The Role of Oral Cavity Biofilm on Metallic Biomaterial Surface Destruction–Corrosion and Friction Aspects. *Int. J. Mol. Sci.* **2018**, *19*, 743. [CrossRef]
23. Bergstrand, F.; Twetman, S. A review on prevention and treatment of post-orthodontic white spot Lesions—Evidence-based methods and emerging technologies. *Open Dent. J.* **2011**, *5*, 158–162. [CrossRef] [PubMed]
24. Sandra, C.D.; Maria, D.; Ingrid, M.D.; Vanessa, H.D.; Katia, V.R.; Leandro, C. Preventing and Arresting the Appearance of White Spot Lesions around the Bracket by applying Fluoride Varnish: A Systematic Review. *Dentistry* **2018**, *8*, 1–7. [CrossRef]
25. Aghoutan, H.; Alami, S.; Quars, F.E.; Diouny, S.; Bourzgui, F. White Spots Lesions in Orthodontic Treatment and Fluoride—Clinical Evidence. In *Emerging Trends in Oral Health Sciences and Dentistry*, 1st ed.; Virdi, M., Ed.; IntechOpen Limited: London, UK, 2015. [CrossRef]
26. Sonesson, M.; Brechter, A.; Abdulraheem, S.; Lindman, R.; Twetman, S.S. Fluoride varnish for the prevention of white spot lesions during orthodontic treatment with fixed appliances: A randomized controlled trial. *Eur. J. Orthod.* **2019**, 1–5. [CrossRef] [PubMed]
27. Höchli, D.; Hersberger-Zurfluh, M.; Papageorgiou, S.N.; Eliades, T. Interventions for orthodontically induced white spot lesions: A systematic review and meta-analysis. *Eur. J. Orthod.* **2017**, *39*, 122–133. [CrossRef] [PubMed]
28. Paula, A.B.; Fernandes, A.R.; Coelho, A.; Marto, C.M.; Marques-Ferreira, M.; Caramelo, F.; Vale, F.; Carrilho, E. Therapies for White Spot Lesions-A Systematic Review. *J. Evid. Based Dent. Pract.* **2017**, *17*, 23–38. [CrossRef]
29. Beyth, N.; Yudovin-Farber, I.; Perez-Davidi, M.; Domb, A.J.; Weiss, E.I. Polyethyleneimine nanoparticles incorporated into resin composite cause cell death and trigger biofilm stress in vivo. *Proc. Natl. Acad. Sci. USA* **2010**, *107*, 22038–22043. [CrossRef]

30. Zaltsman, N.; Shvero, N.D.; Polak, D.; Weiss, E.I.; Beyth, N. Antibacterial orthodontic adhesive incorporating polyethyleneimine nanoparticles. *Oral Health Prev. Dent.* **2017**, *15*, 245–250. [CrossRef]
31. Varon-Shahar, E.; Sharon, E.; Zabrovsky, A.; Houri-Haddad, Y.; Beyth, N. Orthodontic cements and adhesives: A possible solution to streptococcus mutans outgrowth adjacent to orthodontic appliances. *Oral Health Prev. Dent.* **2019**, *17*, 49–56. [CrossRef]
32. Imazato, S.; Ehara, A.; Torii, M.; Ebisu, S. Antibacterial activity of dentine primer containing MDPB after curing. *J. Dent.* **1998**, *26*, 267–271. [CrossRef]
33. Oz, A.Z.; Oz, A.A.; Yazicioglu, S.; Sancaktar, O. Effectiveness of an antibacterial primer used with adhesive-coated brackets on enamel demineralization around brackets: An in vivo study. *Prog. Orthod.* **2019**, *20*, 15. [CrossRef] [PubMed]
34. Feng, X.; Zhang, N.; Xu, H.H.K.; Weir, M.D.; Melo, M.A.S.; Bai, Y.; Zhang, K. Novel orthodontic cement containing dimethylaminohexadecyl methacrylate with strong antibacterial capability. *Dent. Mater. J.* **2017**, *36*, 669–676. [CrossRef] [PubMed]
35. Wang, X.; Zhang, N.; Wang, B.; Park, S.R.; Weir, M.D.; Xu, H.H.K.; Bai, Y. Novel self-etching and antibacterial orthodontic adhesive containing dimethylaminohexadecyl methacrylate to inhibit enamel demineralization. *Dent. Mater. J.* **2018**, *37*, 555–561. [CrossRef]
36. Cheng, L.; Weir, M.D.; Zhang, K.; Arola, D.D.; Zhou, X.; Xu, H.H. Dental primer and adhesive containing a new antibacterial quaternary ammonium monomer dimethylaminododecyl methacrylate. *J. Dent.* **2013**, *41*, 345–355. [CrossRef] [PubMed]
37. Melo, M.A.S.; Wu, J.; Weir, M.D.; Xu, H.H.K. Novel antibacterial orthodontic cement containing quaternary ammonium monomer dimethylaminododecyl methacrylate. *J. Dent.* **2014**, *42*, 1193–1201. [CrossRef]
38. Wang, S.; Zhang, K.; Zhou, X.; Xu, N.; Xu, H.H.K.; Weir, M.D.; Ge, Y.; Wang, S.; Li, M.; Li, Y.; et al. Antibacterial effect of dental adhesive containing dimethylaminododecyl methacrylate on the development of Streptococcus mutans biofilm. *Int. J. Mol. Sci.* **2014**, *15*, 12791–12806. [CrossRef]
39. Huang, L.; Xiao, Y.H.; Xing, X.D.; Li, F.; Ma, S.; Qi, L.L.; Chen, J.H. Antibacterial activity and cytotoxicity of two novel cross-linking antibacterial monomers on oral pathogens. *Arch. Oral Bio* **2011**, *56*, 367–373. [CrossRef]
40. Yu, F.; Dong, Y.; Yu, H.-H.; Lin, P.-T.; Zhang, L.; Sun, X.; Liu, Y.; Xia, Y.-N.; Huang, L.; Chen, J.-H. Antibacterial Activity and Bonding Ability of an Orthodontic Adhesive Containing the Antibacterial Monomer 2-Methacryloxylethyl Hexadecyl Methyl Ammonium Bromide. *Sci. Rep.* **2017**, *7*, 41787. [CrossRef]
41. Jin, X. Method and Antibacterial/Antimicrobial Compositions in Dental Compositions. U.S. Patent 8747831, 2014.
42. Jin, X. Imidazole and Imidazolium Resins and Methods for Preparing Curable Imidazolium Antimicrobial Resins. U.S. Patent RE47512, 2016.
43. De Almeida, C.M.; Da Rosa, W.L.O.; Meereis, C.T.W.; Ribeiro, J.S.; Da Silva, A.F.; Lund, R.G.; De Almeida, S.M. Efficacy of antimicrobial agents incorporated in orthodontic bonding systems: A systematic review and meta-analysis. *J. Orthod.* **2018**, *45*, 79–93. [CrossRef]
44. Garcia, M.T.; Ribosa, I.; Perez, L.; Manresa, A.; Comelles, F. Aggregation Behavior and Antimicrobial Activity of Ester-Functionalized Imidazolium-and Pyridinium-Based Ionic Liquids in Aqueous Solution. *Langmuir* **2013**, *29*, 2536–2545. [CrossRef]
45. Malhotra, S.V.; Kumar, V. A profile of the in vitro anti-tumor activity of imidazolium-based ionic liquids. *Bioorganic Med. Chem. Lett.* **2010**, *20*, 581–585. [CrossRef]
46. Mazzoni, A.; Tjäderhane, L.; Checchi, V.; Di Lenarda, R.; Salo, T.; Tay, F.; Pashley, D.; Breschi, L. Role of dentin MMPs in caries progression and bond stability. *J. Dent. Res.* **2015**, *94*, 241–251. [CrossRef]
47. Sadek, F.T.; Braga, R.R.; Muench, A.; Liu, Y.; Pashley, D.H.; Tay, F.R. Ethanol wet-bonding challenges current anti-degradation strategy. *J. Dent. Res.* **2010**, *89*, 1499–1504. [CrossRef] [PubMed]
48. Gostemeyer, G.; Schwendicke, F. Inhibition of hybrid layer degradation by cavity pretreatment: *Meta-* and trial sequential analysis. *J. Dent.* **2016**, *49*, 14–21. [CrossRef]
49. Ge, Y.; Wang, S.; Zhou, X.; Wang, H.; Xu, H.H.K.; Cheng, L. The use of quaternary ammonium to combat dental caries. *Materials* **2015**, *8*, 3532–3549. [CrossRef] [PubMed]

50. Hwang, G.; Koltisko, B.; Jin, X.; Koo, H. Nonleachable imidazolium-incorporated composite for disruption of bacterial clustering, exopolysaccharide-matrix assembly, and enhanced biofilm removal. *ACS Appl. Mater. Interfaces* **2017**, *9*, 38270–38280. [CrossRef] [PubMed]
51. Xue, Y.; Xiao, H.; Zhang, Y. Antimicrobial Polymeric Materials with Quaternary Ammonium and Phosphonium Salts. *Int. J. Mol. Sci.* **2015**, *16*, 3626–3655. [CrossRef] [PubMed]
52. Riduan, S.N.; Zhang, Y. Imidazolium salts and their polymeric materials for biological applications. *Chem. Soc. Rev.* **2013**, *42*, 9055–9070. [CrossRef]

Publisher's Note: MDPI stays neutral with regard to jurisdictional claims in published maps and institutional affiliations.

 © 2020 by the authors. Licensee MDPI, Basel, Switzerland. This article is an open access article distributed under the terms and conditions of the Creative Commons Attribution (CC BY) license (http://creativecommons.org/licenses/by/4.0/).

Article

Electrospun Scaffolds Containing Silver-Doped Hydroxyapatite with Antimicrobial Properties for Applications in Orthopedic and Dental Bone Surgery

Thomas E. Paterson [1,†], Rui Shi [2,†], Jingjing Tian [3,4,†], Caroline J. Harrison [1], Mailys De Sousa Mendes [5], Paul V. Hatton [1,*], Zhou Li [3,*] and Ilida Ortega [1]

1. School of Clinical Dentistry, University of Sheffield, Shefield 0114, UK; t.paterson@sheffield.ac.uk (T.E.P.); c.j.harrison@sheffield.ac.uk (C.J.H.); i.ortega@sheffield.ac.uk (I.O.)
2. Beijing Laboratory of Biomedical Materials, Institute of Traumatology and Orthopaedics, Beijing Jishuitan Hospital, Beijing 100083, China; shirui@jst-hosp.com.cn
3. Beijing Key Laboratory of Micro-nano Energy and Sensor, Beijing institute of Nanoenergy and Nanosystems, Chinese Academy of Sciences, Beijing 100083, China; tianjingjing@pumch.cn
4. Central Laboratory, Peking Union Medical College Hospital, Peking Union Medical College and Chinese Academy of Medical Sciences, Beijing 100730, China
5. Certara, Simcyp division, Sheffield 0114, UK; mailys.mendes@certara.com
* Correspondence: paul.hatton@sheffield.ac.uk (P.V.H.); zli@binn.cas.cn (Z.L.)
† These authors have contributed equally to the work.

Received: 21 May 2020; Accepted: 28 July 2020; Published: 14 August 2020

Abstract: Preventing the development of osteomyelitis while enhancing bone regeneration is challenging, with relatively little progress to date in translating promising technologies to the clinic. Nanoscale hydroxyapatite (nHA) has been employed as a bone graft substitute, and recent work has shown that it may be modified with silver to introduce antimicrobial activity against known pathogens. The aim of this study was to incorporate silver-doped nHA into electrospun scaffolds for applications in bone repair. Silver-doped nHA was produced using a modified, rapid mixing, wet precipitation method at 2, 5, 10 mol.% silver. The silver-doped nHA was added at 20 wt.% to a polycaprolactone solution for electrospinning. Bacteria studies demonstrated reduced bacterial presence, with *Escherichia coli* and *Staphylococcus aureus* undetectable after 96 h of exposure. Mesenchymal stem cells (MSCs) were used to study both toxicity and osteogenicity of the scaffolds using PrestoBlue® and alkaline phosphatase (ALP) assays. Innovative silver nHA scaffolds significantly reduced *E. coli* and *S. aureus* bacterial populations while maintaining cytocompatibility with mammalian cells and enhancing the differentiation of MSCs into osteoblasts. It was concluded that silver-doped nHA containing scaffolds have the potential to act as an antimicrobial device while supporting bone tissue healing for applications in orthopedic and dental bone surgery.

Keywords: electrospinning; antimicrobial; nano-hydroxyapatite; silver; toxicity; bone regeneration

1. Introduction

Preventing deep bone infections after surgery and enhancing bone regeneration reduces the time required for patient recovery and the costs associated with long hospital stays. Deep bone infections are very challenging to treat once established due to the difficulty of achieving a suitable antibiotic concentration in the affected area and ensuring all bacteria have been removed. Bacteria such as *Staphylococcus aureus* (*S. aureus*) and *Escherichia coli* (*E. coli*) are responsible for causing bone infections post-surgery [1,2], and the application of excessive antibiotics can lead to antimicrobial resistance. Antibacterial poly(methyl methacrylate) beads have been used routinely for treating infections in a reactive manner with additional disadvantages of requiring surgical removal after the treatment ends [3]. Silver has often been used to successfully treat bacterial and fungal infections [4,5] and can be delivered in a localized area. Metallic implants coated with silver have been investigated to prevent infections [6]. However, silver by itself does not enhance wound healing and can have a toxic effect on mammalian cells at higher concentrations [7,8].

To facilitate bone healing, surgeons currently employ hydroxyapatite in granular or paste form to stimulate bone regrowth. Hydroxyapatite is the mineral component of bone and can be manufactured synthetically and has been demonstrated to aid in bone regeneration [9]. Nanoscale hydroxyapatite (nHA) is an increasingly popular biomaterial due to its similarity to the hydroxyapatite in human bone and tooth enamel. Our group recently produced a bioinspired nHA material through a rapid mix preparation, which is slightly calcium deficient to better mimic biological apatite [10]. Using nHA-based medical devices to increase the regeneration rate of the bone after surgery could greatly decrease patient healing times. Unfortunately, despite its success in aiding bone regeneration, hydroxyapatite has no innate antimicrobial properties to prevent infection. Moreover, the implantation of medical devices provides an increased risk of bacterial colonization and eventual biofilm formation if left untreated [11]. Once a biofilm has been established it is very difficult to treat using antibiotics and even surgical intervention involves a high risk of reoccurrence [12].

There is growing interest in using a combination of silver and hydroxyapatite to produce a material with both antimicrobial/antifungal [13,14] and bone-regenerating properties [15–18]. Jiang et al. [19] produced foamed polyurethane scaffolds containing up to 40 wt.% silver phosphate particles (Ag_3PO_4) and nHA. These mold-formed scaffolds were porous and eluted silver for up to 22 days without toxic effects on the human osteosarcoma cell line MG63. Saravanan et al. [20] produced scaffolds using the freeze-drying technique incorporating nano-hydroxyapatite with nanoparticles of silver. Wilcock et al. [21] developed a novel, rapid mixing, wet precipitation method of producing bioinspired, nanoscale hydroxyapatite crystals doped with silver for antimicrobial action. A degradable scaffold could be placed against existing implants or used to hold packed bone substitutes (e.g., bone chips) into a bone defect.

The biomaterials' research community has been keenly interested in using electrospinning as a containment system for ceramic-derived materials including hydroxyapatite [22]. Electrospinning is a technique for producing fibrous scaffolds that roughly resemble the body's extracellular matrix. These can be made porous to allow cell thoroughfare or nanoscale to act as a barrier to cells, fulfilling a wide variety of applications. Ciraldo et al. [23] investigated three different manufacturing approaches using mesoporous glasses containing silver. These were glass coatings prepared using electrospraying, sol-gel methods, and directly spinning with polycaprolactone (PCL). For electrospinning, the authors were limited to 5% mesoporous glasses in PCL but did not investigate antimicrobial action. However, a recent paper by the same group demonstrated that the mesoporous glasses loaded with silver were antimicrobial [24]. Anjaneyulu et al. [25] manufactured doped hydroxyapatite through the sol-gel method and electrospun with poly(vinyl alcohol). Schneider et al. [26] electrospun tricalcium phosphate nanoparticles into poly(lactic-co-glycolic acid) (PLGA) scaffolds, looking at a single concentration of silver, but did not investigate mammalian toxicity of the scaffold. Furtos et al. [27] electrospun a PCL scaffold to contain both nHA and amoxicillin, which they tested against four bacterial strains and found that increasing the nHA content reduced the loading efficiency of the antibiotic.

A variety of polymers and calcium compounds have been used for producing these antimicrobial scaffolds. Different manufacturing technologies have also been employed to produce these as scaffolds. A relatively small number of the above publications have reported testing their material with primary mesenchymal stem cells (MSCs), which are key to understanding bone-healing mechanisms. Few of the published reports have investigated microbiological testing beyond agar diffusion assays with no time course elution antimicrobial testing.

Here we present the fabrication of novel, dual-function scaffolds that combine silver to impart antimicrobial properties and hydroxyapatite to enhance bone regeneration. Our laboratory previously developed a silver-doped nano-hydroxyapatite material formed by the wet precipitation method [21].

Here we show the incorporation of this innovative material into an electrospun scaffold alongside a thorough investigation of metal ion release, antimicrobial activity against clinically relevant bacterial strains, and in vitro biocompatibility using cultured mammalian MSCs. This research has provided an important proof-of-concept study that has the potential to underpin the development of a new generation of dual-action medical devices to address the growing challenges of deep bone infection and antimicrobial resistance.

2. Materials and Methods

2.1. Silver Nano-Hydroxyapatite ($Ca_{10}(PO_4)_6(OH)_2$) Manufacture via Wet Precipitation

This was based on our previously published methods [10,21]. In brief, calcium hydroxide (Sigma Aldrich, Dorset, UK) (50 mmol) was suspended in 500 mL deionized water (dH_2O) into which 0, 1, 2.5, or 5 mmol silver nitrate (Sigma Aldrich, UK) (corresponding to 0, 2, 5, or 10 mol%, respectively) was added and stirred at 400 rpm for 1 h on a hotplate (Fisher Scientific, Loughborough, UK) at 90 °C. Phosphoric acid (Sigma Aldrich, UK) (30 mmol) was dissolved in 250 mL dH_2O, poured into the calcium and silver preparation, and stirred for a further hour. The suspension was left to settle overnight, after which the supernatant was poured off and the silver-doped nHA suspension was washed with dH_2O (3 × 500 mL). The silver-doped nHA suspensions were dried at 60 °C and ground in an agate mortar and pestle.

2.2. Imaging the Silver Nano-Hydroxyapatite and Electrospun Fibers Using SEM

Dried nHA powder or a 5 × 5 mm section of electrospun scaffold was attached to a carbon tab and sputter gold coated and then imaged using a TESCAN VEGA3 SEM (Tescan, Brno, Czech Republic) with an accelerating voltage of 15 kV.

2.3. Imaging the Silver Nano-Hydroxyapatite Using Transmission Electron Microscopy (TEM) and Energy-Dispersive X-Ray Spectroscopy (EDX)

The nHA powder (0.1 g) was dispersed in ethanol in an ultrasonic bath (ThermoFisher, Shanghai, China) for 1 h and a drop was placed onto a copper mounting grid with carbon film and allowed to dry under ambient conditions. Samples were then imaged on a transmission electron microscope (Tecnai G2 F30, ThermoFisher, Shanghai, China) under standard imaging and EDX detecting elements Ca, P, Ag, and O. EDX settings used were 20 kV, takeoff 14.8, amp time 7 µs, and 126.5 eV resolution.

2.4. Analyzing Crystal Phases Present Using X-Ray Diffraction (XRD)

Phase analysis was performed on powdered samples using XRD (STOE IP, Darmstadt, Germany) with Cu Kα λ = 1.5418 Å radiation to determine the crystal structure of the materials. The diffractometer was operated at 40 kV and 35 mA, with a 2θ range of 20–50°. The following The International Centre for Diffraction Data Powder Diffraction (ICDD-PDF) cards were used for phase identification: 9-432 hydroxyapatite, 6-505 silver phosphate.

2.5. Electrospinning Polycaprolactone (PCL) and Incorporating the Ag–nHA

Polycaprolactone (average 80,000 Mn, Sigma, UK) was dissolved in a specific solvent and then propelled by an electric field onto a collector where polymer fibers formed as the solvent evaporated. Adding both the silver-doped and undoped nHA powder to the solution at 20 wt.% allowed it to be incorporated into the electrospun scaffold. PCL (1.5 g) was dissolved into a solvent mixture (90% Dichloromethane [DCM], 10% Dimethylformamide [DMF]) (Sigma Aldrich, UK) and stirred using a magnetic stirrer bar (5 mm bar, Sigma Aldrich, UK) for 2 h. The nHA powders were incorporated at 20 wt.% of the PCL component and added into the dissolved PCL solution. The polymer-ceramic-solvent mixtures were placed in an ultrasonic bath for 30 min, before stirring using a magnetic stirrer bar for 30 min to ensure even distribution of powder in the solution. A 1-mL Luer-Lok syringe (BD, EU) was loaded with 1 mL of the PCL/nHA solution and this was electrospun using a custom electrospinning rig described previously [28]. In total, 2 mL of solution at 1 mL/hr was used to fabricate each scaffold at 17 kV with a spinning gap between the needle and collector of 20 cm.

2.6. Accelerated Degradation Study of Electrospun Scaffolds and Release Profile

To investigate the rate at which nHA was released during scaffold degradation, an accelerated degradation study was performed over 30 days. Samples were weighed and added to a 5-mL solution of 0.1 M NaOH (Sigma Aldrich, UK) in dH$_2$O and placed in an oven at 37 °C. At each time point (days 0, 1, 7, 14, 21, and 28) material was also weighed and 0.5 mL of solution was removed, which was replaced with fresh solution. The solutions obtained were diluted to 5 mL using dH$_2$O.

2.7. Inductively Coupled Plasma (ICP) Measurement of Material and Degradation Product

To ascertain the presence and abundance of various elements within the sample, ICP was utilized. Solid samples (e.g., 0.25 g–0.5 g) were weighed into a glass tube and 9 mL of concentrated hydrochloric acid (Sigma Aldrich, UK) added and 3 mL of concentrated nitric acid (Sigma Aldrich, UK) added. The samples were then placed into the heating block and the temperature was gradually raised to 150 °C and maintained at this temperature for 30 min. After removal from the block and cooling, the samples were diluted to 50 mL with 1% nitric acid. Samples were diluted down to 1 in 100 for analysis. Samples were analyzed using an ICP-OES (Spectro-Ciros-Vision Optical Emission Spectrometer (Spectro, Kleve, Germany) using a procedure with a detection limit of at least 10 ppm.

2.8. Scaffold Preparation for Cell Culture

Disks of each electrospun scaffold were prepared and sterilized prior to culture with cells. Using a 13 mm cork borer, disks (13 mm) of each scaffold were punched out of the primary collector sheet. On the day of use, the culture scaffolds were disinfected with 70% ethanol (Sigma Aldrich, UK) in dH$_2$O for 30 min and then washed three times in dH$_2$O for 5 min.

2.9. Contact Bacterial Toxicity Test

Antimicrobial activity was investigated using a classical diffusion agar assay. Isolates of *E. coli* and *S. aureus* were purchased from the China Center of Industrial Culture Collection (CICC, Beijing, China): *E. coli*, CICC23657 and *S. aureus*, CICC10384. The bacterial suspensions at stationary phase were firstly obtained by culture in LB (Luria-Bertani) medium overnight at 37 °C in a 120-rpm shaker. Then 1 mL of the *E. coli* and *S. aureus* suspensions were pipetted from the overnight phase into another 100-mL flask containing 50 mL of fresh LB. The mixture was then cultured at 37 °C for another 6 h to obtain bacterial suspensions at exponential growth phase. After that, the bacterial suspension was centrifuged at 8000 rpm, washed, and resuspended in sterilized physiological saline to reach a concentration of 1×10^7 colony forming units (CFU)/mL. Then, 0.1 mL of the bacteria suspension with 1×10^7 CFU/mL was pipetted and plated on the agar plate uniformly. Three disks of the electrospun material (12 mm diameter) were placed on each Petri dish. These were incubated at 37 °C for 24 h. The zone of inhibition ring around the electrospun scaffolds was measured.

2.10. Antimicrobial Diffusion Toxicity Test

A disk of the electrospun scaffold (12 mm diameter) was immersed into the bacteria suspension (10 mL) containing 1 mL 10^7 CFU/mL and 9 mL sterile physiological saline in a 15-mL falcon tube. The falcon tubes were placed on a carousel (IKA, Staufen, Germany) and rotated 360° in the z axis at 30 rpm in an incubator at 37 °C. At the time points 3, 18, 24, 48, 72, and 96 h, 200 µL of phosphate buffered saline (PBS) was removed and serially diluted to 10^{-7} before 100 µL of diluted solutions were added to an agar plate and incubated for 24 h before imaging. The CFUs on plates were counted using the dilution, which was the highest with all bacterial colonies visible as single entities.

2.11. Rat Mesenchymal Stem Cell (MSC) Isolation and Passage

MSCs were isolated from four 4–5-week-old male Wistar rat femurs following the method previously described in literature [29,30]. Briefly, the femur was removed from the rats, opened at both ends, and flushed with media. This was then cultured to collect adherent cells, cells from all rats were intermixed, and cells were passaged twice before use. Cells were cultured in minimum essential medium (αMEM) supplemented with 10 U/mL penicillin (Sigma Aldrich, UK), 0.1 mg/mL streptomycin (Sigma Aldrich, UK), 20 mM alanyl-glutamine (Sigma Aldrich, UK), and 10% v/v fetal bovine serum (FBS; Biosera, Heathfield, UK). Cells were passaged 1:5 using trypsin-EDTA (ethylenediaminetetraacetic acid, Sigma Aldrich) and used below passage 10 for toxicity studies and below passage 5 for differentiation studies.

2.12. Noncontact Mammalian Cell Toxicity Study

Toxicity of elution from the electrospun scaffolds was measured by suspending the material over on cultured cells for 24 h. Both primary rat MSCs and the established fibroblast 3T3 cell line were used to test this material. Both cell lines were treated the same in the experiment apart from the rat MSCs were cultured in αMEM as described above and the 3T3 cells were cultured in high-glucose Dulbecco's Modified Eagle Medium (DMEM) with 10% FBS, both supplemented with 20 mM alanyl-glutamine (Sigma Aldrich, UK) and 0.1 mg/mL of streptomycin (Sigma Aldrich, UK). Cells were trypsinized and added to a 24-well plate (ThermoFisher, Loughborough, UK) at 30,000 cells per well in 1 mL of respective media. The cells were cultured for 24 h in an incubator at 37 °C and 5% CO_2. Media was replaced and a Transwell® insert (Millicell Hanging Cell Culture Insert, polyethylene terephthalate (PET) 0.4 µm, 24-well) was added to each well containing cells. In triplicate, the following were added to the Transwell® inserts: No scaffold (positive), 1 mL 70% industrial methylated spirit (IMS) (negative), 0% Ag nHA scaffold, 5% Ag nHA scaffold, and 10% Ag nHA scaffold. One mL of media was added to each Transwell® insert and plates were incubated for 48 h at 37 °C and 5% CO_2. After 48 h the cells within the wells were imaged and metabolic activity was assessed using PrestoBlue™ (ThermoFisher, UK). Media was removed and 700 µL of PrestoBlue™ (10% in complete media) was added to each well and the well plate was incubated for two hours. From the wells, 200 µL of PrestoBlue™ was added in triplicate to a 96-well plate. The fluorescence of the solutions was measured using a plate reader (FLx 800 Bio-Tek Instruments; Thermo Fisher Scientific, UK) with an excitation wavelength of 535 nm and an emission wavelength of 590 nm. The experiment was repeated four times (N = 4, n = 3).

2.13. Contact Biocompatibility Study on Rat MSCs

A direct-contact biocompatibility study was used to assess the toxicity of the scaffolds on cells over 21 days. Rat MSCs were used for this study. Disks of each scaffold were added in triplicate to a 24-well plate and 1 mL of media was added to each for 30 min (including a tissue culture plastic [TCP] control). Cells were trypsinized, suspended in media, and then 30,000 cells were added to each well. Well plates were agitated to aid even cell-seeding across the scaffold. Well plates were incubated for 24 h before cell metabolic activity was measured using PrestoBlue™. The scaffolds were transferred to a new well plate before PrestoBlue™ to avoid interference of cells growing on the tissue culture plastic surrounding the scaffold. PrestoBlue™ was measured at 4, 7, 14, and 21 days in culture. N = 2, n = 3.

2.14. Live and Dead Cell Staining

Cells were stained with two dyes to identify if the cell was alive or if the cell scaffold was disrupted (dead). Cells were stained using the LIVE/DEAD™ Viability/Cytotoxicity Kit (Thermofisher, UK) as per the instructions. In brief, cells were washed in PBS and then Calcien and Ethidium homodimer-1 (both at 5 µM concentration) were incubated with cells in media for 30 min. Samples were imaged using a fluorescent microscope (Axioplan 2, Zeiss, Jena, Germany) using the built-in filters for FITC (calcein) and TRITC (Ethidium homodimer-1).

2.15. Total Sample DNA and Alkaline Phosphatase Production Measurements

Alkaline phosphatase (ALP) is an early indicator of MSC differentiation to osteoblasts and can be measured using an ALP detection test. ALP was normalized to DNA to give a reading of alkaline phosphatase activity per cell regardless of cell number. Media was removed and samples washed three times with PBS. Cell digestion buffer was made up using deionized water and 1% triton X-100 (Sigma, UK). Samples were incubated at 4 °C for 1 h in the cell digestion buffer. The samples were then subjected to two freeze–thaw cycles (−20 °C to room temperature), which included solution agitation by pipetting when thawed to ensure thorough lysis of cells. After the second freeze–thaw cycle, an alkaline phosphatase diethanolamine detection kit was used to assess the alkaline phosphatase activity of the cultures. Then, 40 µL of cell culture lysate solution was placed in a 96-well plate in triplicate with 210 µL alkaline phosphatase reaction buffer. A standard curve of para-nitrophenyl (pNP) product was prepared with concentrations of 250, 100, 50, 10, and 0 nmole.ml^{-1} pNP in the alkaline phosphate reaction buffer. Then, 250 µL of the standard solutions were added in triplicate to the 96-well plate. The para-nitrophenylphosphate (pNPP) substrate solution was prepared by dissolving 6.6 mg pNPP per 1 mL of dH$_2$O. Then, 10 µL substrate was then added to each well using a multichannel pipette and the plate was incubated in the dark at 37 °C. The plate was monitored for color change and, at an appropriate time point, the absorbance at 405 nm was read, ensuring that the readings were within the range provided by the standard solutions. The amount of pNP in the sample wells was then calculated using the standard curve and used as a measure of ALP activity.

After a further freeze–thaw cycle, Quant-iT™ PicoGreen® dsDNA assay (ThermoFisher, UK) was used in accordance with the manufacturer's instructions to quantify the DNA content of the cultures. In detail, 50 µL of cell culture lysate was placed in a 96-well plate in triplicate along with 50 µL dH$_2$O. A standard curve of DNA was also prepared in triplicate with final concentrations of 1000, 500, 100, 50, 10, 5, and 0 ng.mL^{-1} when 100 µL of PicoGreen reagent was added to the 100 µL of standard solution in each well. Then, 100 µL of freshly made PicoGreen reagent was added to all the wells using a multichannel pipette. The plate was incubated in the dark for 4 min, after which the fluorescence was read using an excitation wavelength of 485 nm and an emission wavelength of 528 nm. The standard curve was plotted and used to convert the fluorescence units of the wells into the DNA content of the original cultures.

2.16. Statistical Methodology

Statistical analysis was conducted using ANOVA run with multiple comparisons using GraphPad Prism 8 (GraphPad Software, Inc., San Diego, CA, USA) with significance indicated with * when $p < 0.05$. No demarcation of lower p values was used as all were tested with $p = 0.05$.

3. Results

3.1. Analysis of Prepared Nanoscale Hydroxyapatite

Nano-hydroxyapatite was prepared via the rapid mixing technique previously published [10]. Using TEM, the nanoscale hydroxyapatite particles were visualized and contained crystalline and amorphous regions, typically less than 100 nm in size. Silver was found within the nanoscale hydroxyapatite particles, with the visible formation of nanoparticles of silver. Both materials showed the peaks associated with hydroxyapatite in XRD patterns. XRD showed the presence of silver phosphate peaks but only for the 10 mol.% silver-containing material (Figure 1a). Despite this, in both materials silver nanoparticles were visible under TEM; EDX confirmed that the darker regions found under the TEM contained silver. See Supplementary Figure S1 for EDX mapping of all elements found in sample Figure 1e. The size distribution of particles' diameter between nHA and nHA with 10 mol.% silver substituted in was found to be similar, with an average of 37 and 40 nm, respectively (Figure 1f).

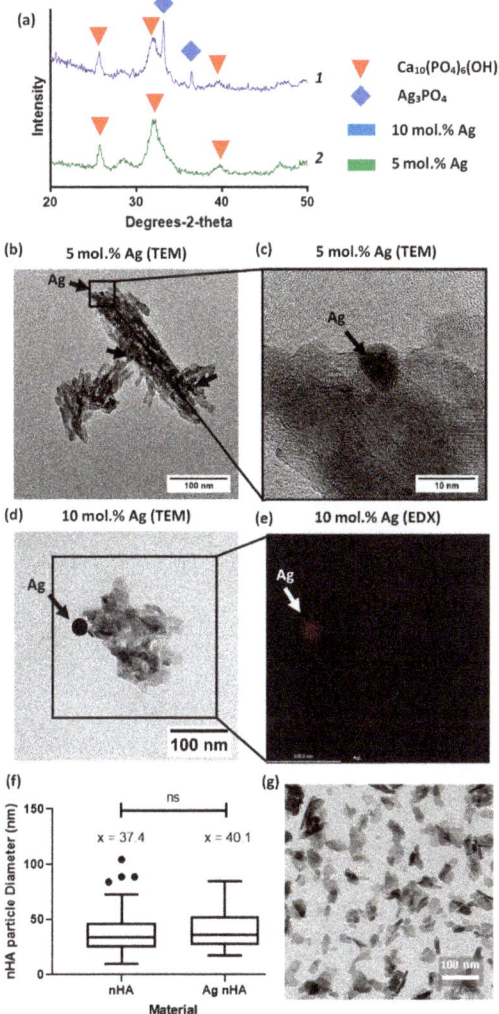

Figure 1. Evaluation of silver-containing nanoscale hydroxyapatite (**a**) XRD patterns of 5 and 10 mol.% silver doped nanoscale hydroxyapatite (Ag nHA), identifying peaks for hydroxyapatite $Ca_{10}(PO_4)_6(OH)_2$ and silver phosphate (Ag_3PO_4) phases within the sample. Silver phosphate was only identified in the 10 mol.% silver-containing material. Plot line 1 (blue) is 10 mol.% Ag nHA and plot line 2 (green) is 5 mol.% Ag nHA; (**b**) TEM image of nHA sample containing 5 mol.% silver. Silver deposits visible and highlighted by black arrows; (**c**) TEM at higher magnification on one of the silver deposit areas; (**d**) TEM image of nHA sample containing 10 mol.% silver. Larger silver deposit noted by black arrow; (**e**) Spatial EDX map of location of silver, mapped over the same location as that boxed in (**d**); (**f**) Particle diameter comparison between nHA and nHA with 10 mol.% silver shown as a Tukey box plot, ns = not significant; (**g**) TEM micrograph of nHA nanoscale particles.

3.2. Ensuring Encapsulation of nHA by Electrospun Fibers

Both silver and nondoped nHA were electrospun into polycaprolactone (PCL) scaffolds with a thickness of between 150 and 200 µm. SEM images showed incorporation of large particles in the fibers that were not observed in plain PCL scaffolds (Figure 2a). When digested, ICP analysis showed that plain PCL had very low levels of elements found within Ag-nHA (Figure 2b). The electrospun nHA showed the presence of calcium and phosphorus but not silver. Both electrospun scaffolds with silver-containing nHA showed the presence of silver alongside calcium and phosphorus (Figure 2b). This demonstrated that nHA was successfully incorporated into the electrospun fibers.

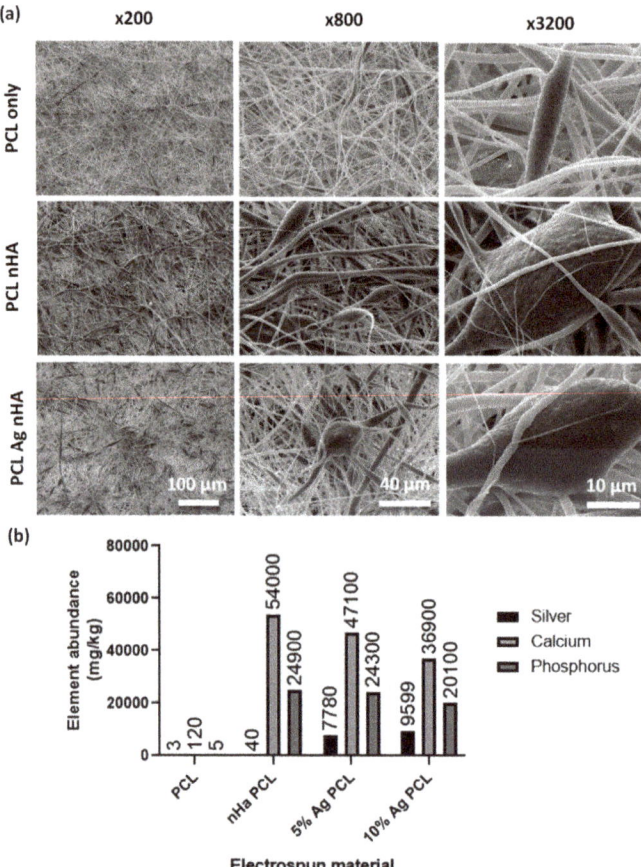

Figure 2. Scaffold fibers containing nHA (**a**) SEM micrographs of electrospun scaffold. Top: Polycaprolactone (PCL) scaffold. Middle: PCL scaffold containing nHA. Bottom: PCL scaffold containing 10 mol.% Ag nHA. Different magnifications were used to visualize encapsulation of nHA particles into the electrospun scaffold; (**b**) Inductively coupled plasma (ICP) analysis of dissolved electrospun scaffolds showing mg/kg of elements present. All % Ag materials are in mol.%.

3.3. Investigation of Silver Release and Scaffold Loss in Accelerated Degradation Study

An accelerated degradation assay was performed using sodium hydroxide at 37 °C to understand the degradation profile of the electrospun scaffolds and to observe release profiles (Figure 3). All scaffolds degraded at the same rate with the presence of nHA having no apparent effect on degradation profile (Figure 3a). Cumulative release of silver was observed for the full 30 days (Figure 3b). The plot of release rate (Figure 3c) indicated that in the early period of days 1 to 5 the release was from surface-bound origins, whereas from days 7 onwards the release was from silver within the polymer that was then released through degradation. Mass loss from degradation was nonexistent in the 1- to

7-day period and then rapidly degraded from there, which reinforced the conclusions drawn from the silver release rate.

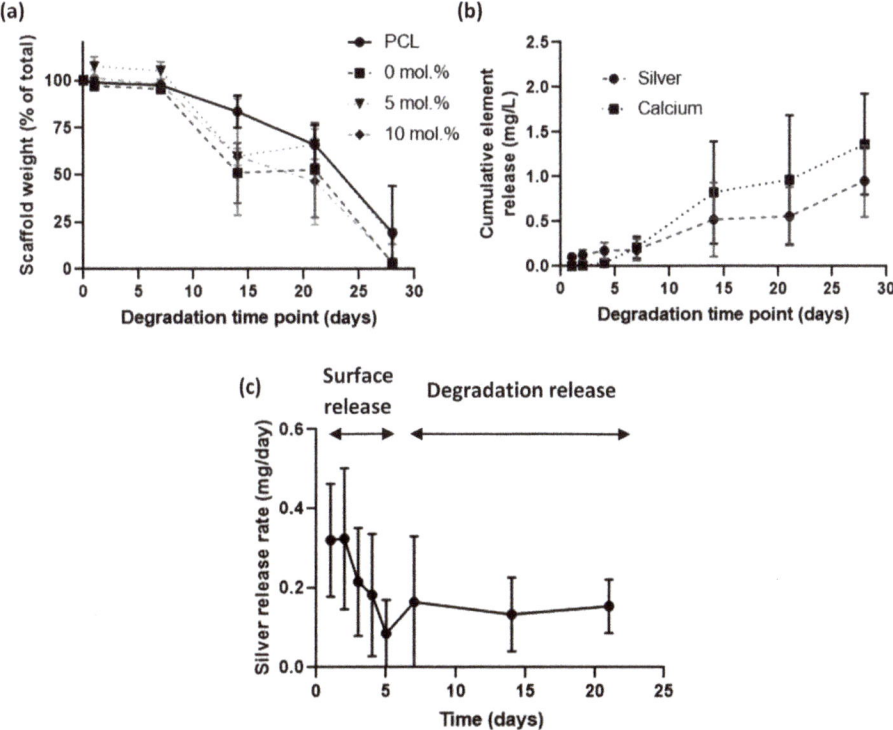

Figure 3. Accelerated degradation study of electrospun scaffolds containing nHA, over a 28-day period in 0.1 M sodium hydroxide. (**a**) Remaining weight over degradation time of different mol.% silver scaffolds; (**b**) Release profile of silver from an electrospun scaffold containing 10 mol.% Ag nHA over a 28-day period; (**c**) Plot of the silver release rate per day. All % Ag materials are in mol.%.

3.4. Antimicrobial Impact of Scaffolds

Both silver-containing scaffolds showed an antibacterial effect on both agar diffusion and on bacterial suspensions. PCL-only scaffolds and those with undoped nHA showed no innate antimicrobial action. Scaffolds containing silver of any concentration showed an antimicrobial response on agar diffusion (Figure 4a,b). Bacterial suspension cultures in PBS exposed to scaffolds for 24 h demonstrated the antimicrobial properties of the scaffolds containing silver. The 10 mol.% silver nHA was studied in a time course assay against bacteria and this showed a significant reduction in bacteria over time (Figure 4e,f). The scaffold reduced the viable bacteria count to undetectable levels by 48 h for *E. coli* and 96 h for *S. aureus*.

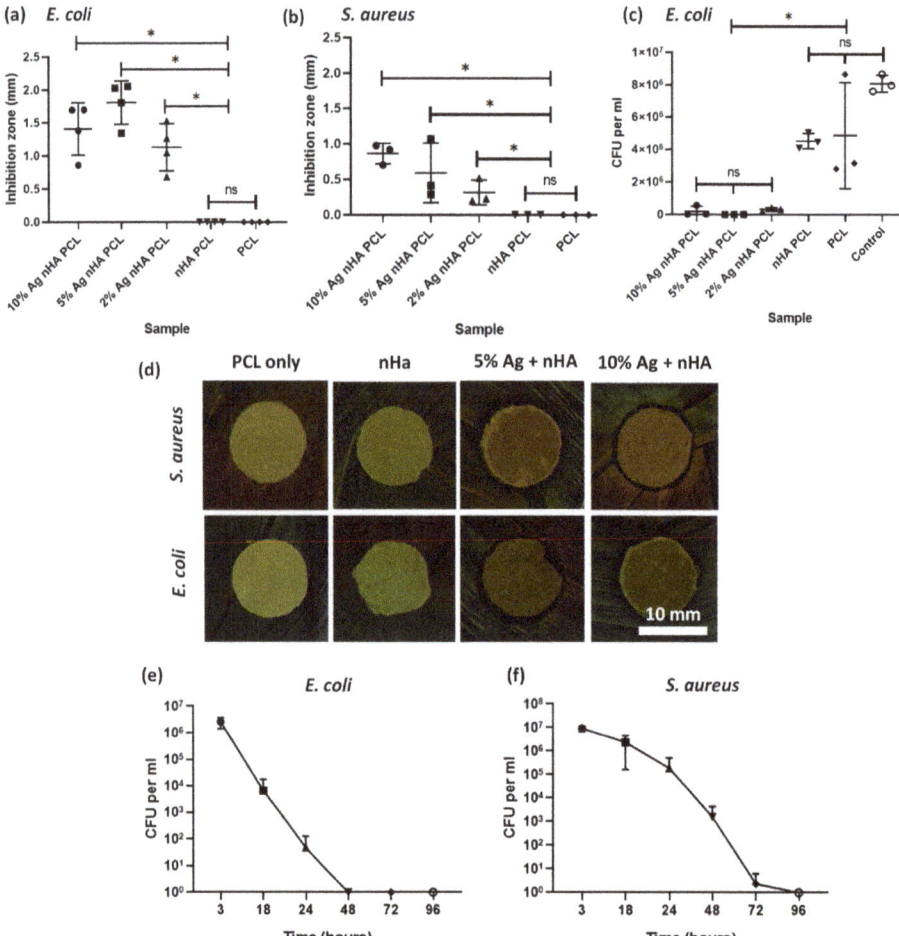

Figure 4. Antibacterial studies on *E. coli* and *S. aureus* bacteria using the electrospun scaffold. (**a**) Agar diffusion test using disks of the electrospun samples against *E. coli* bacteria. Graph displaying individual data plots with a line demarking the mean value; (**b**) Agar diffusion test using disks of the electrospun samples against *S. aureus*; (**c**) Disks of electrospun samples suspended in PBS containing bacteria, measuring viable colonies after 24 h; (**d**) Optical images of the area of inhibition ring surrounding a 12-mm disk of scaffold containing either 10 mol.% silver or plain PCL and tested against *S. aureus* and *E. coli*; (**e**) Bacteria count after *E. coli* are suspended in PBS and exposed to a disk of electrospun 10 mol.% silver nHA for 3, 18, 24, 48, 72, and 96 h; (**f**) Same as (**e**) but using *S. aureus* bacteria. All graphs show mean ± standard deviation. All % Ag materials are in mol.%.

3.5. Antimicrobial Impact of Scaffolds

Indirect toxicity assays found that higher silver concentrations were more toxic to both cell types while a longer, 21-day direct growth experiment showed a reduced toxic impact of these high silver samples on cells over time. When tested using fibroblast 3T3 cells, all scaffolds had a significantly higher viability than the negative (cytotoxic) control (Figure 5a). However, the scaffold containing 10 mol.% silver was also significantly different to the cell-only positive control, indicating some toxic impact. When the materials were tested on primary rat MSCs the toxic impact was more pronounced. The scaffold with 10 mol.% silver was significantly different to the cell-only positive control, but no statistical difference was found with the negative cell death control (Figure 5b). All the remaining materials were statistically different to the negative cell death controls. A longer, 21-day direct growth experiment found that cell metabolic activity for 5 mol.% silver-containing scaffolds were significantly increased compared to the PCL-only scaffolds (Figure 5c). The 10 mol.% silver-containing scaffolds reduced the metabolic activity when compared to 5 mol.% silver but did not show a toxic impact when compared to the PCL-only scaffold. Live cell morphology can be observed from live fluorescent images (Figure 5d) along with the proportion of live to dead cells. Cells appeared to have a more spread out morphology on the nHA silver-containing material than in the PCL-only control.

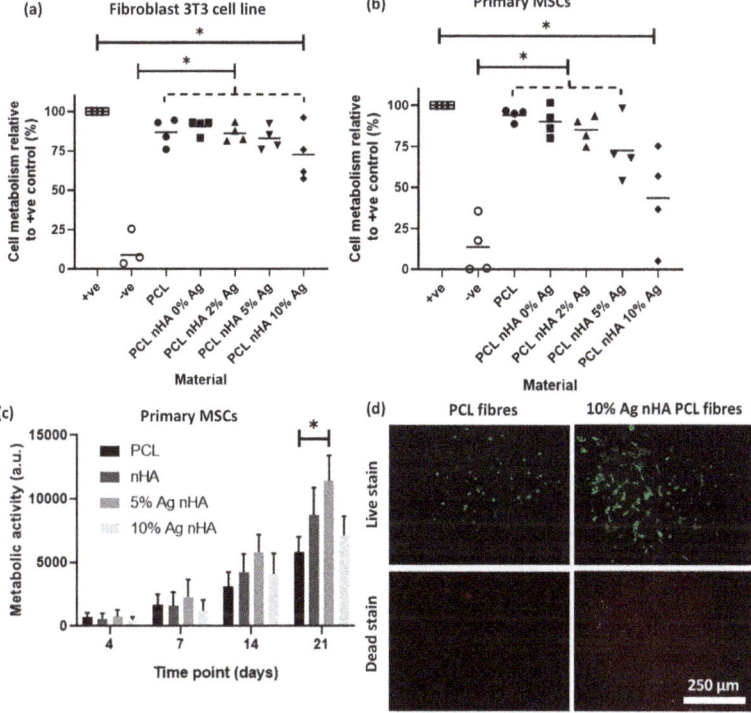

Figure 5. Cell toxicity from the electrospun scaffold with different concentrations of silver nHA assessed by cell metabolic activity via PrestoBlue™ assay. (**a**) A 48-h noncontact toxicity test with the scaffold in the media above a monolayer of 3T3 cells. Graph displaying individual data plots with a line demarking the mean value. Dotted line contains each scaffold under it to be compared to another condition; (**b**) A 48-h noncontact toxicity test with the scaffold in the media above a monolayer of primary rat MSCs. Graph displaying individual data plots with a line demarking the mean value. Dotted line without star contains each scaffold to be compare to another condition; (**c**) Cell metabolism after 21 days of direct culture with the cells seeded on top of the scaffold using metabolic activity at each time point. Statistics (ANOVA) conducted on the day 21 measurements; (**d**) Live/dead staining of MSCs cultured on electrospun scaffolds after 21 days.

3.6. MSC Osteogenic Differentiation Study by ALP Quantification

The impact of silver on the differentiation of primary MSCs down the osteoblast linage was measured after 21 days in culture by quantifying ALP content per unit of DNA (Figure 6). ANOVA showed a significant difference with 5 mol.% and the lower silver concentrations in regard to ALP production. Scaffolds with 10 mol.% silver are not significantly different to 5 mol.% silver scaffolds.

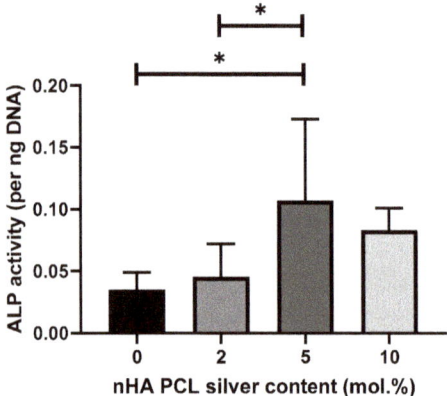

Figure 6. Cell differentiation potential of scaffolds containing different silver contents regarding ALP presence per unit of DNA after 21 days in culture with primary MSCs. Cells were cultured on top of the scaffolds. Graph displaying mean value ± standard deviation. N = 2, n = 3.

4. Discussion

Electrospun scaffolds with incorporated silver nano-hydroxyapatite are nontoxic to cells at a concentration that was antimicrobial, making them suitable for use as antimicrobial bone implants. In 2017, Anjaneyulu et al. [25] published a study looking into nanoscale electrospun fibers containing silver-doped hydroxyapatite and its hemocompatibility. Our study took a different direction, producing scaffolds with fibers to allow ingrowth of cells with larger pore sizes having positive impact on osteoblast cell growth [31]. Our study also included an in-depth microbiology study looking at both contact and noncontact tests. Investigating the impact of the scaffolds on bone regeneration via analyzing the toxicity on primary MSCs, for relevant cell toxicity, and 3T3 fibroblasts allowed toxicity comparison between labs.

The rapid manufacturing of nHA pioneered by Wilcock et al. [10] was replicated to produce silver-doped nHA powder. The characterization of nanoscale hydroxyapatite indicated it was comparable to that produced in previous studies. We were able to go one step further by confirming via TEM XRD that the denser (black) areas on the TEM contained silver. This had been hypothesized by the previous paper but not directly confirmed [21]. Under XRD we confirmed that the material was both hydroxyapatite and that it contained silver. TEM analysis showed that the nHA was in a nanoscale particulate form. The lack of an XRD-measurable silver phosphate phase in the 5 mol.% Ag nHA demonstrated that the silver had been incorporated into the nHA, either by substitution of silver into the HA crystal lattice for calcium or by doping of the silver ions onto the HA structure (i.e., not replacing the ions of the HA crystal). This result was expected for our method as our previous reports demonstrated a pure hydroxyapatite phase precipitated at 5 mol.% using this reaction [21].

The silver-doped nHA was successfully electrospun into PCL scaffolds. PCL is a popular material used in bone-related applications as its long-term degradation profile is suited to the slower healing rate of bone compared to soft tissues. Incorporating silver-containing tricalcium phosphate was reported by Schneider et al. [26], who found it effective against *E. coli* but did not test against the more common bone infection-related pathogen *S. aureus*. Jiang et al. [19] assessed antibacterial activity using both of these bacteria with a scaffold formed by in situ foaming, which allowed inclusion of silver/nHA up to 40% of total weight. This technique does not produce scaffolds, however, instead producing a monolithic block that would need to be carved to dimensions required during theater. Anjaneyulu et al. [25] investigated using both bacteria types and manufactured formed scaffolds but did not assess bone cell biocompatibility.

Our scaffolds degraded in an accelerated degradation study and showed a two-phase silver release system. Initially, a rapid burst of silver was released, decreasing in rate until day 7. During the same initial period of degradation, the polymer had not lost any mass. This rapid silver release is likely to arise from surface- (or near surface-) bound material. After the initial 7 days, the material lost mass and silver was released for the remainder of the experiment. Rothstein et al. described this in a model of an initial surface erosion system, which then transitions into bulk erosion [32]. The combination of the two allows for a rapid release when first implanted to clear immediate microorganisms and then a slower sustained release of the remainder of the implant life to prevent colonization from opportunistic bacteria, which may later attach to the device through a hematogenous route. Liu et al. [15] used electrochemical deposition of silver onto their electrospun fibers to provide an initial burst of silver over two days instead of a continual release. This would help prevent potential buildup of silver toxicity but would not prevent later infections from occurring. A continuous release of calcium and phosphorus was detected using ICP, which has been widely reported to stimulate bone regeneration [33]. An implanted material can act as a desirable location for bacteria to colonize with an infection requiring far fewer bacteria to start than in a normal tissue [34]. A continuous silver release would prevent opportunistic bacteria from infecting the implant at a later date. Due to the limited nature of our degradation method we cannot report on the exact amount of silver released per unit of time in vivo, as physiological conditions would impose a more complicated environment that may impact silver release. Despite not knowing the exact amount released, we showed that it is sufficient to prevent bacteria growth in non-accelerated studies.

Antimicrobial tests using both Gram-positive and Gram-negative bacteria showed that an antimicrobial response was found for 2, 5, and 10 mol.% incorporated silver-doped nHA. In a contact agar diffusion assay, both the silver-containing materials were found to produce an area of inhibition, whereas hydroxyapatite by itself does not [35]. Looking at the reduction of bacteria in solution, both bacteria were reduced beyond detection within 96 h. In our study we found silver was far more effective against *S. aureus* than against the *E. coli* bacteria. This was unexpected, as the thicker wall of the Gram-positive *S. aureus* is arguably better defended against the uptake of silver particles than thinner Gram-negative cell walls [36]. Resistance to silver can be acquired by bacteria [37], although it is a relatively broad spectrum medicant. This is significant, as *S. aureus* is among the most common pathogens found in bone infections [1,2]. Recent studies have shown that treating the motile *S. aureus* in an infected 3D tissue model is more challenging than treating a monolayer bacteria assay [38]. Therefore, this would be the logical progression for the testing of this material. One side effect of silver is that it can also be toxic to mammalian cells at elevated concentrations, which we investigated with our scaffolds.

The toxicity of the scaffolds was related to both the silver concentration and on the cell type being tested. Against the standard cell line of mouse fibroblast 3T3 cells there was a small indication of toxicity with scaffolds containing 10 mol.% silver. When using a primary rat MSCs there was clear toxicity detected with this highest concentration of silver. This shows the importance of not just using a single cell line to determine toxicity of a material but also of the correct choice of relevant cells. For a more in-depth review of the different modes of action of silver nanoparticles on various cell types, see Zhang et al. [39]. Using 3T3 cells is a useful method of comparing relative toxicities between different laboratories as they are well known and utilized in international standards for cytotoxicity (ISO 10993-5). The primary MSCs are more difficult to standardize between labs and are, therefore, harder to use to compare toxicity. They do give a more realistic insight on what levels of toxicity might be harmful in the environment the medical device is intended for. From the results presented, 5 mol.% silver-doped material at 20 wt. mol.% to PCL had an antimicrobial impact without being toxic to mammalian MSCs.

Using a longer study of 21 days, the 10 mol.% silver nHA scaffolds maintained a viable and growing cell population on par with the plain PCL material. Hydroxyapatite appeared to boost the metabolic activity of the MSCs by day 21 with the combination of 5 mol.% silver and hydroxyapatite, resulting in even higher levels of cell activity. Increasing the amount of silver to 10 mol.% did not increase the trend of increased proliferation. The initial high release rate from the material was potentially too toxic for the cells early on but, once the release level had fallen to a consistent 50 mol.%, surviving cells would then have been able to proliferate. After this period, the metabolic activity of cell on the 10 mol.% silver nHA scaffold increased at a similar growth rate to other materials. This is highly advantageous in a wound-healing application as it would allow an initially high burst to clear bacteria, which then lowers to allow the mammalian cells to begin the healing process.

Hydroxyapatite is widely reported to increase MSC differentiation into osteoblasts [40–42] so we focused on investigating the impact of silver on this process showing increased ALP production. Huge disparity can be found in the literature on whether silver has a significant impact on differentiation of MSCs. Some authors find that silver nanoparticles have no impact on MSC differentiation [43,44]. Other papers report on an impact of silver nanoparticles on MSC differentiation down the osteogenic lineage [45,46]. Conversely, some authors argue that silver particles can cause adipogenic differentiation [47] of MSCs, whereas others have found no impact at all on adipogenic differentiation [48]. One aspect that varies between these papers is the size of the silver nanoparticles, which might be a reason between the widely different reporting of the impact of silver on differentiation. Powers et al. found impact on neurodevelopment from silver on zebrafish PC12 cells, which was very dependent on both silver composition and silver nanoparticle size [49]. Another variation between papers is the wide range of different MSCs that are tested for differentiation, from both different tissue and animal origins. As even MSCs from different tissue origins vary in behavior, it is not surprising that different conclusions have been reported by multiple studies [50]. We found that 5 mol.% silver had the highest ALP production at day 21, although there was no significant difference to 10 mol.% silver. It is possible that either diminishing returns or an already reached optimum concentration of silver is the reason for no further increase at 10 mol.% silver. Another explanation is that the slightly more toxic impact of 10 mol.% silver offsets any additional osteogenic stimulation. Our study demonstrated that silver-doped nHA incorporated into PCL scaffolds may have a positive effect regarding MSC differentiation along an osteoblastic lineage due to increased levels of ALP production.

Limitations to the study include no mapping of elements within the electrospun membrane containing nHA. However, Figure 2b does show evidence that the material is present within the electrospun membranes but the precise location of the nHA is not known. Another limitation is that, while using bacteria typical of infections in bone, neither of the isolates was taken from a patient bone infection. Another limitation on the microbiology side is the simple nature of the tests used; it should be considered that (along with other disparities) osteomyelitis is associated with a hypoxic environment, which was not reflected by our experiments. Further work is needed to address these issues, including the use of a bone infection model to more accurately reproduce the conditions found in vivo. In a recent study we found that a material with antimicrobial impact on planktonic bacteria and simple biofilm models was then not effective on *P. aeruginosa* in tissue-engineered skin models, believed to be due to the bacteria motility [24]. This is something we intend to investigate in future work.

This research found that silver nHA containing scaffolds has the potential to act as an antimicrobial scaffold while stimulating bone tissue regeneration. The scaffolds degrade slowly and release a continuous amount of silver into the local environment over time. They are nontoxic to mammalian cells at a concentration that is toxic to both Gram-negative and Gram-positive bacterial strains. Investigating the impact of silver on MSC differentiation has suggested that higher values of silver promote MSC differentiation.

5. Conclusions

Silver-doped nanoscale hydroxyapatite may be incorporated into electrospun PCL scaffolds to fabricate a device that is both nontoxic to cultured mammalian cells and capable of inhibiting the growth of pathogenic microorganisms. The scaffold demonstrated antibacterial activity against both Gram-positive and Gram-negative bacterial strains commonly found in bone infections. The material was shown to release silver in an initial burst followed by a more sustained bulk degradation release. Silver appeared to enhance MSC differentiation down an osteogenic path when used in nontoxic concentrations. Silver-doped nHA containing scaffolds exhibited antimicrobial activity while stimulating bone tissue regeneration, which shows promise as a dual-action medical device to address unmet clinical needs in orthopedic and dental bone surgery.

Supplementary Materials: The following are available online at http://www.mdpi.com/2079-4983/11/3/58/s1, Figure S1: EDX elemental mapping of 10 Ag mol.% samples detecting oxygen, phosphorus, calcium and silver. TEM image of the analysed material is also shown for comparison along with an overlay image of all elements in a single frame.

Author Contributions: Conceptualization, T.E.P., R.S., P.V.H., and I.O.; Formal analysis, M.d.S.M.; Funding acquisition, T.E.P, R.S., and P.V.H.; Investigation, T.E.P. and J.T.; Methodology, T.E.P., J.T., and C.J.H.; Resources, P.V.H., Z.L., and I.O.; Supervision, R.S., Z.L., and I.O.; Validation, C.J.H.; Writing—original draft, T.E.P. and J.T.; Writing—review & editing, R.S., C.J.H., P.V.H., Z.L., and I.O. All authors have read and agreed to the published version of the manuscript.

Funding: This work was funded by MeDe Innovation (the UK EPSRC Centre for Innovative Manufacturing in Medical Devices, grant number EP/K029592/1). UK-China lab placement was made possible by Newton Trust funding, with further support from the National Natural Science Foundation of China (51673029 and 61875015), Beijing Talent Fund (2016000012113ZK34), and Beijing Municipal Health Commission (PXM2018026275000001, BMC2018-4).

Acknowledgments: Mass spectrometry analyses were performed by the biOMICS/chemMS Facility of the Faculty of Science Mass Spectrometry Centre at the University of Sheffield.

Conflicts of Interest: The authors have no competing or conflicted interests to declare.

References

1. Bouza, E.; Muñoz, P. Micro-organisms responsible for osteo-articular infections. *Best Pract. Res. Clin. Rheumatol.* **1999**, *13*, 21–35. [CrossRef] [PubMed]
2. Malone, M.; Bowling, F.L.; Gannass, A.; Jude, E.B.; Boulton, A. Deep wound cultures correlate well with bone biopsy culture in diabetic foot osteomyelitis. *Diabetes Metab Res. Rev.* **2013**, *29*, 546–550. [PubMed]
3. Stone, P.A.; Mousa, A.Y.; Hass, S.M.; Dearing, D.D.; Campbell, J.R., II; Parker, A.; Thompson, S.; AbuRahma, A.F. Antibiotic-loaded polymethylmethacrylate beads for the treatment of extracavitary vascular surgical site infections. *J. Surg.* **2012**, *55*, 1706–1711. [CrossRef] [PubMed]
4. Burduşel, A.-C.; Gherasim, O.; Grumezescu, A.; Mogoantă, L.; Ficai, A.; Andronescu, E. Biomedical applications of silver nanoparticles: An up-to-date overview. *Nanomaterials* **2018**, *8*, 681. [CrossRef]
5. Li, Z.; Tang, H.Y.; Yuan, W.W.; Song, W.; Niu, Y.S.; Yan, L.; Yu, M.; Dai, M.; Feng, S.Y.; Wang, M.H.; et al. Ag nanoparticle-ZnO nanowire hybrid nanostructures as enhanced and robust antimicrobial textiles via a green chemical approach. *Nanotechnology* **2014**, *25*. [CrossRef]
6. Chou, T.G.R.; Petti, C.A.; Szakacs, J.; Bloebaum, R.D. Evaluating antimicrobials and implant materials for infection prevention around transcutaneous osseointegrated implants in a rabbit model. *J. Biomed. Mater. Res. Part A* **2010**, *92*, 942–952. [CrossRef]
7. Ivask, A.; Kurvet, I.; Kasemets, K.; Blinova, I.; Aruoja, V.; Suppi, S.; Vija, H.; Käkinen, A.; Titma, T.; Heinlaan, M. Size-dependent toxicity of silver nanoparticles to bacteria, yeast, algae, crustaceans and mammalian cells in vitro. *PLoS ONE* **2014**, *9*, e102108. [CrossRef]
8. Milić, M.; Leitinger, G.; Pavičić, I.; Zebić Avdičević, M.; Dobrović, S.; Goessler, W.; Vinković Vrček, I. Cellular uptake and toxicity effects of silver nanoparticles in mammalian kidney cells. *J. Appl. Toxicol.* **2015**, *35*, 581–592. [CrossRef]
9. Thorwarth, M.; Schultze-Mosgau, S.; Kessler, P.; Wiltfang, J.; Schlegel, K.A. Bone regeneration in osseous defects using a resorbable nanoparticular hydroxyapatite. *J. Oral Maxillofac. Surg.* **2005**, *63*, 1626–1633. [CrossRef]
10. Wilcock, C.J.; Gentile, P.; Hatton, P.V.; Miller, C.A. Rapid Mix Preparation of Bioinspired Nanoscale Hydroxyapatite for Biomedical Applications. *J. Vis. Exp.* **2017**. [CrossRef]
11. Schierholz, J.; Beuth, J. Implant infections: A haven for opportunistic bacteria. *J. Hosp. Infect.* **2001**, *49*, 87–93. [CrossRef] [PubMed]
12. Zimmerli, W.; Sendi, P. Orthopaedic biofilm infections. *APMIS* **2017**, *125*, 353–364. [CrossRef] [PubMed]

13. Ciobanu, C.S.; Iconaru, S.L.; Le Coustumer, P.; Constantin, L.V.; Predoi, D. Antibacterial activity of silver-doped hydroxyapatite nanoparticles against gram-positive and gram-negative bacteria. *Nanoscale Res. Lett.* **2012**, *7*, 324. [CrossRef] [PubMed]
14. Predoi, D.; Iconaru, S.L.; Predoi, M.V. Bioceramic layers with antifungal properties. *Coatings* **2018**, *8*, 276. [CrossRef]
15. Liu, F.; Wang, X.; Chen, T.; Zhang, N.; Wei, Q.; Tian, J.; Wang, Y.; Ma, C.; Lu, Y. Hydroxyapatite/silver electrospun fibers for anti-infection and osteoinduction. *J. Adv. Res.* **2020**, *21*, 91–102. [CrossRef]
16. Singh, R.P.; Singh, M.; Verma, G.; Shukla, S.; Singh, S.; Singh, S. Structural Analysis of Silver Doped Hydroxyapatite Nanopowders by Rietveld Refinement. *Trans. Indian Inst. Met.* **2017**, *70*, 1973–1980. [CrossRef]
17. Ciobanu, C.S.; Iconaru, S.L.; Chifiriuc, M.C.; Costescu, A.; Le Coustumer, P.; Predoi, D. Synthesis and Antimicrobial Activity of Silver-Doped Hydroxyapatite Nanoparticles. *BioMed. Res. Int.* **2013**, *2013*, 916218. [CrossRef]
18. Jelínek, M.; Weiserová, M.; Kocourek, T.; Zezulová, M.; Strnad, J. Biomedical properties of laser prepared silver-doped hydroxyapatite. *Laser Phys.* **2011**, *21*, 1265–1269. [CrossRef]
19. Jiang, J.; Li, L.; Li, K.; Li, G.; You, F.; Zuo, Y.; Li, Y.; Li, J. Antibacterial nanohydroxyapatite/polyurethane composite scaffolds with silver phosphate particles for bone regeneration. *J. Biomater. Sci. Polym. Ed.* **2016**, *27*, 1584–1598. [CrossRef]
20. Saravanan, S.; Nethala, S.; Pattnaik, S.; Tripathi, A.; Moorthi, A.; Selvamurugan, N. Preparation, characterization and antimicrobial activity of a bio-composite scaffold containing chitosan/nano-hydroxyapatite/nano-silver for bone tissue engineering. *Int. J. Biol. Macromol.* **2011**, *49*, 188–193. [CrossRef]
21. Wilcock, C.; Stafford, G.; Miller, C.; Ryabenkova, Y.; Fatima, M.; Gentile, P.; Möbus, G.; Hatton, P. Preparation and antibacterial properties of silver-doped nanoscale hydroxyapatite pastes for bone repair and augmentation. *J. Biomed. Nanotechnol.* **2017**, *13*, 1168–1176. [CrossRef] [PubMed]
22. Zhang, S.; Jiang, G.; Prabhakaran, M.P.; Qin, X.; Ramakrishna, S. Evaluation of electrospun biomimetic substrate surface-decorated with nanohydroxyapatite precipitation for osteoblasts behavior. *Mater. Sci. Eng. C* **2017**, *79*, 687–696. [CrossRef] [PubMed]
23. Ciraldo, F.; Liverani, L.; Gritsch, L.; Goldmann, W.; Boccaccini, A. Synthesis and characterization of silver-doped mesoporous bioactive glass and its applications in conjunction with electrospinning. *Materials* **2018**, *11*, 692. [CrossRef] [PubMed]
24. Zheng, K.; Balasubramanian, P.; Paterson, T.E.; Stein, R.; MacNeil, S.; Fiorilli, S.; Vitale-Brovarone, C.; Shepherd, J.; Boccaccini, A.R. Ag modified mesoporous bioactive glass nanoparticles for enhanced antibacterial activity in 3D infected skin model. *Mater. Sci. Eng. C* **2019**, *103*, 109764. [CrossRef]
25. Anjaneyulu, U.; Priyadarshini, B.; Grace, A.N.; Vijayalakshmi, U. Fabrication and characterization of Ag doped hydroxyapatite-polyvinyl alcohol composite nanofibers and its in vitro biological evaluations for bone tissue engineering applications. *J. Sol Gel Sci. Technol.* **2017**, *81*, 750–761. [CrossRef]
26. Schneider, O.D.; Loher, S.; Brunner, T.J.; Schmidlin, P.; Stark, W.J. Flexible, silver containing nanocomposites for the repair of bone defects: Antimicrobial effect against E. coli infection and comparison to tetracycline containing scaffolds. *J. Mater. Chem.* **2008**, *18*, 2679–2684. [CrossRef]
27. Furtos, G.; Rivero, G.; Rapuntean, S.; Abraham, G.A. Amoxicillin-loaded electrospun nanocomposite membranes for dental applications. *J. Biomed. Mater. Res. Part B Appl. Biomater.* **2017**, *105*, 966–976. [CrossRef]
28. Paterson, T.E.; Beal, S.N.; Santocildes-Romero, M.E.; Sidambe, A.T.; Hatton, P.V.; Asencio, I.O. Selective laser melting–enabled electrospinning: Introducing complexity within electrospun membranes. *Proc. Inst. Mech. Eng. Part h J. Eng. Med.* **2017**, *231*, 565–574. [CrossRef]
29. Maniatopoulos, C.; Sodek, J.; Melcher, A.H. Bone formation in vitro by stromal cells obtained from bone marrow of young adult rats. *Cell Tissue Res.* **1988**, *254*, 317–330. [CrossRef]
30. Santocildes-Romero, M.E.; Crawford, A.; Hatton, P.V.; Goodchild, R.L.; Reaney, I.M.; Miller, C.A. The osteogenic response of mesenchymal stromal cells to strontium-substituted bioactive glasses. *J. Tissue Eng. Regen. Med.* **2015**, *9*, 619–631. [CrossRef]
31. Oh, S.H.; Park, I.K.; Kim, J.M.; Lee, J.H. In vitro and in vivo characteristics of PCL scaffolds with pore size gradient fabricated by a centrifugation method. *Biomaterials* **2007**, *28*, 1664–1671. [CrossRef] [PubMed]
32. Rothstein, S.N.; Federspiel, W.J.; Little, S.R. A unified mathematical model for the prediction of controlled release from surface and bulk eroding polymer matrices. *Biomaterials* **2009**, *30*, 1657–1664. [CrossRef] [PubMed]

33. LeGeros, R.Z. Calcium phosphate-based osteoinductive materials. *Chem. Rev.* **2008**, *108*, 4742–4753. [CrossRef] [PubMed]
34. Costerton, J.; Montanaro, L.; Arciola, C.R. Biofilm in implant infections: Its production and regulation. *Int. J. Artif. Organs* **2005**, *28*, 1062–1068. [CrossRef]
35. Kim, T.; Feng, Q.L.; Kim, J.; Wu, J.; Wang, H.; Chen, G.; Cui, F. Antimicrobial effects of metal ions (Ag+, Cu2+, Zn2+) in hydroxyapatite. *J. Mater. Sci. Mater. Med.* **1998**, *9*, 129–134. [CrossRef]
36. Feng, Q.L.; Wu, J.; Chen, G.; Cui, F.; Kim, T.; Kim, J. A mechanistic study of the antibacterial effect of silver ions on Escherichia coli and Staphylococcus aureus. *J. Biomed. Mater. Res.* **2000**, *52*, 662–668. [CrossRef]
37. Li, X.-Z.; Nikaido, H.; Williams, K.E. Silver-resistant mutants of Escherichia coli display active efflux of Ag+ and are deficient in porins. *J. Bacteriol.* **1997**, *179*, 6127–6132. [CrossRef]
38. Paterson, T.E.; Bari, A.; Bullock, A.J.; Turner, R.; Montalbano, G.; Fiorilli, S.; Vitale-Brovarone, C.; MacNeil, S.; Shepherd, J. Multifunctional Copper-Containing Mesoporous Glass Nanoparticles as Antibacterial and Proangiogenic Agents for Chronic Wounds. *Front. Bioeng. Biotechnol.* **2020**, *8*, 246. [CrossRef]
39. Zhang, X.-F.; Shen, W.; Gurunathan, S. Silver Nanoparticle-Mediated Cellular Responses in Various Cell Lines: An in Vitro Model. *Int. J. Mol Sci.* **2016**, *17*, 1603. [CrossRef]
40. Venugopal, J.; Rajeswari, R.; Shayanti, M.; Low, S.; Bongso, A.; Giri Dev, V.R.; Deepika, G.; Choon, A.T.; Ramakrishna, S. Electrosprayed hydroxyapatite on polymer nanofibers to differentiate mesenchymal stem cells to osteogenesis. *J. Biomater. Sci. Polym. Ed.* **2013**, *24*, 170–184. [CrossRef]
41. Guo, H.; Su, J.; Wei, J.; Kong, H.; Liu, C. Biocompatibility and osteogenicity of degradable Ca-deficient hydroxyapatite scaffolds from calcium phosphate cement for bone tissue engineering. *Acta Biomater.* **2009**, *5*, 268–278. [CrossRef] [PubMed]
42. Lin, L.; Chow, K.L.; Leng, Y. Study of hydroxyapatite osteoinductivity with an osteogenic differentiation of mesenchymal stem cells. *J. Biomed. Mater. Res. Part A* **2009**, *89*, 326–335. [CrossRef] [PubMed]
43. Liu, X.; He, W.; Fang, Z.; Kienzle, A.; Feng, Q. Influence of silver nanoparticles on osteogenic differentiation of human mesenchymal stem cells. *J. Biomed. Nanotechnol.* **2014**, *10*, 1277–1285. [CrossRef] [PubMed]
44. Samberg, M.E.; Loboa, E.G.; Oldenburg, S.J.; Monteiro-Riviere, N.A. Silver nanoparticles do not influence stem cell differentiation but cause minimal toxicity. *Nanomedicine* **2012**, *7*, 1197–1209. [CrossRef]
45. Qin, H.; Zhu, C.; An, Z.; Jiang, Y.; Zhao, Y.; Wang, J.; Liu, X.; Hui, B.; Zhang, X.; Wang, Y. Silver nanoparticles promote osteogenic differentiation of human urine-derived stem cells at noncytotoxic concentrations. *Int. J. Nanomed.* **2014**, *9*, 2469. [CrossRef]
46. Zhang, R.; Lee, P.; Lui, V.C.; Chen, Y.; Liu, X.; Lok, C.N.; To, M.; Yeung, K.W.; Wong, K.K. Silver nanoparticles promote osteogenesis of mesenchymal stem cells and improve bone fracture healing in osteogenesis mechanism mouse model. *Nanomed. Nanotechnol. Biol. Med.* **2015**, *11*, 1949–1959. [CrossRef]
47. Sengstock, C.; Diendorf, J.; Epple, M.; Schildhauer, T.A.; Köller, M. Effect of silver nanoparticles on human mesenchymal stem cell differentiation. *Beilstein J. Nanotechnol.* **2014**, *5*, 2058–2069. [CrossRef]
48. He, W.; Kienzle, A.; Liu, X.; Müller, W.E.; Elkhooly, T.A.; Feng, Q. In vitro effect of 30 nm silver nanoparticles on adipogenic differentiation of human mesenchymal stem cells. *J. Biomed. Nanotechnol.* **2016**, *12*, 525–535. [CrossRef]
49. Powers, C.M.; Badireddy, A.R.; Ryde, I.T.; Seidler, F.J.; Slotkin, T.A. Silver nanoparticles compromise neurodevelopment in PC12 cells: Critical contributions of silver ion, particle size, coating, and composition. *Environ. Health Perspect.* **2010**, *119*, 37–44. [CrossRef]
50. Ribeiro, A.; Laranjeira, P.; Mendes, S.; Velada, I.; Leite, C.; Andrade, P.; Santos, F.; Henriques, A.; Grãos, M.; Cardoso, C.M.P.; et al. Mesenchymal stem cells from umbilical cord matrix, adipose tissue and bone marrow exhibit different capability to suppress peripheral blood B, natural killer and T cells. *Stem Cell Res. Ther.* **2013**, *4*, 125. [CrossRef]

© 2020 by the authors. Licensee MDPI, Basel, Switzerland. This article is an open access article distributed under the terms and conditions of the Creative Commons Attribution (CC BY) license (http://creativecommons.org/licenses/by/4.0/).

Article

Effect of Novel Antibacterial Composites on Bacterial Biofilms

Rayan B. Yaghmoor [1,2], Wendy Xia [3], Paul Ashley [4], Elaine Allan [5] and Anne M. Young [3,*]

[1] Department of Biomaterials and Tissue Engineering/Department of Microbial Diseases, UCL Eastman Dental Institute, London, NW3 2QG, UK; rayan.yaghmoor.18@ucl.ac.uk
[2] Department of Restorative Dentistry, Umm Al-Qura University, College of Dental Medicine, Makkah 24381, Saudi Arabia
[3] Department of Biomaterials and Tissue Engineering, UCL Eastman Dental Institute, London NW3 2QG, UK; wendy.xia2011@yahoo.co.uk
[4] Unit of Paediatric Dentistry, UCL Eastman Dental Institute, London WC1E 6DE, UK; p.ashley@ucl.ac.uk
[5] Department of Microbial Diseases, UCL Eastman Dental Institute, London NW3 2QG, UK; e.allan@ucl.ac.uk
* Correspondence: anne.young@ucl.ac.uk

Received: 30 June 2020; Accepted: 27 July 2020; Published: 1 August 2020

Abstract: Continuing cariogenic bacterial growth demineralizing dentine beneath a composite filling is the most common cause of tooth restoration failure. Novel composites with antibacterial polylysine (PLS) (0, 4, 6, or 8 wt%) in its filler phase were therefore produced. Remineralising monocalcium phosphate was also included at double the PLS weight. Antibacterial studies involved set composite disc placement in 1% sucrose-supplemented broth containing *Streptococcus mutans* (UA159). Relative surface bacterial biofilm mass ($n = 4$) after 24 h was determined by crystal violet-binding. Live/dead bacteria and biofilm thickness ($n = 3$) were assessed using confocal laser scanning microscopy (CLSM). To understand results and model possible in vivo benefits, cumulative PLS release from discs into water ($n = 3$) was determined by a ninhydrin assay. Results showed biofilm mass and thickness decreased linearly by 28% and 33%, respectively, upon increasing PLS from 0% to 8%. With 4, 6, and 8 wt% PLS, respectively, biofilm dead bacterial percentages and PLS release at 24 h were 20%, 60%, and 80% and 85, 163, and 241 µg/disc. Furthermore, initial PLS release was proportional to the square root of time and levelled after 1, 2, and 3 months at 13%, 28%, and 42%. This suggested diffusion controlled release from water-exposed composite surface layers of 65, 140, and 210 µm thickness, respectively. In conclusion, increasing PLS release initially in any gaps under the restoration to kill residual bacteria or longer-term following composite/tooth interface damage might help prevent recurrent caries.

Keywords: dental composite; antibacterial; antibiofilm; polylysine

1. Introduction

Dental caries is a main public health issue worldwide [1]. In Europe, it affects almost 100% of adults and 20–90% of 6-year-old children [2]. Mercury silver amalgam was for nearly 200 years the most commonly used permanent direct dental filling material in the EU. European legislation, however, in July 2018, aligned with the Minamata convention on mercury reduction to ban the use of dental amalgam for children under 15 years old and pregnant or breastfeeding women. Complete dental amalgam phase out is planned for 2030 [3]. Mercury concerns in combination with growing demand for more conservative and aesthetic restorations have made the resin composite the new main direct long-term restorative of choice [4,5]. Composites, however, have higher failure rate with the most common cause being recurrent (secondary) caries [5–8]. This is largely due to the tendency of resin composite to accumulate bacterial biofilms more than amalgam, which is not only responsible for recurrent caries but also periodontal problems [9].

Dental plaque biofilms consist of microorganisms in an exopolymer matrix (EPM) [10]. These biofilms are three dimensional, spatially-organized microbial communities that show properties as a unit [11]. Metagenomics studies revealed that up to 19,000 bacterial species could be found in a mature dental plaque biofilm [12,13]. Shift in composition of the dental biofilm towards more acidogenic and aciduric bacteria leads to caries. Dentinal caries lesions contain primarily acid-producing and Gram-positive species, with *Streptococcus mutans* being particularly high [14,15]. Their lowering of pH enhances solubility and dissolution of apatite in enamel [16]. Subsequent demineralization of dentine activates enzymes that, over time, slowly hydrolyze remaining collagen [16].

Previously, tooth restoration involved removal of the soft, highly-infected as well as the underlying harder, acid-affected dentine through drilling. Tooth restoration became a major problem during the global SARS-Cov-2 pandemic in 2020, as drilling increases aerosol production that could spread the viral infection [17,18]. Additionally, acid-affected dentine is remineralizable [19]. Furthermore, current recommendations for management of cavitated caries advocate controlling the lesion [19]. This could be better achieved through minimally invasive treatment and bacterial inactivation instead of total demineralized dentine removal [19]. This would reduce the need for drilling, which both dentists and patients consider a major benefit [20].

Bacterial inactivation by composite restorations can be achieved by cavity sealing reducing nutrient ingress [19]. Composite shrinkage during placement or long-term cyclical loading, however, can damage this seal. Residual bacteria or bacterial leakage from surface plaque with sugars from food would then enable biofilm formation in gaps at the composite/tooth interface. Immediate release of antibacterial agents in response to interface damage could potentially reduce this risk. In combination with remineralising agents to seal any gaps, this could provide the time required for the tooth to self-heal.

(Poly-ε-lysine) (ε-PL) (PLS) is a naturally created antibacterial homopolypeptide of 25–30 L-lysine residues that is stable in basic and acidic environments as well as at high temperatures [21,22]. Moreover, it is biodegradable, water-soluble, edible, non-toxic, and has broad-spectrum antibacterial properties [21,23]. Polylysine has therefore recently been incorporated into resin composites with remineralising agents [22,24]. Chemical setting kinetics, volumetric stability, polylysine release kinetics, remineralising features, mechanical properties, and wear have already been investigated [22,24]. One recent study has also studied material effects on bacteria [25]. This previous investigation, however, was restricted to formulations with low PLS concentrations and planktonic *S. mutans* with none of the sucrose required for biofilm maturation [25]. Biofilms, in general, have been shown to be more resistant to antimicrobial agents compared to the same bacterium growing in planktonic culture [26].

The aim of this new study was to evaluate if higher levels of PLS and remineralising agents within composites have the potential to prevent residual bacteria in cavities depositing and forming biofilms early on the composite surface. This is achieved through investigation of new composites incubated with *S. mutans* and sucrose. PLS release kinetics are also provided to help explain the observations.

2. Materials and Methods

2.1. Composite Components

Resin-composite formulations were prepared as in previous studies [25] but with higher antibacterial and remineralising agents by combining two phases: liquid and powder. Liquid components included urethane dimethacrylate (UDMA) as the main base monomer and the initiator camphorquinone (CQ). Both were from DMG (Hamburg, Germany). Poly (propylene glycol) dimethacrylate (PPGDMA) was added to improve fluidity and polymerization kinetics. This and the adhesion promoting monomer, 4-methacryloyloxyethyl trimellitic anhydride (4-META), were from Polysciences (PA, Warrington, UK) [25].

The filler consisted primarily of silane-treated radiopaque strontium aluminosilicate glass filler of 7 and 0.7 µm from DMG. These were combined with silica nano-glass particles (Azelis, Hertford, UK). Monocalcium phosphate monohydrate (MCPM) (Himed, Old Bethpage, NY, USA) and polylysine (PLS) (Handary, Brussels, Belgium) were added as remineralising and antibacterial agents, respectively. Particle SEM images are provided in a previous work [24].

2.2. Paste and Solid Sample Disc Composition and Preparation

Four semi-fluid paste formulations were prepared using a powder to liquid weight ratio of 3:1. The clear liquid phase consisted of UDMA (72 wt%), PPGDMA, (24 wt%), 4-META (3 wt%), and CQ (1 wt%). This monomer system polymerizes effectively upon light activation without need of an amine activator [24]. Then, 7 µm glass, 0.7 µm glass, and silica were combined in the weight ratio 6:3:1 [24]. This maximized particle packing and provided flow and material handling characteristics approved by clinical focus groups. MCPM (0, 8, 12, and 16 wt%) and PLS (0, 4, 6, and 8 wt%) were added to this powder with the MCPM:PLS weight ratio fixed at 2:1.

Clear liquids were prepared by completely dissolving solid 4-META and CQ in PPGDMA followed by UDMA addition. The powder components were mixed first at 3500 rpm for 10 s using a Speedmixer (DAC600.2 CM51, Synergy Devices Ltd., High Wycombe, UK). Monomers and powders were then mixed at 3500 rpm for 40 s.

To prepare solid disc specimens (1 mm thick, 9.4 mm diameter, and mass 0.13 g), pastes were pressed within metal circlips sandwiched between two acetate sheets then light cured. Each surface was cured using a light-emitting diode (LED) with a wavelength of 450–470 nm and power of 1100–1300 mW/cm^2 (Demi Plus, Kerr Dental, Bioggio, Switzerland) in contact with the acetate sheet with 4 overlapping irradiation cycles according to ISO:4049 [27]. Then, 40 s per cycle was used to ensure maximum room temperature monomer conversion of 72% [24]. After curing, composite discs were removed from the moulds, and excess composite was trimmed.

2.3. Biofilm Formation on Composite Discs

S. mutans (UA159) was cultured on brain heart infusion (BHI) Agar (CM1136 by OXOID, Thermo Fisher Scientific, Loughborough, UK) at 37 °C in an atmosphere of air enriched with 5% CO_2 for 72 h. A single colony was inoculated into 10 mL BHI broth (CM1135 by OXOID) and allowed to grow for 18 h at 37 °C in air enriched with 5% CO_2. This was diluted to obtain a bacterial density of 8×10^7 CFU/mL, which was confirmed by colony forming unit (CFU) counting and calibrated optical density (caused by absorbance and light scatter) at 595 nm (OD_{595}). The OD_{595} was measured using a spectrophotometer (Biochrom WPA CO8000, Cambridge, UK).

Specimen discs were placed randomly in 24 well plates (Greiner bio-one, Cellstar, Kremsmunster, Austria) and sterilized through top and bottom surface exposure to UV light for 40 min. Biofilms were grown on each composite disc by adding 25 µL of 8×10^7 CFU/mL inoculum (containing 2×10^6 CFU) with 1 mL BHI broth (OXOID) containing 1% (w/v) sucrose. The well plate was incubated at 37 °C statically in air enriched with 5% CO_2. After 24 h, the medium was carefully removed. The resultant biofilms were washed twice with 1 mL of sterile phosphate buffered saline (PBS) and fixed by exposure to 1 mL methanol during 15 min.

2.4. Crystal Violet (CV) Biofilm Assay

A crystal violet (CV) assay was used to evaluate relative biofilm mass on composite surfaces [28]. Biofilms were produced on 4 separate days; 3 composite discs per formula were tested each day requiring 12 discs per formula in total.

The procedure involved addition of 1 mL of aqueous 0.1% CV (Pro-Lab Diagnostics, Bromborough, UK) to each disc for 5 min then washing with 1 mL of PBS twice. After drying, bound CV was solubilized in 1 mL of 30% aqueous acetic acid. Once fully dissolved, the resultant CV solutions were diluted to enable absorbance (OD_{595}) determination, and the OD_{595} of the undiluted solutions were calculated and reported.

2.5. Live/Dead and Biofilm Thickness Assessment

Live/dead assay and biofilm thickness were assessed using confocal laser scanning microscopy (CLSM, Lasersharp 2000) (Radiance 2100, Biorad, Hercules, CA, USA) with a 40× magnification wet lens. Biofilms were prepared and examined on different days in 3 independent assays with one disc per formulation each day.

Then, 20 µL of LIVE/DEAD stain (Viability Kit, Thermofisher Scientific, Loughborough, UK) and 10 mL of PBS were added to the biofilms on discs within wells. Green and red images of live versus dead bacteria were obtained from near the biofilm centre (260 × 260 µm^2). In the following, red and green images are shown overlapped. For analysis, the individual red or green images were converted to black (background) and white (stained bacteria), and the percentage area that was white was determined using ImageJ software. The ratio of live/dead bacteria was obtained by dividing the percentage of white areas from the green versus red images; this was then converted to percentages. Additionally, movement of the microscope stage in the z direction enabled determination of biofilm thickness. Average results at three random points within the image area were obtained. Normalised thickness was obtained by dividing biofilm thickness by that for the control with no PLS on each given day.

2.6. Polylysine Release

To quantify kinetics of PLS release, 6 composite discs of each formulation of known total mass were stored in plastic pots containing 5 mL of deionized water at room temperature (RT). A total of 2 discs per pot enabled sufficient release for ease of detection and a sample repetition number of $n = 3$. The discs were transferred to new plastic containers with fresh 5 mL of deionized water at the following time points: 6 h, 24 h, 3 days, and at 1, 2, 3, 4, 5, 6, 7, 8, 9, 10, 12, 14, and 18 weeks.

The concentration of PLS that had leached into the deionized water was determined using a ninhydrin assay that was modified from previously published work [29]. It involved adding 1 mL of 0.5% ethanolic Ninhydrin (Sigma-Aldrich, Gillingham, UK) solution to boiling tubes containing 4 mL of sample storage solution or aqueous polylysine solutions of known concentration. After vortex-mixing for 10 s, 1.5 g (± 0.1) of marble chips (BDH, Poole, England, UK) were added to each tube. All tubes were covered with aluminium foil and placed in a boiling water bath to accelerate a ninhydrin/PLS reaction. After 15 min, reactions were quenched by rapid cooling to RT.

Following sedimentation, supernatant liquids were analysed using a spectrophotometer (Thermo Spectronic Unicam UV 500 Spectrophotometer, Loughborough, England with Vision pro™ software). The absorbance was recorded between 400–700 nm to ensure curves were consistent and free of background scattering (for example due to residual non sedimented marble chips). Absorbances at 570 nm (A_{570}) for known polylysine concentrations of 0, 4, 10, 16, 20, 40, 60, 80, and 100 ppm were used to establish a calibration graph (A_{570} = 0.023/ppm, Pearson correlation coefficient, RSQ = 0.96). This was then used to determine polylysine concentration in each storage solution and cumulative release in µg/disc versus time. Cumulative PLS (%) was also calculated using:

$$\% \text{ PLS release} = \frac{100[\sum_0^t R_t]}{w_c} \qquad (1)$$

where R_t is the amount of released PLS (g) at time t, w_c is the amount of *PLS* (g) in the composite disc.

2.7. Statistical Analysis

The 95% confidence intervals (95%CI) (equal to 2 SD/SQRT(n) where SD and SQRT are standard deviation and square root) are provided as error bars in Figures. Average percentage standard deviations (%SD) were 7%, 9%, and 15% and sample numbers n = 4, 3, and 3 for crystal violet (CV) absorbance, normalised biofilm thickness, and PLS release studies, respectively. Significance ($p < 0.05$) was determined using one-way Analysis of Variance (ANOVA) followed by pairwise comparisons (Kruskal–Wallis test). Excel was used for curve fitting and to calculate the Pearson correlation coefficient (RSQ).

3. Results

3.1. Biofilm Mass

CV adsorption and subsequent absorbance indicated an inverse relationship between the amount of biofilm mass and the concentration of PLS in composite formulations (Figure 1). All PLS-containing formulations (4, 6, and 8 wt%) had biofilms with significantly less CV adsorption than the control formula with no PLS. In addition, the formula containing 8% PLS had significantly less CV adsorption compared to the material containing 4% PLS.

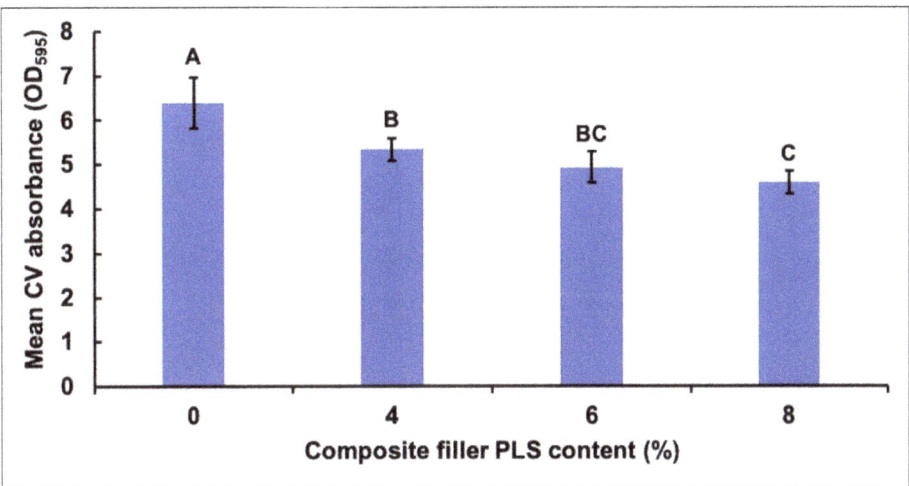

Figure 1. Mean ± (95%CI, n = 4) absorbance (OD$_{595}$) due to crystal violet (CV) adsorbed by biofilms on the four composite formulations with 0, 4, 6, or 8 wt% polylysine (PLS) in filler phase (error bars are for 4 repetitions on different days with an average result of 3 specimens per formulation each day). The absorbance values were 10 times those measured in the spectrometer to account for dilution. Composite formulations with the same uppercase letter/s above the bars were not significantly different at $p < 0.05$.

3.2. Live/Dead S.Mutans

Confocal microscopy images showed primarily dead bacteria with 8% PLS but live bacteria with 0% and 4% PLS levels. The 6% PLS samples had both live and dead bacteria (Figure 2). Quantification of live versus dead stain confirmed that formulae containing 6% and 8% PLS had significantly more dead *S. mutans* and significantly less live *S. mutans* than the 4% and the control formulations (Figure 3). Additionally, live and dead percentages gave linear trends versus concentration between 4% and 8% PLS (Figure 3). Additionally, live and dead percentages gave linear trends versus concentration between 4% and 8% PLS (Figure 3).

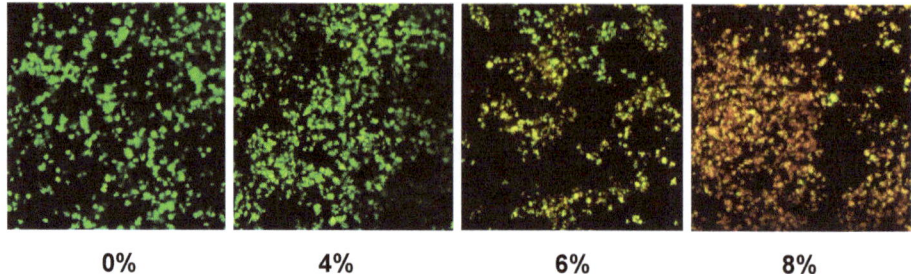

Figure 2. Example confocal microscopy images of live/dead stained biofilms formed on composite formulations containing different percentages of PLS added to the filler phase. Live bacteria stain green whereas dead bacteria are red.

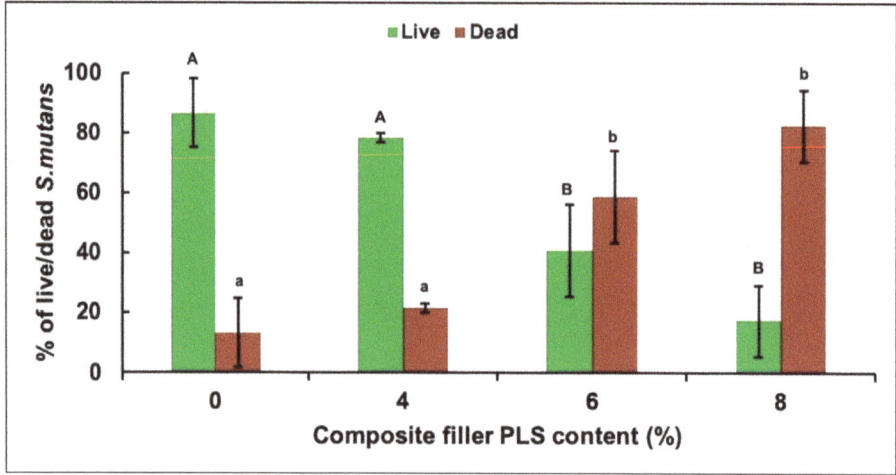

Figure 3. Mean ± (95%CI, $n = 3$) percentage of live and dead *S. mutans* in biofilm formed on discs of the four composite formulations with 0, 4, 6, or 8 wt% PLS in filler phase at 24 h (error bars are for 3 repetitions on different days with 1 specimen per formulation per day). Composite formulations with the same uppercase/lowercase letter/s were not significantly different at $p < 0.05$.

3.3. Biofilm Thickness

Due to the variability in biofilm thickness arising from day to day, the data were normalised by biofilm thickness of the control on the given day. All PLS-containing formulations (4, 6, and 8%) then showed significantly reduced normalised biofilm thickness compared to the control (Figure 4). In addition, the formula containing 8% PLS had significantly less biofilm thickness compared to that containing 4% PLS. Additionally, a linear decline in biofilm thickness was observed with increasing PLS concentration (Figure 4).

When biofilm mass and biofilm thickness were both normalised by results for the control, the data within experimental error overlapped. Additionally, both showed a linear decline in normalised mean with increasing PLS concentration (Figure 4). Linear declines of up to 28% and 33% in the biofilm mass and the biofilm thickness, respectively, were observed with increasing PLS concentration to 8% (Figure 4).

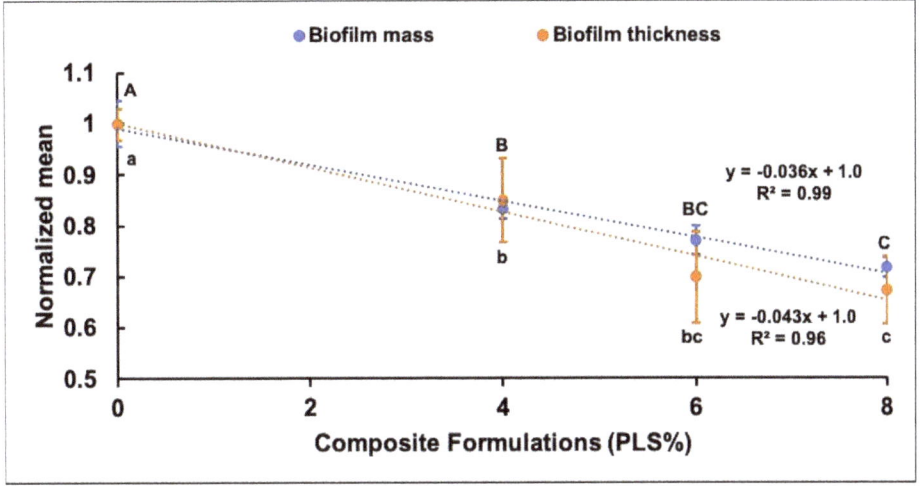

Figure 4. Scatter graph showing the relationship between normalised mean ± (95%CI) biofilm thickness ($n = 3$) and biofilm mass ($n = 4$) (Y-axis) for different composite formulations (X-axis). For biofilm mass and thickness, the formulations with the same uppercase letters above and lowercase letters below the bars, respectively, were not significantly different at $p < 0.05$.

3.4. Polylysine Release

Percentage PLS release was proportional to the square root of time up to 1, 2, or 3 months with 4%, 6%, and 8% PLS, respectively, but then levelled (Figure 5). The overall general linear model revealed a significant difference in the percentage of PLS released between different composite formulations at every time point. Cumulative release in µg/disc at 24 h and final percentage release both increased with raising the PLS filler content (Figure 6). Within error, final percentage release was 60/350 times the 24 h µg/disc release for all formulations (Figure 6). Final percentage releases were 42%, 28%, and 13% with 8%, 6%, and 4% PLS in the filler phase, respectively. Release at 24 h was 241, 163 and 85 µg/disc, respectively (Figure 6). At 6 h, the release was half these values.

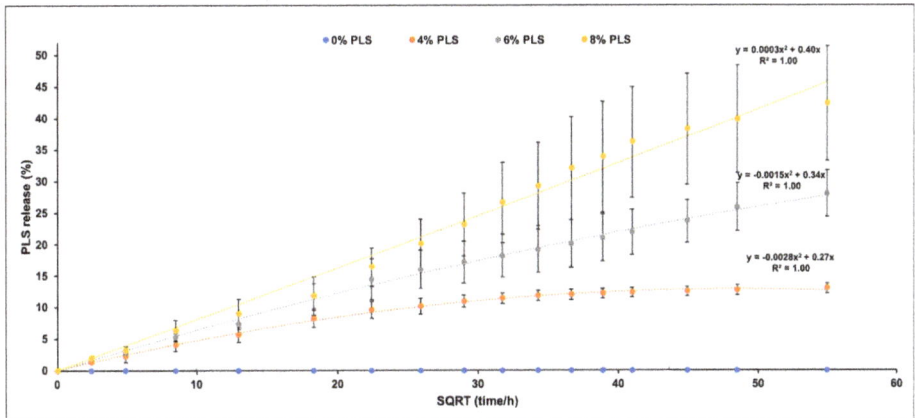

Figure 5. Scatter graph of mean ± (95%CI, $n = 3$) percentage PLS release (Y-axis) versus the square root of time/hours (X-axis) for each composite formula up to 18 weeks (the legend shows the composite filler PLS %). All final release values were significantly different for different PLS levels at $p < 0.05$.

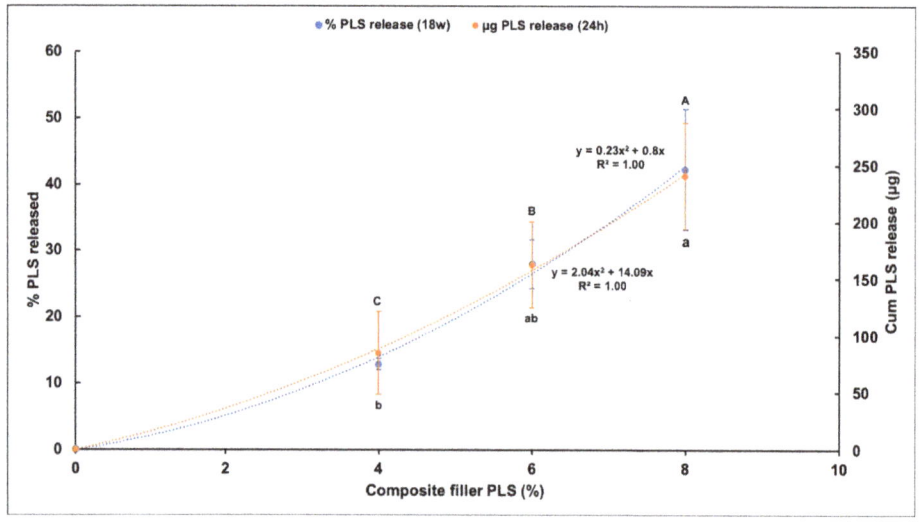

Figure 6. Scatter graph of mean ± (95%CI, $n = 3$) cumulative percentage PLS release at 18 weeks (left Y-axis) and mean ± (95%CI) cumulative PLS release in (µg/disc) at 24 h (right Y-axis) for different composite formulations (X-axis). The formulations with the same uppercase letter/s above or lowercase letter/s below the bars that indicate 18 week and 24 h data, respectively, were not significantly different at $p < 0.05$.

4. Discussion

In the current study, the antibacterial effects of composites formulated with simultaneously increasing levels of antibacterial PLS and remineralising MCPM was evaluated. MCPM/PLS weight ratio was fixed at 2. From relative densities and molecular weights, this gave approximately comparable volumes of the two components and the moles of MCPM and the lysine monomer units in all formulations. Previous studies have shown that, in composites, these may work synergistically at promoting surface mineralizing properties [22]. The aim of this study, however, was to assess how formulations with these two components affect *S. mutans* biofilm deposition and survival on the composite surface. Subsequent polylysine release rates were used to explain the large differences observed between materials with different PLS and MCPM levels.

Polylysine is effective against a wide range of bacteria, although concentrations required tend to be much higher than for conventional antibiotics or agents such as chlorhexidine. A recent study showed that increase in PLS concentration above 250 ppm caused reduction in survival rate, enzymatic activity, and adenosine triphosphate (ATP) levels of *Staphylococcus aureus*. It also increased cell collapse and membrane permeability. Additionally, it destroyed the peptidoglycan component of the cell wall and caused changes in the levels of several metabolites [23]. Another study compared the action of PLS of 750 and 74 ppm on Gram-positive and Gram-negative bacteria, *Listeria innocua*, and *E. coli*, respectively. In both cases, PLS exerted its activity by binding to negatively charged phospholipid head groups in the lipid bilayer of bacterial membranes and destabilized them, thus increasing permeability. In the case of *E. coli*, PLS also binds to and destroys the lipopolysaccharide component of the outer membrane [30].

The minimum inhibitory concentration of PLS against *S. mutans* with the standard initial inoculum level of 5.5×10^5 CFU/mL was found to be 20 µg/mL [31], but this increases with the higher bacterial levels seen in the oral cavity. *S. mutans* is a Gram-positive facultative anaerobic bacterium that was first identified as a cariogenic bacterium in 1924 [32]. A previous longitudinal study reported that 96.6% of patients aged between 6–30 have *S. mutans* present in their oral cavity [33]. In a recent study, composite formulations that released 93 ppm PLS upon the first 24 h of water immersion were able to reduce an initial inoculum level of 8×10^5 CFU/mL down to 10^2 CFU/mL [25]. In this present study, the formulation with 4% PLS with a similar level of PLS release, however, was unable to prevent *S. mutans* biofilm growth. This was likely due to the addition of sucrose enhancing the rate of bacterial growth in addition to enabling production of a protective extracellular matrix (ECM) [34]. The observed linear decline in biofilm mass with increasing PLS suggests it may have been causing a linear decline in the numbers of viable bacteria in the surrounding medium. Alternatively, this effect could have been independent of PLS bactericidal activity and due instead to biofilm matrix disruption caused by PLS [35].

Since CV may stain the ECM in addition to the bacterial cell wall, it provides a measure of the combined live and dead bacteria. In the present study, a live/dead stain was additionally used to determine the fraction of the bacterial population that were live. In regard to the live/dead stain, propidium iodide staining (red) revealed a damaged cell membrane. A statistically significant increase in the percentage of dead *S. mutans* was apparent with the composites containing 6% and 8% PLS compared to the 4% and the control formulations. This study, in conjunction with the previous work [25], suggests that the concentrations of PLS and MCPM in the filler need to be increased significantly to be effective when sucrose is added.

Considerable variation in biofilm thickness detected by confocal microscopy was observed between experiments, although this was not apparent in the biofilm mass measured by CV staining. This may reflect differences in biofilm structural organization between experiments. Nonetheless, when the data were normalised to the control in individual experiments, a consistent decline in both biofilm mass and thickness with increasing PLS concentration was evident.

The most widely used antibacterial agent in dental composites has been chlorhexidine (CHX). Previous studies, however, showed development of antibacterial resistance toward CHX [36] as well as severe [37] and sometimes fatal hypersensitivity reactions [38]. Furthermore, release from hydrophobic composites [39,40] can be restricted and can require addition of hydrophilic components to promote water sorption, which decreases strength. Whilst, in standard tests, minimum inhibitory concentrations of PLS are generally much higher (20 ppm) [31] than those of CHX (<1 ppm), the new studies suggest its very much higher aqueous solubility enables faster release from the composite surface.

In order to quantify the kinetics of PLS release from each formulation, a ninhydrin assay was used. Ninhydrin is used for the detection of ammonia, primary or secondary amines, amino acids, peptides, and proteins [41]. The reaction between NH_2-group present in protein, peptide, or amino acid with ninhydrin (originally yellow color) produces a coloured ninhydrin chromophore (deep purple colour) (Ruhemann's purple λ_{max} 570 nm) [41]. PLS is one of the peptides that can be detected by the ninhydrin reaction. Under appropriate conditions, the intensity of the purple colour is proportional to the NH_2-group (PLS concentration). Ninhydrin reaction is extremely sensitive to low concentrations of protein, peptide, or amino acid [41].

Unlike previous studies [22], there was no initial burst release of PLS in this new study. This may have been a consequence of higher filler content in the earlier work (80 wt%) leading to particles making more direct contact with the surface. In the current study, the percentage of PLS released was instead proportional to the square root of time. This is expected from the Higuchi equation for drug release from thin layers [42].

$$\frac{P_t}{P_\infty} = \frac{4}{L}\sqrt{\frac{Dt}{\pi}} \tag{2}$$

P_t and P_∞ are the percentage *PLS release* at time t and at infinity, respectively. D is a diffusion coefficient and L the thickness of the layer top and bottom of the sample from which the drug is released. If all the PLS is released from this layer but none is released from the remaining material bulk, layer thickness would be expected to be given by:

$$L = \frac{P_\infty h}{100} \tag{3}$$

where h is half the sample thickness and is equal to 0.5 mm. As maximum percentage polylysine releases are 13%, 28%, and 42%, the top and the bottom surface layers depleted of PLS are expected from this expression to be 65, 140 and 210 µm with 4%, 6%, and 8% PLS in the filler, respectively. These are approximately 2, 4, and 6 times the diameters of the PLS and the MCPM particles. The increase in thickness is likely due to higher levels of these hydrophilic components upon their release, generating pores in the surface that enable PLS release from deeper within the specimens. Doubling the concentration of PLS in the filler therefore more than doubles the release from the composite.

Combining Equations (2) and (3) gives:

$$P_t = \frac{400}{h}\sqrt{\frac{Dt}{\pi}} \tag{4}$$

At 24 h, this becomes:

$$P_{24} = \frac{2000}{h}\sqrt{\frac{D}{\pi}} \tag{5}$$

Additionally, from Figure 6:

$$P_{24} = 100\frac{w_{24}}{w_c} = \frac{[14+2c]}{100mf} \tag{6}$$

w_{24} and w_c are the weight of PLS released at 24 h in micrograms and total PLS in the composite disc. m, c, and f are mass of disc in grams (0.13 g), percentage of PLS in the filler, and filler fraction (0.75) in the composite. This equation shows that, upon doubling the PLS concentration in the filler, the early release in micrograms more than doubles. h, m, and f are all constants. Comparing Equations (5) and (6) therefore suggests that, upon increasing c, the diffusion coefficient for PLS increases, enabling faster release. This may be a consequence of higher c increasing PLS, MCPM, and water sorption. Further studies would be required to separate the effects of PLS and MCPM variables. The results, however, show that doubling both PLS and MCPM filler content more than doubles early PLS release mass.

When the composite is in direct contact with the dentine, release of PLS would be restricted. Release could occur, however, upon composite/dentine interface damage creating a water filled gap. These gaps may occur due poor bonding technique generated through composite shrinkage upon placement or formed at a later time due to cyclic stress on the bond. Rapid release of PLS into this gap could help to prevent biofilm formation initiating recurrent caries. The caries could, at early times after restoration, arise due to residual bacteria, or, if the gap is larger than a few microns, may be initiated by bacteria penetrating from the oral cavity. Additional diffusion of nutrients from the oral cavity would accelerate the bacterial growth within the gap.

If release of PLS is sufficiently fast to enable the minimum inhibitory concentration to be reached in the medium before exponential bacterial growth, then biofilm formation should be prevented. Alternatively, as occurred in this study, if release is not sufficiently fast to prevent biofilm initiation, longer-term release may kill the bacteria that are adherent on the composite surface. This study suggests that higher MCPM and PLS are required to achieve this mechanism of biofilm destruction. This antibacterial action would enable reduced risk with partial caries removal and possibly allow less affected dental hard tissue removal. In addition, it might increase the survival rate of the restoration and decrease the prevalence of recurrent caries. Excessive release of surface PLS, however, would have a negative effect on the top surface roughness and hardness. High levels could additionally reduce mechanical properties. This, however, could be addressed through use of the composite with high levels of MCPM and PLS beneath the control composite. This would likely block release of PLS from the top surface into the oral cavity.

5. Conclusions

Composites were prepared with PLS added to the filler phase at levels of 0, 4, 6, and 8 wt% and with monocalcium phosphate added simultaneously at double these levels. Upon incubating set composite discs with *S. mutans* and sucrose, a linear decline in surface biofilm mass and thickness occurred with increasing PLS level. Conversely, an abrupt increase in dead bacteria within the biofilm was observed with PLS levels above 4 wt%. Release studies suggest this was a consequence of 24 h PLS release needing to be above a critical level to kill bacteria within the surface biofilms. Early PLS release from the composite was proportional to the square root of time, as expected for a diffusion controlled process. Release levelled below 100%, suggesting it was from surface layers, which increased in thickness with raising PLS level. Upon damage of the composite restoration/tooth interface, either through polymerization shrinkage during placement or upon cyclic loading, a new composite surface was generated that was in contact with fluid. With rapid and sufficient PLS release from this new surface, risk of recurrent caries within the microgaps formed could potentially be reduced.

Author Contributions: Conceptualization, W.X., P.A., E.A., and A.M.Y.; Data curation, R.B.Y., W.X., and A.M.Y.; Formal analysis, R.B.Y. and A.M.Y.; Funding acquisition, R.B.Y., E.A., and A.M.Y.; Investigation, R.B.Y.; Methodology, W.X., E.A., and A.M.Y.; Project administration, E.A. and A.M.Y.; Resources, P.A., E.A., and A.M.Y.; Software, R.B.Y.; Supervision, P.A., E.A., and A.M.Y.; Validation, R.B.Y., W.X., P.A., E.A., and A.M.Y.; Visualization, R.B.Y.; Writing—original draft, R.B.Y.; Writing—review & editing, P.A., E.A., and A.M.Y. All authors have read and agreed to the published version of the manuscript.

Funding: W.X., A.Y., and P.A. were funded by National Institute for Health Research (Invention for Innovation (i4 i, http://www.nihr.ac.uk/funding/invention-for-innovation.htm)); optimisation and commercial manufacture of tooth-coloured composite dental-fillings with added poly-antimicrobial (PAM) and remineralising calcium phosphate (CaP), (II-LB-0214-20002), UK EPSRC (Engineering and Physical Sciences Research Council, https://www.epsrc.ac.uk) (EPI/I02234/1); and Welcome Trust (Wtissf 3: Institutional Strategic Support Fund (Issf) Third Tranche, www.wellcome.ac.uk/Funding/WTP057769.htm), (ISSF/FHCE/0079, The National Institute for Health Research, University College London Hospitals, Biomedical Research Centre www.uclhospitals.brc.nihr.ac.uk. The funders had no role in study design, data collection and analysis, or interpretation, decision to publish, or preparation of the manuscript.

Acknowledgments: Authors would like to acknowledge Nicola J. Mordan for providing technical support.

Conflicts of Interest: The corresponding author Anne Young has, with Paul Ashley, had funding from NIHR and EPSRC. This funding supported the salary of Wendy Xia. In the future, the corresponding author and inventor on the above patents (Anne Young) may receive royalties if a commercial product is produced. The team is currently working with Schottlander Dental Company to aid CE marking of a similar product to those in the publication.

Patents: The work is covered by the following licensed patent families: Formulations and composites with reactive fillers (US8252851 B2, EP2066703B1, US20100069469, WO2008037991A1), and Formulations and materials with cationic polymers (PCT/GB2014/052349, WO2015015212 A1, EP3027164A1, US20160184190).

References

1. WHO. Sugars and Dental Caries. Available online: https://apps.who.int/iris/bitstream/handle/10665/259413/WHO-NMH-NHD-17.12-eng.pdf;jsessionid=32E391ADF801ABA40F4C08F06FAB6382?sequence=1 (accessed on 29 June 2020).
2. WHO. Data and Statistics. Available online: http://www.euro.who.int/en/health-topics/disease-prevention/oral-health/data-and-statistics (accessed on 29 June 2020).
3. Bourguignon, D. Briefing EU Legislation in Progress Mercury Aligning EU legislation with Minamata. Available online: http://www.europarl.europa.eu/RegData/etudes/BRIE/2017/595887/EPRS_BRI%282017%29595887_EN.pdf (accessed on 29 June 2020).
4. Heintze, S.D.; Rousson, V. Clinical effectiveness of direct class II restorations-a meta-analysis. *J. Adhes. Dent.* **2012**, *14*, 407–431. [CrossRef] [PubMed]
5. Sunnegardh-Gronberg, K.; van Dijken, J.W.; Funegard, U.; Lindberg, A.; Nilsson, M. Selection of dental materials and longevity of replaced restorations in Public Dental Health clinics in northern Sweden. *J. Dent.* **2009**, *37*, 673–678. [CrossRef] [PubMed]

6. Beck, F.; Lettner, S.; Graf, A.; Bitriol, B.; Dumitrescu, N.; Bauer, P.; Moritz, A.; Schedle, A. Survival of direct resin restorations in posterior teeth within a 19-year period (1996–2015): A meta-analysis of prospective studies. *Dent. Mater.* **2015**, *31*, 958–985. [CrossRef] [PubMed]
7. Alvanforoush, N.; Palamara, J.; Wong, R.H.; Burrow, M.F. Comparison between published clinical success of direct resin composite restorations in vital posterior teeth in 1995–2005 and 2006–2016 periods. *Aust. Dent. J.* **2017**, *62*, 132–145. [CrossRef] [PubMed]
8. Kopperud, S.E.; Tveit, A.B.; Gaarden, T.; Sandvik, L.; Espelid, I. Longevity of posterior dental restorations and reasons for failure. *Eur. J. Oral. Sci.* **2012**, *120*, 539–548. [CrossRef] [PubMed]
9. Bourbia, M.; Ma, D.; Cvitkovitch, D.G.; Santerre, J.P.; Finer, Y. Cariogenic bacteria degrade dental resin composites and adhesives. *J. Dent. Res.* **2013**, *92*, 989–994. [CrossRef]
10. Marsh, P.D. Dental plaque as a microbial biofilm. *Caries Res.* **2004**, *38*, 204–211. [CrossRef]
11. Seneviratne, C.J.; Zhang, C.F.; Samaranayake, L.P. Dental plaque biofilm in oral health and disease. *Chin. J. Dent. Res.* **2011**, *14*, 87–94.
12. Keijser, B.J.; Zaura, E.; Huse, S.M.; van der Vossen, J.M.; Schuren, F.H.; Montijn, R.C.; ten Cate, J.M.; Crielaard, W. Pyrosequencing analysis of the oral microflora of healthy adults. *J. Dent. Res.* **2008**, *87*, 1016–1020. [CrossRef]
13. Dewhirst, F.E.; Chen, T.; Izard, J.; Paster, B.J.; Tanner, A.C.; Yu, W.H.; Lakshmanan, A.; Wade, W.G. The human oral microbiome. *J. Bacteriol.* **2010**, *192*, 5002–5017. [CrossRef]
14. Aas, J.A.; Griffen, A.L.; Dardis, S.R.; Lee, A.M.; Olsen, I.; Dewhirst, F.E.; Leys, E.J.; Paster, B.J. Bacteria of dental caries in primary and permanent teeth in children and young adults. *J. Clin. Microbiol.* **2008**, *46*, 1407–1417. [CrossRef] [PubMed]
15. Munson, M.A.; Banerjee, A.; Watson, T.F.; Wade, W.G. Molecular analysis of the microflora associated with dental caries. *J. Clin. Microbiol.* **2004**, *42*, 3023–3029. [CrossRef] [PubMed]
16. Takahashi, N.; Nyvad, B. Ecological Hypothesis of Dentin and Root Caries. *Caries Res.* **2016**, *50*, 422–431. [CrossRef] [PubMed]
17. Ge, Z.Y.; Yang, L.M.; Xia, J.J.; Fu, X.H.; Zhang, Y.Z. Possible aerosol transmission of COVID-19 and special precautions in dentistry. *J. Zhejiang Univ. Sci. B* **2020**, *21*, 361–368. [CrossRef]
18. Odeh, N.D.; Babkair, H.; Abu-Hammad, S.; Borzangy, S.; Abu-Hammad, A.; Abu-Hammad, O. COVID-19: Present and Future Challenges for Dental Practice. *Int. J. Environ. Res. Public Health* **2020**, *17*, 3151. [CrossRef]
19. Schwendicke, F.; Frencken, J.E.; Bjorndal, L.; Maltz, M.; Manton, D.J.; Ricketts, D.; Van Landuyt, K.; Banerjee, A.; Campus, G.; Domejean, S.; et al. Managing Carious Lesions: Consensus Recommendations on Carious Tissue Removal. *Adv. Dent. Res.* **2016**, *28*, 58–67. [CrossRef]
20. Frencken, J.E.; Leal, S.C.; Navarro, M.F. Twenty-five-year atraumatic restorative treatment (ART) approach: A comprehensive overview. *Clin. Oral Investig.* **2012**, *16*, 1337–1346. [CrossRef]
21. Ye, R.; Xu, H.; Wan, C.; Peng, S.; Wang, L.; Xu, H.; Aguilar, Z.P.; Xiong, Y.; Zeng, Z.; Wei, H. Antibacterial activity and mechanism of action of ε-poly-L-lysine. *Biochem. Biophys. Res. Commun.* **2013**, *439*, 148–153. [CrossRef]
22. Panpisut, P.; Liaqat, S.; Zacharaki, E.; Xia, W.; Petridis, H.; Young, A.M. Dental Composites with Calcium/Strontium Phosphates and Polylysine. *PLoS ONE* **2016**, *11*, e0164653. [CrossRef]
23. Tan, Z.; Shi, Y.; Xing, B.; Hou, Y.; Cui, J.; Jia, S. The antimicrobial effects and mechanism of ε-poly-lysine against Staphylococcus aureus. *Bioresour. Bioprocess.* **2019**, *6*, 11. [CrossRef]
24. Kangwankai, K.; Sani, S.; Panpisut, P.; Xia, W.; Ashley, P.; Petridis, H.; Young, A.M. Monomer conversion, dimensional stability, strength, modulus, surface apatite precipitation and wear of novel, reactive calcium phosphate and polylysine-containing dental composites. *PLoS ONE* **2017**, *12*, e0187757. [CrossRef] [PubMed]
25. Nikolaos, N.N.; Allan, E.; Xia, W.; Ashley, P.F.; Young, A.M. Early polylysine release from dental composites and its effects on planktonic Streptococcus mutans growth. *J. Funct. Biomater.* **2020**, *11*, 53. [CrossRef]
26. Davies, D. Understanding biofilm resistance to antibacterial agents. *Nat. Rev. Drug. Discov.* **2003**, *2*, 114–122. [CrossRef] [PubMed]
27. ISO 4049:2019 Dentistry—Polymer-Based Restorative Materials. Available online: https://www.iso.org/standard/67596.html (accessed on 30 July 2020).
28. Astasov-Frauenhoffer, M.; Glauser, S.; Fischer, J.; Schmidli, F.; Waltimo, T.; Rohr, N. Biofilm formation on restorative materials and resin composite cements. *Dent. Mater.* **2018**, *34*, 1702–1709. [CrossRef] [PubMed]

29. Amrita. Quantitative Estimation of Amino Acids by Ninhydrin. Available online: http://vlab.amrita.edu/?sub=3&brch=63&sim=156&cnt=2 (accessed on 29 June 2020).
30. Hyldgaard, M.; Mygind, T.; Vad, B.S.; Stenvang, M.; Otzen, D.E.; Meyer, R.L. The antimicrobial mechanism of action of epsilon-poly-l-lysine. *Appl. Environ. Microbiol.* **2014**, *80*, 7758–7770. [CrossRef]
31. Badaoui Najjar, M.; Kashtanov, D.; Chikindas, M.L. Natural Antimicrobials epsilon-Poly-L-lysine and Nisin A for Control of Oral Microflora. *Probiotics Antimicrob. Proteins* **2009**, *1*, 143. [CrossRef]
32. Clarke, J.K. On the Bacterial Factor in the Ætiology of Dental Caries. *Br. J. Exp. Pathol.* **1924**, *5*, 141–147.
33. Oda, Y.; Hayashi, F.; Okada, M. Longitudinal study of dental caries incidence associated with Streptococcus mutans and Streptococcus sobrinus in patients with intellectual disabilities. *BMC Oral Health* **2015**, *15*, 102. [CrossRef]
34. Limoli, D.H.; Jones, C.J.; Wozniak, D.J. Bacterial Extracellular Polysaccharides in Biofilm Formation and Function. *Microbiol. Spectr.* **2015**, *3*. [CrossRef]
35. Feldman, M.; Sionov, R.; Smoum, R.; Mechoulam, R.; Ginsburg, I.; Steinberg, D. Comparative Evaluation of Combinatory Interaction between Endocannabinoid System Compounds and Poly-L-lysine against Streptococcus mutans Growth and Biofilm Formation. *Biomed. Res. Int.* **2020**, *2020*, 7258380. [CrossRef]
36. Kulik, E.M.; Waltimo, T.; Weiger, R.; Schweizer, I.; Lenkeit, K.; Filipuzzi-Jenny, E.; Walter, C. Development of resistance of mutans streptococci and Porphyromonas gingivalis to chlorhexidine digluconate and amine fluoride/stannous fluoride-containing mouthrinses, in vitro. *Clin. Oral Investig.* **2015**, *19*, 1547–1553. [CrossRef] [PubMed]
37. Calogiuri, G.; Di Leo, E.; Trautmann, A.; Nettis, E.; Ferrannini, A.; Vacca, A. Chlorhexidine hypersensitivity: A critical and updated review. *J. Allergy Ther.* **2013**, *4*, 10.4172.
38. Pemberton, M.N.; Gibson, J. Chlorhexidine and hypersensitivity reactions in dentistry. *Br. Dent. J.* **2012**, *213*, 547–550. [CrossRef]
39. Aljabo, A.; Abou Neel, E.A.; Knowles, J.C.; Young, A.M. Development of dental composites with reactive fillers that promote precipitation of antibacterial-hydroxyapatite layers. *Mater. Sci. Eng. C Mater. Biol. Appl.* **2016**, *60*, 285–292. [CrossRef] [PubMed]
40. Zhang, J.F.; Wu, R.; Fan, Y.; Liao, S.; Wang, Y.; Wen, Z.T.; Xu, X. Antibacterial Dental Composites with Chlorhexidine and Mesoporous Silica. *J. Dent. Res.* **2014**, *93*, 1283–1289. [CrossRef]
41. Friedman, M. Applications of the ninhydrin reaction for analysis of amino acids, peptides, and proteins to agricultural and biomedical sciences. *J. Agric. Food Chem.* **2004**, *52*, 385–406. [CrossRef]
42. Siepmann, J.; Peppas, N.A. Higuchi equation: Derivation, applications, use and misuse. *Int. J. Pharm.* **2011**, *418*, 6–12. [CrossRef]

© 2020 by the authors. Licensee MDPI, Basel, Switzerland. This article is an open access article distributed under the terms and conditions of the Creative Commons Attribution (CC BY) license (http://creativecommons.org/licenses/by/4.0/).

Review

Use of Protein Repellents to Enhance the Antimicrobial Functionality of Quaternary Ammonium Containing Dental Materials

Leopoldo Torres Jr and Diane R. Bienek *

ADA Science & Research Institute, LLC, Innovative & Technology Research, Frederick, MD 21704, USA; torresl@ada.org
* Correspondence: bienekd@ada.org; Tel.: +1-301-694-2999

Received: 29 June 2020; Accepted: 23 July 2020; Published: 1 August 2020

Abstract: An advancement in preventing secondary caries has been the incorporation of quaternary ammonium containing (QAC) compounds into a composite resin mixture. The permanent positive charge on the monomers allows for electrostatic-based killing of bacteria. Spontaneous adsorption of salivary proteins onto restorations dampens the antimicrobial capabilities of QAC compounds. Protein-repellent monomers can work with QAC restorations to achieve the technology's full potential. We discuss the theory behind macromolecular adsorption, direct and indirect characterization methods, and advances of protein repellent dental materials. The translation of protein adsorption to microbial colonization is covered, and the concerns and fallbacks of the state-of-the-art protein-resistant monomers are addressed. Last, we present new and exciting avenues for protein repellent monomer design that have yet to be explored in dental materials.

Keywords: protein repellent; restorations; zwitterionic polymers; dental materials; antimicrobial; antifouling

1. Introduction

Tooth decay, also known as dental caries, is one of the most prevalent infections globally that afflicts both the developed and developing world. It affects young and old at large percentages, while being a preventable disease. Due to the high incidence of patient affliction, the economic toll is large in the US, approximately $442 billion [1]. To combat caries, clinicians remove the decayed tissue and replace with restorative materials. These composite materials consist of a stiff polymeric matrix (e.g., bisphenol A-glycidyl methacrylate (BisGMA) [2], urethane dimethacrylate [3], or methacryl polyhedral oligomeric silsesquioxane [4,5]) and inorganic filler components (i.e., amorphous calcium phosphate nanoparticles [6], borosilicate microparticles [7], and hydroxyapatite [8]). Additional problems arise from secondary caries, subsequent infections beneath or in the micro-cracks of the composite [9]. This occurs up to 44% for all adult patients and could be mitigated by antimicrobial technologies [10].

Integration of polymerizable antimicrobial methacrylates in dental resin offers the benefit of providing lasting antimicrobial activity, while being chemically stable. Specifically, quaternary ammonium containing (QAC) monomers have been incorporated into dental resins, to enable contact-killing of microorganisms. This concept was first introduced in 1993 by Imazato et al. [11]. Briefly, a quaternary ammonium compound, 12-methacryloyloxydodecylpyridinium bromide (MDPB), was incorporated into a resin, to formulate an antimicrobial composite [12]. Reports have indicated that QAC compounds destroy bacterial cell membrane integrity and eventually lead to cell death [13–15]. The design of these monomers has been heavily studied to optimize antimicrobial capabilities and elucidate the mechanism of bacterial killing [16]. For example, Li and coworkers synthesized QAC monomers with varying carbon lengths, following the positive quaternary amine, to enhance the insertion of the dangling monomer into the *Streptococcus mutans* membrane [17]. Another strategy has been to investigate monomers with varying degrees of flexibility, for improved incorporation into the bacterial membrane. The effect of alkyl chain length on antimicrobial properties of monomethacrylate monomers suggested a rise in antibacterial activity with the increasing alkyl chain [18,19]. However, the antimicrobial efficacy–structure relationships are not strictly linear. For instance, the longer chain length of novel adhesive methacrylate dental monomers had a less marked effect on reducing *S. mutans* biofilms [20]. Furthermore, when drawing conclusions about chain length, it is important to consider that antimicrobial functionality is also affected by molecular mass, spacer rigidity, hydrophobicity, charge density, and charge distribution [21,22]. Since the first Imazato and coworkers' QAC resin manuscript, advances have been made in synthesizing QAC monomers with dual functionality, such as increased shear bond strength [20,23,24] and silane-coupling capabilities [25–27]. Similarly, other antimicrobial approaches have been studied in dentistry, including metallic nanoparticles with inherent bactericidal properties [28] and are not discussed in this review.

While previous studies on QAC dental materials are significant, their impact has been dampened by the reduction in antimicrobial efficacy due to protein adsorption [29]. Salivary proteins form a thin coating onto the enamel surface, called the pellicle [30]. The pellicle allows for the attachment of early colonizing bacteria [31]. Many bacterial species possess surface structures (i.e., fimbriae and fibrils), which facilitate their attachment [32]. In the early stages of biofilm formation, planktonic bacteria directly attach to surfaces or indirectly bind to other bacteria that have already colonized [33]. Acidic and high-molecular-weight mucin fractions, acidic proline-rich proteins, and a multidomain glycosylated protein of the salivary pellicle are reported to bind bacteria and adhere to the non-native surface, to support biofilm attachment at the composite–adhesive–tooth interfaces [34]. The proteins and pathogens (and their interactions) attributing to failure at the composite–tooth interfaces have been presented in an elegant review [35]. After attachment, the bacterial cells proliferate, form microcolonies, and mature. Generally, biofilms can be morphologically heterogeneous 3D structures with the shape affected by spatiotemporal stress [36]. The 3D biofilm structures can be interspersed with bacteria-free channels used as diffusion pathways [32]. Dental caries and periodontal disease are a net result of the cross-talk between pathogenic dental-plaque biofilm and the host-tissue response [36]. While clinical examination and X-rays are commonly used to diagnose oral disease, advancement of salivary biomarkers and metaproteomic analyses of the oral microbiota may be exploited for future diagnosis of opportunistic and infectious disease [37].

To improve the antimicrobial properties of QAC monomers, protein-repellent functionality should be incorporated to prevent the extent of adsorption that current restorations experience (Figure 1). To accomplish this, the dental-material community has implemented approaches from the surface-science and blood-contacting material literature. Much of the recent literature has focused on 2-methacryloyloxyethyl phosphorylcholine (MPC), a commercially available and U.S. Food and Drug Administration-cleared zwitterionic polymer with well-studied protein-repellent capability.

The objective of this review is to focus on (1) the theoretical and practical considerations of protein adsorption; (2) methods to quantify protein adsorption; and (3) protein repelling functionality in dental restoratives and mouthwash technologies. Moreover, we also identify challenges with commonly used protein repellants and consider the potential of developing novel monomers.

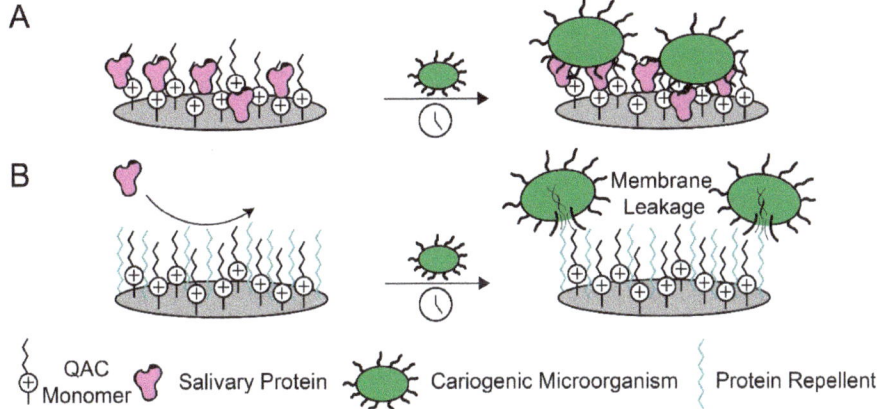

Figure 1. Dual functional dental materials enable contact-killing of cariogenic microorganisms by repelling proteins and disrupting bacterial membranes via charged interactions. (**A**) Salivary proteins adsorb to quaternary ammonium-containing (QAC) monomers, inhibiting their long-term antimicrobial properties. (**B**) Protein-repellent molecules work with QAC monomers to disrupt the formation of biofilms.

2. Theoretical and Practical Considerations for Protein Adsorption

Protein adsorption is a spontaneous process arising from a contribution of electrostatic and hydrophobic interactions, or hydrogen bonding [38]. Net charges on proteins can electrostatically bond with surfaces that are oppositely charged. This process can be reversed as the pH is altered, owing to the proteins pKa [39]. Additionally, hydrophobic regions of proteins can unravel (face outward) to bond with hydrophobic surfaces, minimizing the interactions between water and salivary proteins and between the surface and salivary proteins [40]. Lastly, both proteins and dental materials can participate in hydrogen bonding with each other if that is more favorable than surface solvation interactions [41].

Understanding the mechanisms of adsorption has led to the development of design principles for protein-repellent materials. The Whitesides group in 1993 reported a method for fabricating self-assembled monolayers (SAMs) as a tool to study protein adsorption [42]. These SAMs consisted of densely packed alkylthiol molecules that aligned parallel with one another on the surface of gold substrates. The tail end of the alkylthiol molecules were terminated with a functional group that could be covalently bonded to a chemical group of interest. Surface Plasmon Resonance (SPR) was coupled with a SAMs device to quantify the adsorption, observed as a change in frequency of the plasmon resonance. Their group subsequently followed their initial article with a survey of chemical structure relationship on adsorption, for which they reported four distinct rules for effective protein-repellent surfaces: The surface exposed monomers should be (1) hydrophilic; (2) contain hydrogen bond acceptors; (3) have no hydrogen bond donors; and (4) be electrically neutral [43–45]. Poly(ethylene glycol) (PEG)-based polymers [46] and many zwitterions [47] fall into this set of criteria and have been widely used for coatings in blood-contacting materials and implants.

In practice, additional considerations need to be taken into account to ensure effective protein repelling. The findings of the Whitesides' group can only translate if a critical density of antifouling monomers is on the surface of the material. Surface-coating treatment of protein-repellent monomers will not be as effective in dentistry, due to the formation of micro-cracks in composite materials produced during polymerization shrinkage [3]. These cracks are susceptible to bacterial colonization and consequently secondary caries, which are largely responsible for the failure of composite fillings [10]. It is imperative that dental materials contain protein-repellent molecules on their surface and in the bulk without (or minor) effect on the mechanical properties. In the same vein, researchers have developed hydrogels with protein-repellent monomers incorporated in the bulk and surface [48–50]. Last, the QAC monomers have functional groups that inherently adsorb proteins more readily. Specifically, the charged quaternary amine will participate in electrostatic interactions with proteins, leading toward adsorption. The long alkyl chain in many QAC monomers will also aid in hydrophobic–hydrophobic interactions, causing unfavorable adsorption. Therefore, dental materials require sufficient coverage of protein-repellent monomers to minimize attractive forces between proteins and QAC monomers.

3. Characterization Methods for Quantifying Protein Adsorption

Several techniques and assays have been developed to study the degree of protein coating on material surfaces (Figure 2). These methods range in sensitivity and each have trade-offs and should be considered for studying the adsorption of protein on dental materials. A comprehensive guide to characterization techniques for protein adsorption can be found in a previous article [51].

The most sensitive techniques (able to detect 1 ng to 1 µg of protein) commonly utilized are SPR or quartz crystal microbalance (QCM). Both analyze the surface of a small (<5 × 5 mm) substrate that is functionalized with a protein repellent of interest. Samples are prepared by coating or chemically functionalizing the surface with the protein repellent of interest. By preparing a thin sample, the interactions between a protein solution and protein-repellent monomer can be probed. In a typical SPR detection apparatus, a thin gold-coated slide is coated with the monomer of interest [42,52]. The slide is mounted, glass side down, onto a prism, and the functionalized side is used as part of a microfluidic channel. A light source with narrow range emission is projected through the prism and glass slide and reflected off the thin gold layer. A detector collects the angle of the reflected light, which corresponds to the index of refraction of the functionalized gold layer. When a protein solution flows onto the functionalized surface, the index of refraction increases, causing the angle of the reflected light to change. This change can then be used to calculate the mass of protein on the surface of the substrate.

QCM devices implement a piezoelectric functionalized substrate [53–55]. When a current is applied to the substrate, the material vibrates at a frequency proportional to its mass. The piezoelectric material surface is coated or chemically bonded with a material of interest, to probe how much protein adsorbs to the material surface. These substrates are incorporated into microfluidic devices that flow protein solutions onto the substrate surface and are coupled with real-time sensing. When proteins adsorb onto the substrate, the frequency at which it vibrates changes. This change in frequency is converted to the mass of protein adsorbed to the substrate. QCM can be used to yield dynamic properties of adsorbed proteins, such as revealing changes to the salivary pellicle on hydroxyapatite surfaces when various detergents are flowed over the substrates [56]. In addition to monitoring protein adsorption, QCM has been used to probe the formation of biofilms and bacterial death in a clinically relevant microorganism model [57]. SPR and QCM techniques are useful for protein interactions with high-density surfaces and are best for studying low adsorption, as they saturate with milligram quantities of protein.

Dental composite materials exhibit polymerization stress, causing the composite to crack when curing [3]. These cracks expose the bulk, leaving sites for adsorption without a surface-modified layer. A more practical method for detection of protein adsorption is through colorimetric analysis of protein solution surrounding a dental material. Biochemical assays such as the bicinchoninic acid (BCA), Lowry protein, or Bradford assay utilize reagents that alter their visible-light absorbance upon reacting with a protein in solution. In this regard, three protocol variations have been widely used by the dental materials community. A material sample is submerged in a protein solution for a predetermined time. (1) The material is rinsed with saline solution, to remove non-adsorbed proteins, and the adsorbed protein is removed by rinsing the material with a sodium dodecyl sulfate (SDS) solution. This SDS solution is then used in a colorimetric assay [58]. While this method is the most used in the dental material literature, a comprehensive investigation determined that SDS rinsing does not adequately remove adsorbed proteins [59]. This may lead researchers to conclude a protein-repelling capacity that is inaccurate. (2) In lieu of rinsing the material with SDS to remove the adsorbed protein, the protein solution and material can be vortexed to remove non-adsorbed proteins, and the protein solution surrounding the material is analyzed. This method yields quantification of the remaining solution compared to the initial starting concentration. (3) A small sample of protein solution is placed on a flat, clear, and polymerized dental resin. After the desired time, the non-adherent protein is removed by rinsing with saline. The material is then submerged in the reagents of a colorimetric assay, allowing the adsorbed proteins to react with the assay reagents. The optical density at the assays absorbance wavelength is then performed to quantify the adsorbed proteins [60,61].

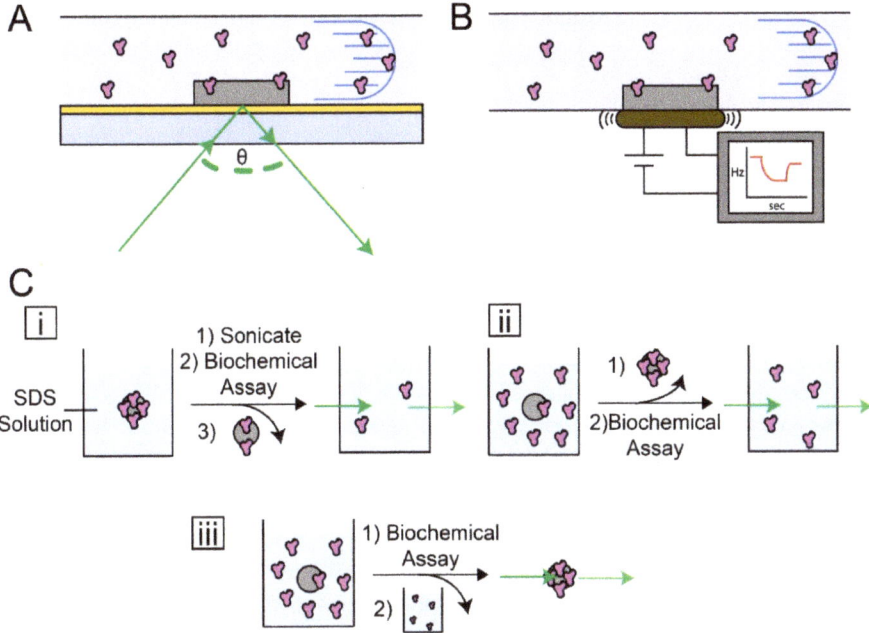

Figure 2. Characterization methods for quantifying protein adsorption on dental materials. (A) A general Surface Plasmon Resonance (SPR) device setup. Protein solution is flowed through a microfluidic channel and onto the material of interest bonded to a thin gold layer. Light is projected through a prism and onto the gold layer to discern protein–material interactions. (B) A general quartz crystal microbalance (QCM) device setup. The material of interest is fabricated onto a piezoelectric sensor. As protein accumulates onto the material, the vibration frequency changes. (C) Colorimetric methods for quantifying protein adsorption. (i) A disk with protein adsorbed to the surface is placed into a sodium dodecyl sulfate (SDS) buffer, to remove the protein. This solution is then analyzed by using an amino acid colorimetric reactive dye. (ii) A material of interest is placed in a protein solution of known concentration. After some time, the material is removed, and the remaining solution is analyzed. (iii) A material is placed in protein solution and is removed from the solution after a desired time point. The material with adsorbed protein is placed in a solution with colorimetric reagents, and the optical density is measured.

Topographical features and visualization of adsorption are important in understanding the growth of biofilm formation on dental materials. Looking toward the future, dental materials researchers should explore atomic force microscopy as a characterization tool. This technique probes the surface of materials, using a cantilever tip (100 nm–100 μm) and a laser to gauge the position of the tip, producing geometric information of a material [62]. Few dental material groups have explored the use of atomic force microscopy. to visualize protein adsorption on the surface [63]. Information that could be useful to researchers include the homogeneity of the adsorption layer, thickness of the adsorption layer, and force required to break bonds between the material surface and adsorbed proteins.

4. Dental Materials with Protein-Repellent Functionality

MPC is a methacrylate zwitterionic polymer that contains a negatively charged phosphorylcholine and a positively charged quaternary ammonium head. It is the most investigated molecule for protein-repelling dental materials [47,64,65]. It blends well into BisGMA/triethylene glycol methacrylate (TEGMA) and other hydrophilic resins. It was first introduced in the dental material literature by the Xu group in 2015 as an additive to a 50:50 BisGMA and TEGMA (BT) resin with dimethylaminohexyadecyl methacrylate (DMAHDM) and barium boroaluminosilicate filler [58]. The authors found that, out of various compositions investigated, a 3% (wt/wt) MPC and 1.5% (wt/wt) DMAHDM composite demonstrated nine times less bovine serum albumin (BSA) adsorption (~1 µg/cm^2), compared to the commercial control. For this composition, the flexural strength decreased from 100 to 80 MPa and elastic modulus from 6.7 to 6.5 MPa, compared to the resin without MPC. Oral biofilms, derived from the saliva of human donors, were cultured on the composites, to assess the antimicrobial capabilities of the MPC composites, with and without DMAHDM. Compared to the commercial control, a 3% MPC composite exhibited an order of magnitude lower total microorganisms, while the 3% MPC + 1.5% DMAHDM exhibited a three order of magnitude decrease of total microorganisms. This result demonstrated that the full potential of QAC resins could be realized with the addition of MPC.

The Xu lab continued their efforts into investigating the capability of MPC in many facets of materials in dentistry. A subsequent article optimized the addition of MPC into BT resins to identify a formulation with high protein repellency, without compromising the mechanical properties of the material [66]. They demonstrated that the flexural strength and elastic modulus suffer with materials containing 4.5% MPC and above. Specifically, the 4.5% MPC composite exhibited a flexural strength decrease to ~60 MPa compared to ~85 MPa in the case of the control. Moreover, the elastic modulus decreased to 5 MPa, compared to 6 MPa, as it was with the control. In a BSA adsorption assay, the researchers reported that a 3% MPC sample decreased the amount adsorbed by 85%, as compared to a polymerized BT sample. This same formulation also exhibited eight times fewer total microorganisms compared to the BT control.

MPC was then incorporated into a dental primer, to establish whether it could be useful as a restoration [67]. In conjunction with DMAHDM, MPC mixed into a commercial dental bond primer (3M Scotchbond Multi-Purpose Adhesive and Primer). A 7.5% MPC composite demonstrated less than 1 µg/cm^2 of protein adsorption, nearly 10 times less than the control. The same formulation had a comparable dentin shear bond strength, at ~27 MPa, compared to the control, at ~33 MPa. The degree of conversion was minimally impacted in all the formulations tested, indicating that MPC blends well with conventional dental resins. Last, the 7.5% MPC formulation was the most effective at reducing the total microbial count, by four orders of magnitude lower than the primer control. A follow-up investigation with amorphous calcium phosphate nanoparticles as the filler revealed that the shear bond strength decreased from 30 to 22 MPa [68]. The addition of the particles did decrease the shear bond strength to 25 from 30 MPa for a filler content of 30%, but did not alter the protein-repelling effects or antimicrobial efficacy. Even in long-term water aging of 180 days, MPC composites demonstrated closely similar protein-repelling and antimicrobial efficacy, likely due to the high degree of conversion [69].

To probe whether MPC inhibited or enhanced the release of calcium and phosphate ions from ACP particles, a 1:1 mixture of ethoxylated bisphenol A dimethacrylate (EBPADMA) and pyromellitic dianhydride glycerol dimetrocralte (PMGDM) was used as the resin matrix, abbreviated to EBPM [70]. This resin formulation was found to allow for the release of calcium and phosphorous ions when used with a ACP particle filler [71]. The combination of resin, MPC, and DMAHDM did not alter the protein-repelling properties of a 3% MPC formulation. In a four-organism biofilm challenge, a 3% MPC

+ 3% DMAHDM composite inhibited the colony count of *Porphyromonas gingivalis* and *Aggregatibacter actinomycetemcomitans* by four orders of magnitude and *Prevotella intermedia* and *Fusobacterium nucleatum* by three orders of magnitude, as compared to the resin alone [70]. The hydrophilicity of MPC caused more swelling in the composites, leading to a higher release of calcium and phosphorous ions, compared to the formulation without MPC [72]. By altering the amount of MPC, the amount of ions released could be tuned [73].

Poly(methyl methacrylate) (PMMA) is a common biomaterial routinely used in dentures and can be a breeding ground for oral microbes due to the heavy coating of salivary proteins it endures. MPC was incorporated into a methyl methacrylate monomer mixture and thermally cured. A 3% MPC formulation was enough to substantially decrease the amount of BSA adsorbed to ~2.0 µg/cm^2 compared to bare PMMA (~12 µg/cm^2) [74]. Computational modeling of surface interactions of MPC grafted onto PMMA revealed that MPC forms a tight hydration layer and a network of hydrogen bonding between adjacent MPC chains (in high-density grafting), which inhibits the adsorption of proteins and the anchoring of bacteria to the material surface [75].

More recently, there has been an effort to understand how QAC composites affect the microbial composition in biofilm models. MPC in QAC composites were shown to be more effective at decreasing the microbial growth of a single species biofilm of *P. gingivalis* [76]. As the microbial diversity increased, the composite was less effective at decreasing growth. Ultimately, the composite decreased the total microbe count by three orders of magnitude, compared to the resin control. It is well-documented that genetic information is shared between microbes in biofilm communities that act as a defense against chemical agents [77]. These composites also have shown to decrease the *S. mutans* composition in biofilms, leaving non-cariogenic species to thrive [78]. A summary of the protein repellent dental material capabilities discussed can be found in Table 1.

5. Mouthwash Coating Technology

A potential solution to repetitive cariogenic bacterial attachment to dental tissue is through the use of oral rinses with safe protein-repellent molecules that bind to enamel, root, or dentin surfaces. Recently, this concept was tested in a small clinical study by evaluating the number of microbes in dental plaque before and after rinsing with a solution of 5% MPC in saline [79]. Twenty patients had oral samples collected via gargle immediately after and 5 h after brushing their teeth. Half of the subjects were given a saline rinse as a control and the other half the MPC treatment. The patients treated with MPC saw a microbial decrease of 45%, compared to the control, through electric counting of the patients' gargle, indicating that protein adsorption was lessened. The number of fusobacteria, a mediator of bacterial aggregation and plague formation, was inhibited by this treatment. While no chemical modification of the oral environment was mentioned, a more biologically compatible material may be necessary for frequent rinses. In a separate study, a self-assembly approach was used to coat the oral cavity with lysozyme aggregated particles tethered to PEG to repel proteins [53]. Lysozyme was reduced with tris(2-carboxyethyl)phosphine (TCEP) to induce aggregation. These particles have a high tendency to physically bond to many material surfaces, including dentin and enamel surfaces. The particles densely pack at the surface, allowing for the formation of a tight hydration shell around the outward facing PEG molecules, yielding an effective protein repellent monolayer. The authors successfully tested their coatings against BSA, concanavalin A, fibronectin, saliva, fetal bovine serum, milk, egg whites, and various polysaccharides. Importantly, the authors found they can overcome the potential of esterase degradation by incorporating both positive and negative charges into the PEG molecules to induce a zwitterion effect and increase protein repellency of esterases. These approaches are novel and justify further investigations to determine feasibility and effect on the biofilm formation.

Table 1. Summary of bovine serum albumin (BSA) adsorption values for references in this review.

Protein Repellent Compound	Bulk Material	Filler	Adsorption Value (ng/cm^2)	Quantification Method	Reference
3% MPC (w/w)	25.5% 1:1 BisGMA/TEGDMA	70% Barium boroaluminosilicate	1240	SDS removal + BCA Assay	[58]
3% MPC	27% 1:1 BisGMA/TEGDMA	70% Barium boroaluminosilicate	960	SDS removal + BCA Assay	[66]
7.5% MPC	75% 1:1 Scotchbond Multi-Purpose Primer and Adhesive	15% Amorphous calcium phosphate	321	SDS removal + BCA Assay	[68]
3% MPC	25.5% 50:50 BisGMA/TEGDMA	70% Barium boroaluminosilicate	972 (with 180 days water aging)	SDS removal + BCA Assay	[69]
3% MPC	24% 1:1 EBPM	20% Amorphous calcium phosphate; 50% barium boroaluminosilicate	1200	SDS removal + BCA Assay	[70]
3% MPC	44.5% PMGDM, 39.5% EBPADMA, 10% 2-hydroxyethyl methacrylate, 5% BisGMA	30% Amorphous calcium phosphate	416	SDS removal + BCA Assay	[72]
3% MPC	47.75% Nature Cryl^{TM} liquid	47.75% Nature Cryl^{TM} powder	2150	SDS removal + BCA Assay	[74]
3% MPC	24% 1:1 EBPM	20% NACP, 50% barium boroaluminosilicate	1000	SDS removal + BCA Assay	[76]
PEG	Self-assembled PEG lysozyme	N/A	8	QCM	[53]
33% trimethylamine N-oxide Zwitterionic Hydrogel	N/A	N/A	3 *	SPR	[80]
9% Poly(carboxybetaine acrylamide) Zwitterionic Hydrogel	N/A	N/A	4.3 *	SPR	[49]

* Adsorption values for human serum, not BSA.

6. Limitations of Existing Technologies

While MPC has gained attention in the dental material literature, it has several shortcomings: (1) MPC contains an ester group connecting the zwitterionic component and the polymerizable methacrylate group. It is well-documented that ester bonds can be cleaved by esterases in saliva via hydrolysis [81]. (2) To achieve low levels of protein adsorption (ng/cm^2), MPC concentrations ≥ 5% need to be included into the bulk, which has a detrimental effect on the flexural strength, elastic modulus, and hardness. (3) Currently, there is a paucity of information regarding the long-term protein adsorption of dental resins containing MPC [82]. (4) Although, in theory, MPC meets many design criteria for being the superior protein-repellent candidate, many groups have shown that other conventional polymers outperform MPC in various experimental models [83–86]. New zwitterionic polymers are needed to overcome some of the long-term concerns with MPC. In addition, the potential clinical benefit needs to be confirmed, as MPC may have a deleterious effect on the remineralization capacity of restoratives (i.e., possible binding of the re-mineralizing calcium ions by MPC).

Recently, a zwitterionic polymer was synthesized by oxidizing an acrylamide monomer with 50% hydrogen peroxide, resulting in a permanently positive quaternary amine bonded to a permanently negative oxygen atom [80]. The proximity of the two charged atoms forms a tight hydration layer in an aqueous environment, leading to high repellent efficacy (5 ng/cm^2 by SPR). The polymer also exhibited satisfactory cytotoxicity and immunogenicity in a mouse model and should be studied as a candidate for protein repellency in dental materials. A separate group also synthesized amide-based mono- and bi-functional monomers to create hydrogels, which also demonstrated excellent compatibility in vivo and high protein repellency (4.3 ng/cm^2 by SPR) [49,87]. Further investigations are warranted to determine if these types of molecules can produce dental materials with satisfactory mixing with resin monomers, mechanical properties, and degree of conversion.

7. Conclusions

Protein-repellent technology has the potential to decrease the global burden of dental caries. When incorporated into composites alone or with QAC monomers, protein-repellent technology can inhibit the adsorption of proteins onto dental materials and thus slow the formation of biofilm and associated oral diseases. Many research groups have contributed to the understanding of protein adsorption and material design, and it is now time for dental material researchers to make an impact in the clinical setting. A concerted effort should be placed on understanding the long-term effect on the oral microflora with protein-repellent restorations, as it is not ideal to remove "good" bacteria from the oral cavity. New, more design-driven monomers should be explored to enhance the stability and protein repellency of QAC restorations, to allow for prolonged antimicrobial properties. Last, the dental material field should unify in using reliable protein-adsorption protocols, to ensure consistent comparisons across different material platforms.

Author Contributions: L.T.J. and D.R.B. contributed to the preparation/review of this manuscript. All authors have read and agreed to the published version of the manuscript.

Funding: This study was supported in part by the National Institute for Dental and Craniofacial Research (grants R01 DE26122 and DE026122-04S1) and the American Dental Association.

Conflicts of Interest: The authors declare no conflict of interest.

Disclaimer: The sole purpose of identifying certain commercial materials and equipment in this article was to adequately define the experimental protocols. Such identification, in no instance, implies recommendation or endorsement by the ADA or ADA Science & Research Institute, LLC, or means that the material/equipment specified is the best available for the purpose.

References

1. Sugars and Dental Caries. Available online: https://www.who.int/news-room/fact-sheets/detail/sugars-and-dental-caries (accessed on 15 June 2020).
2. Venhoven, B.A.M.; de Gee, A.J.; Davidson, C.L. Polymerization contraction and conversion of light-curing BisGMA-based methacrylate resins. *Biomaterials* **1993**, *14*, 871–875. [CrossRef]
3. O'Donnell, J.N.R.; Skrtic, D. Degree of Vinyl Conversion, Polymerization Shrinkage and Stress Development in Experimental Endodontic Composite. *J. Biomim. Biomater. Tissue Eng.* **2009**, *4*, 1–12. [CrossRef] [PubMed]
4. Wang, J.; Liu, Y.; Yu, J.; Sun, Y.; Xie, W. Study of POSS on the Properties of Novel Inorganic Dental Composite Resin. *Polymers* **2020**, *12*, 478. [CrossRef] [PubMed]
5. Liu, Y.; Wu, X.; Sun, Y.; Xie, W. POSS Dental Nanocomposite Resin: Synthesis, Shrinkage, Double Bond Conversion, Hardness, and Resistance Properties. *Polymers* **2018**, *10*, 369. [CrossRef]
6. Skrtic, D.; Antonucci, J.M.; Eanes, E.D. Amorphous calcium phosphate-based bioactive polymeric composites for mineralized tissue regeneration. *J. Res. Natl. Inst. Stand. Technol.* **2003**, *108*, 167. [CrossRef]
7. Tanaka, J.; Inoue, K.; Masamura, H.; Matsumura, K.; Najai, H.; Inoue, K. The Application of Fluorinated Aromatic Dimethacrylates to Experimental Light-cured Radiopaque Composite Resin, Containing Barium-Borosilicate Glass Filler—A Progress in Nonwaterdegradable Properties. *Dent. Mater. J.* **1993**, *12*, 1–11. [CrossRef]
8. Arcís, R.W.; López-Macipe, A.; Toledano, M.; Osorio, E.; Rodríguez-Clemente, R.; Murtra, J.; Fanovich, M.A.; Pascual, C.D. Mechanical properties of visible light-cured resins reinforced with hydroxyapatite for dental restoration. *Dent. Mater.* **2002**, *18*, 49–57. [CrossRef]

9. Kuper, N.K.; van de Sande, F.H.; Opdam, N.J.M.; Bronkhorst, E.M.; de Soet, J.J.; Cenci, M.S.; Huysmans, M.C.D.J.N.M. Restoration Materials and Secondary Caries Using an In Vitro Biofilm Model. *J. Dent. Res.* **2015**, *94*, 62–68. [CrossRef]
10. Nedeljkovic, I.; Teughels, W.; De Munck, J.; Van Meerbeek, B.; Van Landuyt, K.L. Is secondary caries with composites a material-based problem? *Dent. Mater.* **2015**, *31*, e247–e277. [CrossRef]
11. Imazato, S.; Torri, M.; Tsuchitani, Y. Immobilization of an antibacterial component in composite resin. *Dent. Jpn.* **1993**, *30*, 63–68.
12. Imazato, S.; Torii, M.; Tsuchitani, Y.; McCabe, J.F.; Russell, R.R.B. Incorporation of Bacterial Inhibitor into Resin Composite. *J. Dent. Res.* **1994**. [CrossRef]
13. Gottenbos, B.; van der Mei, H.C.; Klatter, F.; Nieuwenhuis, P.; Busscher, H.J. In vitro and in vivo antimicrobial activity of covalently coupled quaternary ammonium silane coatings on silicone rubber. *Biomaterials* **2002**, *23*, 1417–1423. [CrossRef]
14. Murata, H.; Koepsel, R.R.; Matyjaszewski, K.; Russell, A.J. Permanent, non-leaching antibacterial surfaces—2: How high density cationic surfaces kill bacterial cells. *Biomaterials* **2007**, *28*, 4870–4879. [CrossRef] [PubMed]
15. Lu, G.; Wu, D.; Fu, R. Studies on the synthesis and antibacterial activities of polymeric quaternary ammonium salts from dimethylaminoethyl methacrylate. *React. Funct. Polym.* **2007**, *67*, 355–366. [CrossRef]
16. Jain, A.; Duvvuri, L.S.; Farah, S.; Beyth, N.; Domb, A.J.; Khan, W. Antimicrobial Polymers. *Adv. Healthc. Mater.* **2014**, *3*, 1969–1985. [CrossRef]
17. Li, F.; Weir, M.D.; Xu, H.H.K. Effects of Quaternary Ammonium Chain Length on Antibacterial Bonding Agents. *J. Dent. Res.* **2013**, *92*, 932–938. [CrossRef]
18. Gozzelino, G.; Lisanti, C.; Beneventi, S. Quaternary ammonium monomers for UV crosslinked antibacterial surfaces. *Colloids Surf. A Physicochem. Eng. Asp.* **2013**, *430*, 21–28. [CrossRef]
19. He, J.; Söderling, E.; Österblad, M.; Vallittu, P.K.; Lassila, L.V.J. Synthesis of Methacrylate Monomers with Antibacterial Effects Against S. Mutans. *Molecules* **2011**, *16*, 9755–9763. [CrossRef]
20. Bienek, D.R.; Giuseppetti, A.A.; Okeke, U.C.; Frukhtbeyn, S.A.; Dupree, P.J.; Khajotia, S.S.; Florez, F.L.E.; Hiers, R.D.; Skrtic, D. Antimicrobial, biocompatibility, and physicochemical properties of novel adhesive methacrylate dental monomers. *J. Bioact. Compat. Polym.* **2020**. [CrossRef]
21. Xue, Y.; Xiao, H.; Zhang, Y. Antimicrobial Polymeric Materials with Quaternary Ammonium and Phosphonium Salts. *Int. J. Mol. Sci.* **2015**, *16*, 3626–3655. [CrossRef]
22. Makvandi, P.; Jamaledin, R.; Jabbari, M.; Nikfarjam, N.; Borzacchiello, A. Antibacterial quaternary ammonium compounds in dental materials: A systematic review. *Dent. Mater.* **2018**, *34*, 851–867. [CrossRef] [PubMed]
23. Antonucci, J.M.; Zeiger, D.N.; Tang, K.; Lin-Gibson, S.; Fowler, B.O.; Lin, N.J. Synthesis and characterization of dimethacrylates containing quaternary ammonium functionalities for dental applications. *Dent. Mater.* **2012**, *28*, 219–228. [CrossRef] [PubMed]
24. Bienek, D.; Frukhtbeyn, S.; Giuseppetti, A.; Okeke, U.; Skrtic, D. Antimicrobial Monomers for Polymeric Dental Restoratives: Cytotoxicity and Physicochemical Properties. *JFB* **2018**, *9*, 20. [CrossRef] [PubMed]
25. Daood, U.; Parolia, A.; Elkezza, A.; Yiu, C.K.; Abbott, P.; Matinlinna, J.P.; Fawzy, A.S. An in vitro study of a novel quaternary ammonium silane endodontic irrigant. *Dent. Mater.* **2019**, *35*, 1264–1278. [CrossRef] [PubMed]
26. Bienek, D.R.; Giuseppetti, A.A.; Frukhtbeyn, S.A.; Hiers, R.D.; Esteban Florez, F.L.; Khajotia, S.S.; Skrtic, D. Physicochemical, Mechanical, and Antimicrobial Properties of Novel Dental Polymers Containing Quaternary Ammonium and Trimethoxysilyl Functionalities. *JFB* **2019**, *11*, 1. [CrossRef] [PubMed]
27. Daood, U.; Matinlinna, J.P.; Pichika, M.R.; Mak, K.-K.; Nagendrababu, V.; Fawzy, A.S. A quaternary ammonium silane antimicrobial triggers bacterial membrane and biofilm destruction. *Sci. Rep.* **2020**, *10*, 10970. [CrossRef] [PubMed]
28. Makvandi, P.; Wang, C.; Zare, E.N.; Borzacchiello, A.; Niu, L.; Tay, F.R. Metal-Based Nanomaterials in Biomedical Applications: Antimicrobial Activity and Cytotoxicity Aspects. *Adv. Funct. Mater.* **2020**, *30*, 1910021. [CrossRef]
29. Imazato, S. Bio-active restorative materials with antibacterial effects: New dimension of innovation in restorative dentistry. *Dent. Mater. J.* **2009**, *28*, 11–19. [CrossRef]
30. Li, F.; Weir, M.D.; Fouad, A.F.; Xu, H.H.K. Effect of salivary pellicle on antibacterial activity of novel antibacterial dental adhesives using a dental plaque microcosm biofilm model. *Dent. Mater.* **2014**, *30*, 182–191. [CrossRef]

31. Ten Cate, J.M. Biofilms, a new approach to the microbiology of dental plaque. *Odontology* **2006**, *94*, 1–9. [CrossRef]
32. Saini, R.; Saini, S.; Sharma, S. Biofilm: A dental microbial infection. *J. Nat. Sci. Biol. Med.* **2011**, *2*, 71. [CrossRef]
33. Hojo, K.; Nagaoka, S.; Ohshima, T.; Maeda, N. Bacterial Interactions in Dental Biofilm Development. *J. Dent. Res.* **2009**, *88*, 982–990. [CrossRef] [PubMed]
34. Gibbons, R.J.; Hay, D.I. Adsorbed Salivary Acidic Proline-rich Proteins Contribute to the Adhesion of Streptococcus mutans JBP to Apatitic Surfaces. *J. Dent. Res.* **1989**, *68*, 1303–1307. [CrossRef] [PubMed]
35. Spencer, P.; Ye, Q.; Misra, A.; Goncalves, S.E.P.; Laurence, J.S. Proteins, Pathogens, and Failure at the Composite-Tooth Interface. *J. Dent. Res.* **2014**, *93*, 1243–1249. [CrossRef]
36. Seneviratne, C.J.; Zhang, C.F.; Samaranayake, L.P. Dental plaque biofim in oral health and disease. *Chin. J. Dent. Res.* **2011**, *14*, 87–94.
37. Castagnola, M.; Scarano, E.; Passali, G.C.; Messana, I.; Cabras, T.; Iavarone, F.; Cintio, G.D.; Fiorita, A.; Corso, E.D.; Paludetti, G. Salivary biomarkers and proteomics: Future diagnostic and clinical utilities. *Acta Otorhinolaryngolog. Italica* **2017**, *37*, 94.
38. McPherson, T.B.; Lee, S.J.; Park, K. Analysis of the Prevention of Protein Adsorption by Steric Repulsion Theory. In *Proteins at Interfaces II*; Horbett, T.A., Brash, J.L., Eds.; ACS Symposium Series; American Chemical Society: Washington, DC, USA, 1995; Volume 602, pp. 395–404. ISBN 978-0-8412-3304-1.
39. Müller, C.; Wald, J.; Hoth-Hannig, W.; Umanskaya, N.; Scholz, D.; Hannig, M.; Ziegler, C. Protein adhesion on dental surfaces—A combined surface analytical approach. *Anal. Bioanal. Chem.* **2011**, *400*, 679–689. [CrossRef]
40. Tilton, R.D.; Robertson, C.R.; Gast, A.P. Manipulation of hydrophobic interactions in protein adsorption. *Langmuir* **1991**, *7*, 2710–2718. [CrossRef]
41. Chen, S.; Li, L.; Zhao, C.; Zheng, J. Surface hydration: Principles and applications toward low-fouling/nonfouling biomaterials. *Polymer* **2010**, *51*, 5283–5293. [CrossRef]
42. KL Prime; G Whitesides Self-assembled organic monolayers: Model systems for studying adsorption of proteins at surfaces. *Science* **1991**, *252*, 1164–1167. [CrossRef]
43. Ostuni, E.; Chapman, R.G.; Holmlin, R.E.; Takayama, S.; Whitesides, G.M. A Survey of Structure–Property Relationships of Surfaces that Resist the Adsorption of Protein. *Langmuir* **2001**, *17*, 5605–5620. [CrossRef]
44. Chapman, R.G.; Ostuni, E.; Takayama, S.; Holmlin, R.E.; Yan, L.; Whitesides, G.M. Surveying for Surfaces that Resist the Adsorption of Proteins. *J. Am. Chem. Soc.* **2000**, *122*, 8303–8304. [CrossRef]
45. Holmlin, R.E.; Chen, X.; Chapman, R.G.; Takayama, S.; Whitesides, G.M. Zwitterionic SAMs that Resist Nonspecific Adsorption of Protein from Aqueous Buffer. *Langmuir* **2001**, *17*, 2841–2850. [CrossRef]
46. Bernhard, C.; Roeters, S.J.; Franz, J.; Weidner, T.; Bonn, M.; Gonella, G. Repelling and ordering: The influence of poly(ethylene glycol) on protein adsorption. *Phys. Chem. Chem. Phys.* **2017**, *19*, 28182–28188. [CrossRef]
47. Baggerman, J.; Smulders, M.M.J.; Zuilhof, H. Romantic Surfaces: A Systematic Overview of Stable, Biospecific, and Antifouling Zwitterionic Surfaces. *Langmuir* **2019**, *35*, 1072–1084. [CrossRef] [PubMed]
48. Jain, P.; Hung, H.-C.; Lin, X.; Ma, J.; Zhang, P.; Sun, F.; Wu, K.; Jiang, S. Poly(ectoine) Hydrogels Resist Nonspecific Protein Adsorption. *Langmuir* **2017**, *33*, 11264–11269. [CrossRef] [PubMed]
49. Chou, Y.-N.; Sun, F.; Hung, H.-C.; Jain, P.; Sinclair, A.; Zhang, P.; Bai, T.; Chang, Y.; Wen, T.-C.; Yu, Q.; et al. Ultra-low fouling and high antibody loading zwitterionic hydrogel coatings for sensing and detection in complex media. *Acta Biomater.* **2016**, *40*, 31–37. [CrossRef] [PubMed]
50. Sabaté del Río, J.; Henry, O.Y.F.; Jolly, P.; Ingber, D.E. An antifouling coating that enables affinity-based electrochemical biosensing in complex biological fluids. *Nat. Nanotechnol.* **2019**, *14*, 1143–1149. [CrossRef]
51. Migliorini, E.; Weidenhaupt, M.; Picart, C. Practical guide to characterize biomolecule adsorption on solid surfaces (Review). *Biointerphases* **2018**, *13*, 06D303. [CrossRef]
52. Prabowo, B.; Purwidyantri, A.; Liu, K.-C. Surface Plasmon Resonance Optical Sensor: A Review on Light Source Technology. *Biosensors* **2018**, *8*, 80. [CrossRef]
53. Li, C.; Lu, D.; Deng, J.; Zhang, X.; Yang, P. Amyloid-Like Rapid Surface Modification for Antifouling and In-Depth Remineralization of Dentine Tubules to Treat Dental Hypersensitivity. *Adv. Mater.* **2019**, *31*, 1903973. [CrossRef] [PubMed]

54. Bhakta, S.A.; Evans, E.; Benavidez, T.E.; Garcia, C.D. Protein adsorption onto nanomaterials for the development of biosensors and analytical devices: A review. *Anal. Chim. Acta* **2015**, *872*, 7–25. [CrossRef] [PubMed]
55. Höök, F.; Vörös, J.; Rodahl, M.; Kurrat, R.; Böni, P.; Ramsden, J.J.; Textor, M.; Spencer, N.D.; Tengvall, P.; Gold, J.; et al. A comparative study of protein adsorption on titanium oxide surfaces using in situ ellipsometry, optical waveguide lightmode spectroscopy, and quartz crystal microbalance/dissipation. *Colloids Surf. B Biointerfaces* **2002**, *24*, 155–170. [CrossRef]
56. Ash, A.; Mulholland, F.; Burnett, G.R.; Wilde, P.J. Structural and compositional changes in the salivary pellicle induced upon exposure to SDS and STP. *Biofouling* **2014**, *30*, 1183–1197. [CrossRef]
57. Xu, Z.; Coriand, L.; Loeffler, R.; Geis-Gerstorfer, J.; Zhou, Y.; Scheideler, L.; Fleischer, M.; Gehring, F.K.; Rupp, F. Saliva-coated titanium biosensor detects specific bacterial adhesion and bactericide caused mass loading upon cell death. *Biosens. Bioelectron.* **2019**, *129*, 198–207. [CrossRef] [PubMed]
58. Zhang, N.; Ma, J.; Melo, M.A.S.; Weir, M.D.; Bai, Y.; Xu, H.H.K. Protein-repellent and antibacterial dental composite to inhibit biofilms and caries. *J. Dent.* **2015**, *43*, 225–234. [CrossRef] [PubMed]
59. Kratz, F.; Grass, S.; Umanskaya, N.; Scheibe, C.; Müller-Renno, C.; Davoudi, N.; Hannig, M.; Ziegler, C. Cleaning of biomaterial surfaces: Protein removal by different solvents. *Colloids Surf. B Biointerfaces* **2015**, *128*, 28–35. [CrossRef]
60. Kwon, J.-S.; Lee, M.-J.; Kim, J.-Y.; Kim, D.; Ryu, J.-H.; Jang, S.; Kim, K.-M.; Hwang, C.-J.; Choi, S.-H. Novel anti-biofouling light-curable fluoride varnish containing 2-methacryloyloxyethyl phosphorylcholine to prevent enamel demineralization. *Sci. Rep.* **2019**, *9*. [CrossRef]
61. Lee, M.-J.; Kwon, J.-S.; Kim, J.-Y.; Ryu, J.-H.; Seo, J.-Y.; Jang, S.; Kim, K.-M.; Hwang, C.-J.; Choi, S.-H. Bioactive resin-based composite with surface pre-reacted glass-ionomer filler and zwitterionic material to prevent the formation of multi-species biofilm. *Dent. Mater.* **2019**, *35*, 1331–1341. [CrossRef]
62. Binnig, G.; Quate, C.F.; Gerber, C. Atomic Force Microscope. *Phys. Rev. Lett.* **1986**, *56*, 930–933. [CrossRef]
63. Chen, X.; Davies, M.C.; Roberts, C.J.; Tendler, S.J.B.; Williams, P.M.; Davies, J.; Dawkes, A.C.; Edwards, J.C. Recognition of Protein Adsorption onto Polymer Surfaces by Scanning Force Microscopy and Probe–Surface Adhesion Measurements with Protein-Coated Probes. *Langmuir* **1997**, *13*, 4106–4111. [CrossRef]
64. Cao, L.; Wu, J.; Zhang, Q.; Baras, B.; Bhadila, G.; Li, Y.; Melo, M.A.S.; Weir, M.D.; Bai, Y.; Zhang, N.; et al. Novel Protein-Repellent and Antibacterial Resins and Cements to Inhibit Lesions and Protect Teeth. *Int. J. Polym. Sci.* **2019**, *2019*, 1–11. [CrossRef]
65. Zhang, N.; Zhang, K.; Xie, X.; Dai, Z.; Zhao, Z.; Imazato, S.; Al-Dulaijan, Y.; Al-Qarni, F.; Weir, M.; Reynolds, M.; et al. Nanostructured Polymeric Materials with Protein-Repellent and Anti-Caries Properties for Dental Applications. *Nanomaterials* **2018**, *8*, 393. [CrossRef] [PubMed]
66. Zhang, N.; Chen, C.; Melo, M.A.; Bai, Y.-X.; Cheng, L.; Xu, H.H. A novel protein-repellent dental composite containing 2-methacryloyloxyethyl phosphorylcholine. *Int. J. Oral Sci.* **2015**, *7*, 103–109. [CrossRef]
67. Zhang, N.; Weir, M.D.; Romberg, E.; Bai, Y.; Xu, H.H.K. Development of novel dental adhesive with double benefits of protein-repellent and antibacterial capabilities. *Dent. Mater.* **2015**, *31*, 845–854. [CrossRef]
68. Zhang, N.; Melo, M.A.S.; Chen, C.; Liu, J.; Weir, M.D.; Bai, Y.; Xu, H.H.K. Development of a multifunctional adhesive system for prevention of root caries and secondary caries. *Dent. Mater.* **2015**, *31*, 1119–1131. [CrossRef]
69. Zhang, N.; Zhang, K.; Melo, M.; Weir, M.; Xu, D.; Bai, Y.; Xu, H. Effects of Long-Term Water-Aging on Novel Anti-Biofilm and Protein-Repellent Dental Composite. *IJMS* **2017**, *18*, 186. [CrossRef]
70. Wang, L.; Xie, X.; Imazato, S.; Weir, M.D.; Reynolds, M.A.; Xu, H.H.K. A protein-repellent and antibacterial nanocomposite for Class-V restorations to inhibit periodontitis-related pathogens. *Mater. Sci. Eng. C* **2016**, *67*, 702–710. [CrossRef]
71. Zhang, L.; Weir, M.D.; Chow, L.C.; Antonucci, J.M.; Chen, J.; Xu, H.H.K. Novel rechargeable calcium phosphate dental nanocomposite. *Dent. Mater.* **2016**, *32*, 285–293. [CrossRef]
72. Al-Qarni, F.D.; Tay, F.; Weir, M.D.; Melo, M.A.S.; Sun, J.; Oates, T.W.; Xie, X.; Xu, H.H.K. Protein-repelling adhesive resin containing calcium phosphate nanoparticles with repeated ion-recharge and re-releases. *J. Dent.* **2018**, *78*, 91–99. [CrossRef]
73. Al-Dulaijan, Y.A.; Weir, M.D.; Melo, M.A.S.; Sun, J.; Oates, T.W.; Zhang, K.; Xu, H.H.K. Protein-repellent nanocomposite with rechargeable calcium and phosphate for long-term ion release. *Dent. Mater.* **2018**, *34*, 1735–1747. [CrossRef] [PubMed]

74. Cao, L.; Xie, X.; Wang, B.; Weir, M.D.; Oates, T.W.; Xu, H.H.K.; Zhang, N.; Bai, Y. Protein-repellent and antibacterial effects of a novel polymethyl methacrylate resin. *J. Dent.* **2018**, *79*, 39–45. [CrossRef]
75. Choi, W.; Jin, J.; Park, S.; Kim, J.-Y.; Lee, M.-J.; Sun, H.; Kwon, J.-S.; Lee, H.; Choi, S.-H.; Hong, J. Quantitative Interpretation of Hydration Dynamics Enabled the Fabrication of a Zwitterionic Antifouling Surface. *ACS Appl. Mater. Interfaces* **2020**, *12*, 7951–7965. [CrossRef] [PubMed]
76. Wang, L.; Xie, X.; Qi, M.; Weir, M.D.; Reynolds, M.A.; Li, C.; Zhou, C.; Xu, H.H.K. Effects of single species versus multispecies periodontal biofilms on the antibacterial efficacy of a novel bioactive Class-V nanocomposite. *Dent. Mater.* **2019**, *35*, 847–861. [CrossRef] [PubMed]
77. Chen, H.; Tang, Y.; Weir, M.D.; Lei, L.; Masri, R.; Lynch, C.D.; Oates, T.W.; Zhang, K.; Hu, T.; Xu, H.H.K. Effects of *S. mutans* gene-modification and antibacterial calcium phosphate nanocomposite on secondary caries and marginal enamel hardness. *RSC Adv.* **2019**, *9*, 41672–41683. [CrossRef]
78. Wang, H.; Wang, S.; Cheng, L.; Jiang, Y.; Melo, M.A.S.; Weir, M.D.; Oates, T.W.; Zhou, X.; Xu, H.H.K. Novel dental composite with capability to suppress cariogenic species and promote non-cariogenic species in oral biofilms. *Mater. Sci. Eng. C* **2019**, *94*, 587–596. [CrossRef]
79. Fujiwara, N.; Yumoto, H.; Miyamoto, K.; Hirota, K.; Nakae, H.; Tanaka, S.; Murakami, K.; Kudo, Y.; Ozaki, K.; Miyake, Y. 2-Methacryloyloxyethyl phosphorylcholine (MPC)-polymer suppresses an increase of oral bacteria: A single-blind, crossover clinical trial. *Clin. Oral Investig.* **2019**, *23*, 739–746. [CrossRef]
80. Li, B.; Jain, P.; Ma, J.; Smith, J.K.; Yuan, Z.; Hung, H.-C.; He, Y.; Lin, X.; Wu, K.; Pfaendtner, J.; et al. Trimethylamine N-oxide–derived zwitterionic polymers: A new class of ultralow fouling bioinspired materials. *Sci. Adv.* **2019**, *5*. [CrossRef]
81. Gonzalez-Bonet, A.; Kaufman, G.; Yang, Y.; Wong, C.; Jackson, A.; Huyang, G.; Bowen, R.; Sun, J. Preparation of Dental Resins Resistant to Enzymatic and Hydrolytic Degradation in Oral Environments. *Biomacromolecules* **2015**, *16*, 3381–3388. [CrossRef]
82. Tone, S.; Hasegawa, M.; Puppulin, L.; Pezzotti, G.; Sudo, A. Surface modifications and oxidative degradation in MPC-grafted highly cross-linked polyethylene liners retrieved from short-term total hip arthroplasty. *Acta Biomater.* **2018**, *66*, 157–165. [CrossRef] [PubMed]
83. Van Andel, E.; Lange, S.C.; Pujari, S.P.; Tijhaar, E.J.; Smulders, M.M.J.; Savelkoul, H.F.J.; Zuilhof, H. Systematic Comparison of Zwitterionic and Non-Zwitterionic Antifouling Polymer Brushes on a Bead-Based Platform. *Langmuir* **2019**, *35*, 1181–1191. [CrossRef] [PubMed]
84. Gu, M.; Vegas, A.J.; Anderson, D.G.; Langer, R.S.; Kilduff, J.E.; Belfort, G. Combinatorial synthesis with high throughput discovery of protein-resistant membrane surfaces. *Biomaterials* **2013**, *34*, 6133–6138. [CrossRef] [PubMed]
85. Imbrogno, J.; Williams, M.D.; Belfort, G. A New Combinatorial Method for Synthesizing, Screening, and Discovering Antifouling Surface Chemistries. *ACS Appl. Mater. Interfaces* **2015**, *7*, 2385–2392. [CrossRef] [PubMed]
86. Li, Q.; Imbrogno, J.; Belfort, G.; Wang, X.-L. Making polymeric membranes antifouling via "grafting from" polymerization of zwitterions. *J. Appl. Polym. Sci.* **2015**, *132*. [CrossRef]
87. Zhang, P.; Sun, F.; Tsao, C.; Liu, S.; Jain, P.; Sinclair, A.; Hung, H.-C.; Bai, T.; Wu, K.; Jiang, S. Zwitterionic gel encapsulation promotes protein stability, enhances pharmacokinetics, and reduces immunogenicity. *Proc. Natl. Acad. Sci. USA* **2015**, *112*, 12046–12051. [CrossRef] [PubMed]

© 2020 by the authors. Licensee MDPI, Basel, Switzerland. This article is an open access article distributed under the terms and conditions of the Creative Commons Attribution (CC BY) license (http://creativecommons.org/licenses/by/4.0/).

Article

Myristyltrimethylammonium Bromide (MYTAB) as a Cationic Surface Agent to Inhibit *Streptococcus mutans* Grown over Dental Resins: An In Vitro Study

Paola Andrea Mena Silva [1,2,3], Isadora Martini Garcia [1], Julia Nunes [4], Fernanda Visioli [4], Vicente Castelo Branco Leitune [1], Mary Anne Melo [5,6,*] and Fabrício Mezzomo Collares [1,*]

1. Dental Materials Laboratory, School of Dentistry, Federal University of Rio Grande do Sul, 90035-003 Porto Alegre-RS, Brazil; pao_mena100@hotmail.com (P.A.M.S.); isadora.garcia@ufrgs.br (I.M.G.); vicente.leitune@ufrgs.br (V.C.B.L.)
2. Universidad Central del Ecuador, 170129 Quito, Ecuador
3. Postgraduate Department, Universidad Regional Autónoma de Los Andes, 170129 Quito, Ecuador
4. Oral Pathology Department, School of Dentistry, Federal University of Rio Grande do Sul, 90035-003 Porto Alegre-RS, Brazil; jusnunes@icloud.com (J.N.); fernanda.visioli@ufrgs.br (F.V.)
5. Operative Dentistry Division, General Dentistry Department, University of Maryland School of Dentistry, Baltimore, MD 21201, USA
6. Ph.D. Program in Biomedical Sciences, University of Maryland School of Dentistry, Baltimore, MD 21201, USA
* Correspondence: mmelo@umaryland.edu (M.A.M.); fabricio.collares@ufrgs.br (F.M.C.)

Received: 16 January 2020; Accepted: 11 February 2020; Published: 15 February 2020

Abstract: This in vitro study evaluated the effect of myristyltrimethylammonium bromide (MYTAB) on the physical, chemical, and biological properties of an experimental dental resin. The resin was formulated with dental dimetacrylate monomers and a photoinitiator/co-initiator system. MYTAB was added at 0.5 ($G_{0.5\%}$), 1 ($G_{1\%}$), and 2 ($G_{2\%}$) wt %, and one group remained without MYTAB and was used as the control (G_{Ctrl}). The resins were analyzed for the polymerization kinetics, degree of conversion, ultimate tensile strength (UTS), antibacterial activity against *Streptococcus mutans*, and cytotoxicity against human keratinocytes. Changes in the polymerization kinetics profiling were observed, and the degree of conversion ranged from 57.36% (±2.50%) for $G_{2\%}$ to 61.88% (±1.91%) for $G_{0.5\%}$, without a statistically significant difference among groups ($p > 0.05$). The UTS values ranged from 32.85 (±6.08) MPa for $G_{0.5\%}$ to 35.12 (±5.74) MPa for G_{Ctrl} ($p > 0.05$). MYTAB groups showed antibacterial activity against biofilm formation from 0.5 wt % ($p < 0.05$) and against planktonic bacteria from 1 wt % ($p < 0.05$). The higher the MYTAB concentration, the higher the cytotoxic effect, without differences between G_{Ctrl} e $G_{0.5\%}$ ($p > 0.05$). In conclusion, the addition of 0.5 wt % of MYTAB did not alter the physical and chemical properties of the dental resin and provided antibacterial activity without cytotoxic effect.

Keywords: dental materials; dentistry; anti-bacterial agents; dental caries; biocompatible materials; biofilms; quaternary ammonium compounds

---

1. Introduction

Dental caries-linked bacteria growing in biofilms play a pivotal role in the initial formation and development of carious lesions. The demand for the development of antibacterial surfaces has gained prominence in order to reduce patients' susceptibility to new or repeated diseases [1–5]. In dentistry, dental materials have been developed with antibacterial agents to provide new preventive and treatment dynamics for patients [6–9]. The antibacterial approach for dental materials relies on biological interactions of the bacteria grown over the materials, and the contact with antibacterial agents presents in the surface or is released by the materials [10–13]. In this context, the achievement of surfaces able to reduce caries-linked biofilm formation over dental tissues or restorative structures has gained the attention of dental biomaterial researchers.

The current understanding of the dental caries disease process and the new advances in dental materials promote the preservation of dental structure and the application of minimally invasive techniques [14]. Under this framework, dental resins are the first option for the replacement of dental hard tissues that were lost due to carious lesions [15,16]. Dental resins are materials cured by chemical or physicochemical (via photo-activation) processes, and they present inorganic fillers depending on their purpose. In areas where it is necessary to increase strength (occlusal/chewing surfaces of teeth), a high amount of inorganic fillers is incorporated [17]. Dental resins with low viscosity can also be successfully used to seal the biting surfaces of teeth, decreasing the incidence of caries lesions [18].

The oral environment provides many challenges to the physical and chemical stability of dental resins, such as high humidity, temperature, and pH variations [19,20]. Likewise, acids leached by high acidogenic caries-linked bacteria such as *Streptococcus mutans*, conjointly with a degradative attack of enzymes, inherently present in the saliva, can jeopardize the materials' properties over time [21]. More expressively, acidic attack from high acidogenic caries-linked bacteria is a crucial step in tooth demineralization around the restorations [22]. On this basis, the search for long-lasting strategies challenging biofilm accumulation on dental resins, as well as the consequent caries lesion development around restorations, must be addressed.

Antibacterial surfaces with a bacteria-killing function have shown great promise in biological and biomedical applications, in particular for dental resin-based materials. This approach employs the incorporation of quaternary ammonium compounds (QACs) in the monomeric blend of the resin-based material formulation [9,11,23]. QACs have facile synthesis, antibacterial property, and a lack of a detrimental effect on the mechanical and physical properties when incorporated in concentrations. Typically, QAS exhibits a positive charge, which confers the surface with the ability to attach and kill bacteria efficiently. The killing effect is attributed to their electronically interaction and linking to bacteria membrane and wall, along with their possible diffusion into the cytoplasmic membrane, an increase of osmotic pressure, and the release of some cytoplasmic constituents [24,25].

Myristyltrimethylammonium bromide (MYTAB) is a QAC with an alkyl-chain length of 14 carbons presenting a chemical structure of $C_{17}H_{38}NBr$, a molecular weight of 336.39, and a cationic polar head group. Trimethyl alkylammonium compounds such as the MYTAB series can also strongly affect bacterial membrane properties of a wide range of microorganisms, including changes in electrokinetic potential as well as net surface charge [26]. A recent study has shown promising application for the sealing of dental roots during endodontic treatment with expressive bacterial reduction of the endodontic pathogen [27]. Additionally, the long-chain MYTAB presents the potential to act as a useful endocytosis inhibitor for cell biology, inhibiting different forms of endocytosis in multiple cell systems [26].

The present study aimed to evaluate the potential ability of MYTAB to impair bacterial reduction when incorporated into dental resin at increasing concentrations. Its antibacterial effects against *S. mutans*, a pivotal cariogenic pathogen presented in planktonic and biofilm stages, were assessed. The cytotoxicity effect on human keratinocytes was investigated. The chemical and physical effects of its incorporation on the materials were also explored.

2. Results

Figure 1 shows the results of polymerization kinetics (Figure 1a–c), degree of conversion (Figure 1d), and ultimate tensile strength (UTS) (Figure 1e). The experimental dental resins had different polymerization behavior throughout the 40 s of photoactivation. The degree of conversion per time is shown in Figure 1a. The results (Figure 2d) ranged from 57.36% (±2.50%) for $G_{2\%}$ to 61.88% (±1.91%) for $G_{0.5\%}$, without statistical difference among groups ($p > 0.05$). Figure 1b indicates the results of the polymerization rate per time, showing that the higher the concentration of MYTAB, the higher the delay in achieving the maximum polymerization rate. Moreover, the higher the concentration of MYTAB, the lower the maximum polymerization rate. Figure 1c displays these differences, showing that at the same degree of conversion among groups, the polymerization rate of G_{Ctrl} was higher than $G_{1\%}$ and $G_{2\%}$ and lower than $G_{0.5\%}$. In Figure 1d, the degree of conversion revealed similar behavior for the tested groups ($p > 0.05$). The mechanical property of the dental resins was evaluated under UTS and expressed in MPa, as shown in Figure 1e. The UTS ranged from 32.85 (±6.08) MPa for $G_{0.5\%}$ to 35.12 (±5.74) MPa for G_{Ctrl}. There was no statistical difference among groups ($p > 0.05$).

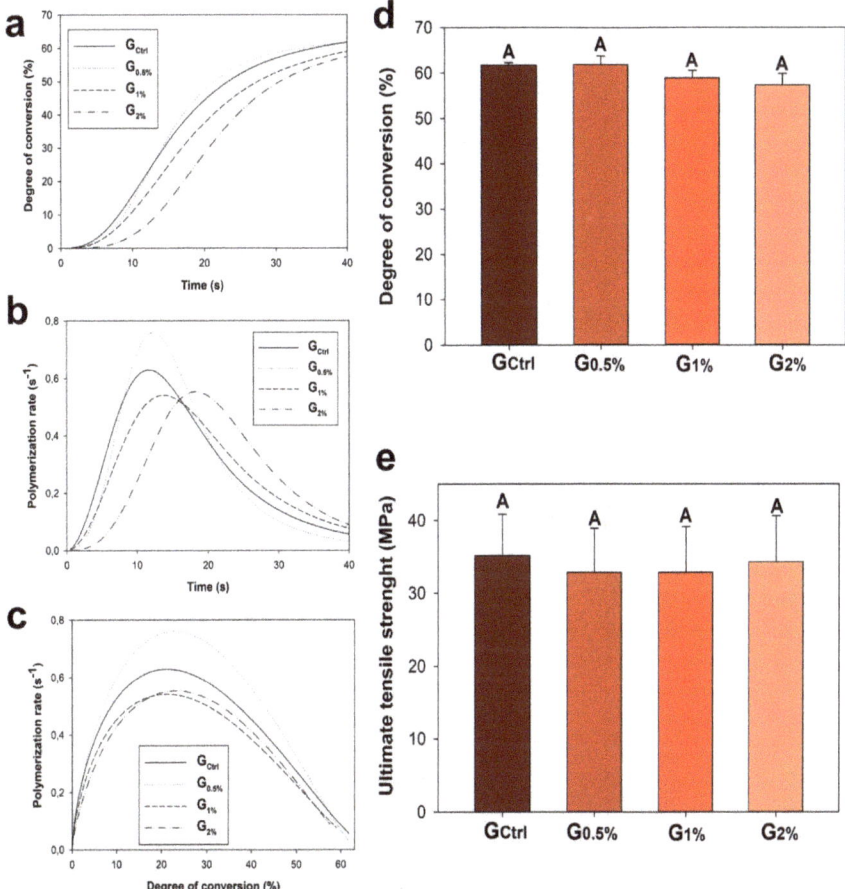

Figure 1. Comprehensive results of polymerization kinetics (**a**–**c**), degree of conversion after 40 s of photoactivation (**d**), and ultimate tensile strength (UTS) (**e**). Same capital letters indicate no statistical difference among groups ($p > 0.05$). The groups had different polymerization kinetics without statistical difference for the degree of conversion and UTS.

Figure 2 shows the results of the antibacterial activity of the experimental dental resins against biofilm formation of *S. mutans* and planktonic *S. mutans*, expressed by the log reduction in colony-forming unit per milliliter. In the microbiological assessment against biofilm formation (Figure 2a), the values ranged from 4.58 (±0.08) log CFU/mL for $G_{2\%}$ to 7.21 (±0.08) log CFU/mL for G_{Ctrl} ($p < 0.05$). A greater bacterial reduction ($p < 0.05$) was observed at the higher concentration of MYTAB incorporated into the dental resin. In the test against planktonic bacteria (Figure 2b), the values ranged from 6.68 (±0.58) log CFU/mL for $G_{2\%}$ to 8.28 (±0.05) log CFU/mL for G_{Ctrl} ($p < 0.05$). The group presenting MYTAB concentration at 2% expressed the highest *S. mutans* bacterial reduction ($p < 0.05$).

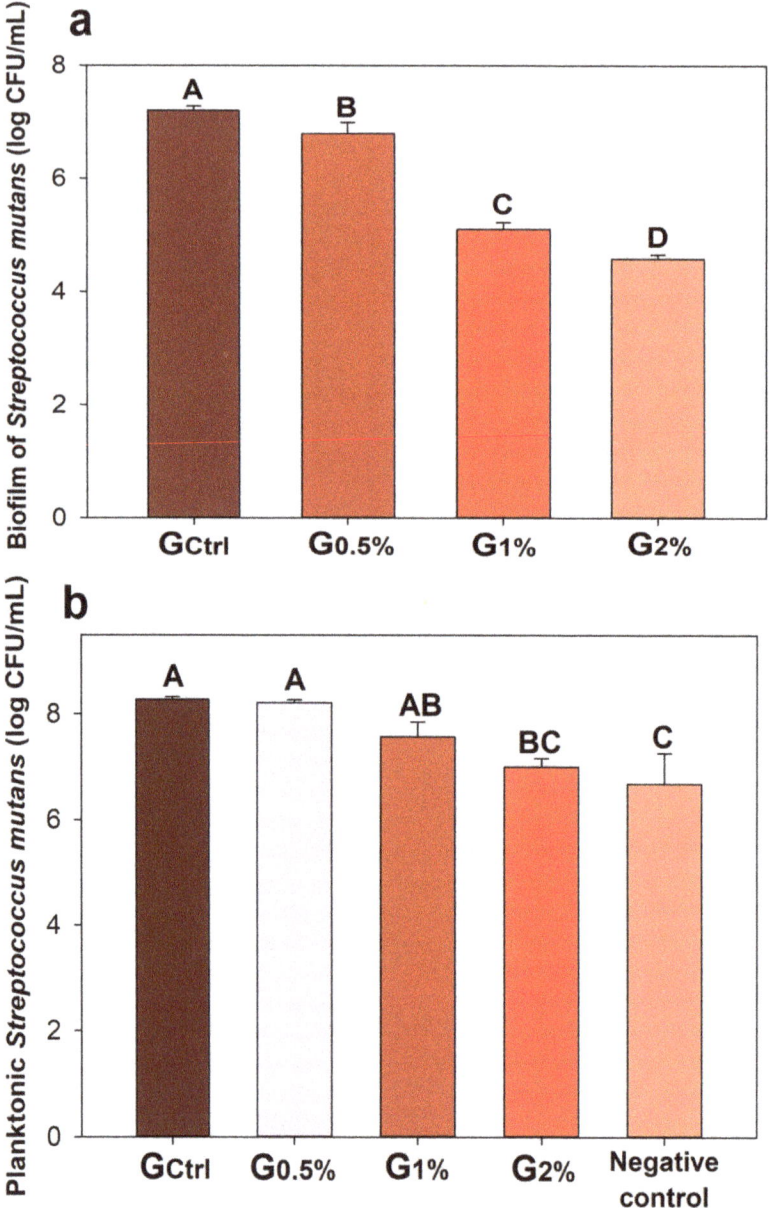

Figure 2. Bacterial colony forming unit counting: (**a**) *Streptococcus mutans* biofilms and (**b**) planktonic *S. mutans* that were in contact with the polymerized samples. Values indicated by different letters indicate statistical differences among the groups ($p < 0.05$).

Figure 3 presents the effects of MYTAB incorporated into a dental resin on normal human keratinocytes (HaCaT) for cytotoxicity assessed by sulforhodamine B (SRB) assay. The percentage of viability of the cells ranged from 45.26% (±14.11%) for $G_{2\%}$ to 110.16% (±14.64%) for G_{Ctrl} ($p < 0.05$). $G_{0.5\%}$ (91.82% ± 12.17%) presented no statistical difference in comparison to G_{Ctrl} ($p > 0.05$) for cytotoxicity against human keratinocytes.

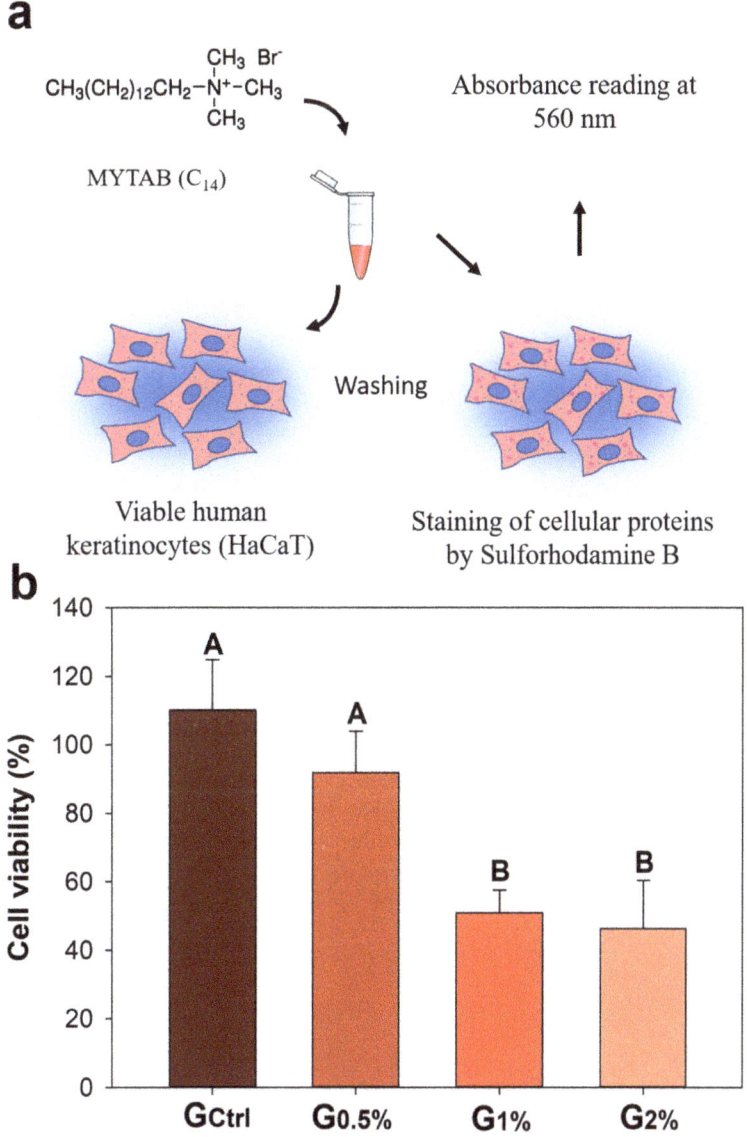

Figure 3. Cytotoxicity evaluation of the experimental dental resins expressed by percentage of cell viability: (**a**) structure of myristyltrimethylammonium bromide (MYTAB) and schematic drawing of sulforhodamine B (SRB) assay; and (**b**) MYTAB cytotoxicity assessed in normal human keratinocytes (HaCaT) line. Different capital letters indicate statistical differences among groups ($p < 0.05$).

3. Discussion

Dental resins are reliable materials to restore teeth, and, when used as pit and fissure sealants, could prevent new caries lesions [18,28,29]. Nevertheless, the formation of caries lesions around dental resins is still a major concern due to biofilm accumulation [22,30]. In order to prevent this issue, we investigated the effect of a cationic organic compound, MYTAB, in the properties of an experimental dental resin. The formulated resin is a suitable material for dental restorative purposes.

The longevity of dental restorative materials is strongly related to material rates of polymerization. High monomer conversion is essential for polymers to achieve reliable properties and stability [31]. The study of their polymerization behavior of modified formulations assists in understanding the effects of the incorporation of the compound on the functional aspect of the restorative material. Here, the formulated experimental dental resins showed different polymerization kinetics depending on the concentration of MYTAB. The effects were more evident with the addition of 1 wt % of MYTAB in the resin (Figure 2 A–C). From this concentration, the polymerization process was delayed, and the groups reached the maximum polymerization rate later compared to G_{Ctrl} and $G_{0.5\%}$.

On the other hand, $G_{0.5\%}$ showed a higher maximum polymerization rate compared to all groups (Figure 2 B). The rationale for this may be attributed to different viscosities among groups [32]. It was previously suggested that cationic surfactants could increase monomer chain mobility and modify polymerization behavior [33]. The lower viscosity and the higher monomer chain mobility for $G_{0.5\%}$ may lead to a higher maximum polymerization rate. Nevertheless, by increasing MYTAB incorporation, the spaces among monomer chains would be increased further, leading to a lower polymerization rate for $G_{1\%}$ and $G_{2\%}$. Despite these events, there was no difference in the degree of conversion among groups, and they achieved high values of conversion, similar to commercial dental resins [34].

Even though all groups presented a reliable degree of conversion, the delay during the polymerization kinetics observed for $G_{1\%}$ and $G_{2\%}$ could induce the formation of a more linear polymer, with lower crosslinking density [35]. Therefore, the mechanical evaluation of the formulated materials was essential in better understanding their performance. The specimens of dental resins were submitted to tensile strength until fracture in a universal testing machine with no statistical difference among groups. The UTS was a promising outcome because the incorporation of antibacterial agents could lead to lower mechanical properties [36]. The dental resins formulated in this study may be used in occlusal surfaces, where repetitive chewing stress is applied. The maintenance of resins' mechanical properties besides the antibacterial activity presented is essential in keeping the material in function.

In chewing sites, the experimental material proposed could be used not only for the prevention of caries lesions [18] but also to treat teeth already affected by the disease [29]. Pits and fissures are sites of difficult hygiene, favoring biofilm accumulation. Dental sealants successfully inhibit dental caries due to their ability to seal the demineralized tissue and form a mechanical barrier, inhibiting bacterial growth, hampering lesion progression, or preventing the demineralization of the sealed area [18,37]. The use of dental sealants in children and adolescents decreases the susceptibility to caries development in occlusal surfaces of permanent molars in comparison to people with no sealed teeth. Even among the people with sealed teeth, there are around 18% who present new lesions over time [18]. The composition of the sealants available in the market does not present antibacterial agents in their composition, which could be a strategy to reduce this percentage.

The parental composite resin formulated had its viscosity manually accessed for friendly use in clinical settings as a sealant. In the previous study with similar QAC, a high antibacterial effect of myristyltrimethylammonium bromide against *Enterococcus faecalis* was observed [27]. In both cases, these QACs have a long alkyl chain, which increases the QAC's hydrophobicity [38]. Consequently, when leached, these agents are more prone to penetrate bacterial walls and membranes [39]. With MYTAB, the higher its concentration in the dental resin, the lower the viability of *S. mutans* in biofilm and in planktonic stages.

For the assessment of antibacterial property, discs of the polymerized dental resins entered into contact with an enriched broth containing *S. mutans* as previously performed [13,33,40–42].

This bacterium is Gram-positive, and it is present in intraoral multispecies biofilms [43]. *S. mutans* group is the main bacteria associated with caries lesion development [44], and they can attach to dental and material surfaces [1,21]. The samples were exposed to planktonic as well as biofilm stages. During the contact with a dental resin containing MYTAB, planktonic bacteria reduction is suggested because of the leaching of some MYTAB molecules for the broth. The lack of assessment of the long-term antibacterial effect of the dental resins is a limitation of this study. Besides this, the quantities of MYTAB released over time from the polymerized resin is not known. The knowledge about this event could be valuable in order to understand the long-term behavior of the material and to assist in predicting the in vivo outcomes. However, the main goal of this study was to investigate the potential adverse effects of the addition of this compound into the composite to impart an antibacterial effect.

The cytotoxicity test was performed against human keratinocytes via the SRB method. Through this method, proteins of viable cells are stained, indicating increased viability when higher optical density is achieved [45]. The higher the incorporation of MYTAB, the higher the cytotoxic effect observed. This result corroborates with previous studies, which showed that cationic compounds with long-alkyl chains lead to high cytotoxic effects [46]. The International Organization for Standardization (ISO) recommends that biomaterials must promote up to 70% of cells' viability in order to not be considered cytotoxic [47]. It is worth mentioning that, despite the lower values of viability found for 1 and 2 wt % of MYTAB, all groups were directly treated with the eluates from the samples. In other words, we did not dilute the agent as other studies do [46], but we did somewhat increase the challenge by using the eluates on human cells during 72 h of contact.

A similar compound to MYTAB, a quaternary ammounium compound called ATAB, has been found to show antibacterial activity against *Enterococcus faecalis*, a relevant bacterium to endodontic infections [48]. The main difference between MYTAB and ATAB is the presence of mixing of QACs with different aliphatic chain lengths for ATAB, which does not occur in the case of MYTAB. In the previous study, ATAB was associated with halloysite nanotubes (HNT) that were incorporated in the sealer without the evaluation of ATAB alone [27], which might have repercussions for physical, chemical, or biological properties. In the same study, there were no cytotoxic effects for pulp cells with the incorporation of 10 wt % of ATAB/ halloysite nanotubes in the sealer, even when the proportion of ATAB/ halloysite nanotubes was 2:1 [27]. Here, MYTAB was not carried out by another system because we aimed to evaluate the QAC itself as a free drug, which may have increased the biological effects. $G_{0.5\%}$ showed no cytotoxicity in comparison to G_{Ctrl}, and both promoted viability higher than 70%. Therefore, the addition of 0.5 wt % of MYTAB may be a promising method for providing antibacterial activity for a dental resin without compromising its physical, chemical, and biological properties. The material here formulated is an exciting approach to be further translated for clinical trials.

4. Materials and Methods

4.1. Study Design and Formulation of Dental Resins

The study design is described in the flowchart presented in Figure 4. All reagents of the analytical grade for in vitro experiments were purchased from Sigma-Aldrich (Sigma-Aldrich Chemical Company, St. Louis, MO, USA) if not otherwise specified. First, a parental resin was formulated with two dimethacrylate monomers: bisphenol A glycol dimethacrylates (BisGMA) and triethylene glycol dimethacrylate (TEGDMA), at the proportion of 1:1. As a photoinitiator/co-initiator system, camphorquinone and ethyl 4-dimethylaminobenzoate were added to the resin at 1 mol % each. Butylated hydroxytoluene was added at 0.01 wt % as a polymerization inhibitor. Calcium tungstate ($CaWO_4$) was added as inorganic filler at 30 wt %. Colloidal silicon dioxide (SiO_2; Aerosil 200, Evonik, Essen, Germany) was incorporated at 0.7 wt % to adjust the resin's viscosity. After being hand-mixed for 5 min, they were sonicated for 180 s and hand-mixed for 5 min.

MYTAB with a purity of > 99% was added to the parental dental resin formulation at an increased double concentration of 0.5, 1, and 2% wt% mass fractions.

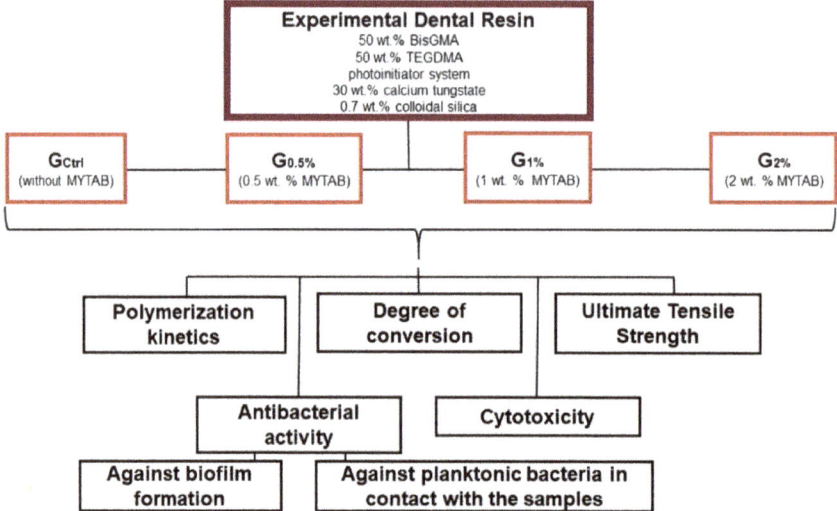

Figure 4. Flowchart of the study design. Experimental dental resins were formulated with different concentrations of a quaternary ammonium compound (MYTAB). Bisphenol A glycol dimethacrylates (BisGMA) and triethylene glycol dimethacrylate (TEGDMA) were used for the monomeric blend. The four experimental groups were evaluated for chemical, physical, and antibacterial properties, alongside their effect against human keratinocytes.

4.2. Polymerization Kinetics and Degree of Conversion

Fourier transform infrared spectroscopy (FTIR, Vertex 70, Bruker Optics, Ettinger, Germany) was used to evaluate the polymerization kinetics and the degree of conversion of the experimental dental resin. For this test, three samples per group were analyzed by dispensing them on the attenuated total reflectance (ATR) device in the polyvinylsiloxane matrix measuring 1 mm thickness and 4 mm in diameter. To perform the photoactivation of each sample, the light-cured unit (Radii Cal, SDI, Australia; 1200 mW/cm^2) was fixed using support to maintain 1 mm between the tip of the light-cured unit and the top of each sample. During the 40 s of photoactivation, two spectra were obtained per second in absorbance mode (10 kHz velocity, 4 cm^{-1} resolution; Opus 6.5 software, Bruker Optics, Ettlingen, Germany) in the range of 4000 to 400 cm^{-1}. The first spectrum obtained was used as "uncured resin dental resin", and the last spectrum as "cured resin dental resin" in the calculation of the degree of conversion. The peak at 1610 cm^{-1} from aromatic carbon–carbon double bond was used as an internal standard, and the peak at 1640 cm^{-1} was used as an aliphatic carbon–carbon double bond to calculate the conversion in percentage (Equation (1)) [31]. The polymerization rate was calculated using the degree of conversion at time t subtracted from the degree of conversion at time $t-1$.

Equation (1):

$$DC\,(\%) = 100 \times \left(\frac{\text{peak height of cured aliphatic C=C/peak height of cured aromatic C=C}}{\text{peak height of uncured aliphatic C=C/peak height of uncured aromatic C=C}} \right) \quad (1)$$

4.3. Ultimate Tensile Strength (UTS)

Ten samples per group with hourglass shape were prepared in a metallic matrix that was 8.0 mm long, 2.0 mm wide, and of 1.0 mm thickness, with a cross-sectional area of ±1 mm^2. Each uncured sample was placed in the mold and photoactivated for 20 s on each side (bottom and top). The prepared samples were stored in distilled water at 37 °C for 24 h. The samples were fixed in jigs with cyanoacrylate resin, and they were submitted under tensile strength in a universal testing machine (EZ-SX Series,

Shimadzu, Kyoto, Japan) at 1 mm/min until fracture. The maximum force value (Newtons, N) achieved was divided by the constriction area of each sample, which was measured with a digital caliper (Mitutoyo, Kawasaki, Kanagawa, Japan) to calculate the maximum value of tensile strength (Equation (2)). The results were expressed in megapascals (MPa).

Equation (2):

$$\textbf{UTS (MPa)} = \left(\frac{Force\ (N)}{Constriction\ area\ (mm^2)}\right) \quad (2)$$

4.4. Antibacterial Activity

Two in vitro assessments were performed to evaluate the antibacterial activity of the experimental dental resin: (1) planktonic bacteria and (2) against biofilm formation on the polymerized samples. The bacteria used in both tests were *S. mutans* (NCTC 10449). Three samples per group were prepared for the biofilm test, and the other three samples per group were used in the test against planktonic bacteria. *S. mutans* were prepared according to a previous study [13], and the initial inoculum used for the tests was assessed by serial dilution method and colony counting, which indicated an inoculum at 7.8×10^7 CFU/mL.

To evaluate the antibacterial activity against biofilm formation, the polymerized samples (1 mm thickness and 4 mm diameter) were fixed on Teflon specimens that were fixed on the cover of a 48-well plate, and this assembly was sterilized with hydrogen peroxide plasma (58%, 48 min, 56 °C) [13,49]. From the initial inoculum, 100 µL was added in each well of a 48-well plate with 900 µL of brain–heart infusion (BHI) broth with 1 wt % of sucrose. All reagents used in the antibacterial activity analysis were purchased from Aldrich Chemical Company (St. Louis, MO, USA). The sterile set of cover and samples was joined with this 48-well plate and kept for 24 h under 37 °C for biofilm formation on the samples. After this period, each sample was detached from the cover and vortexed for 1 min in 1 mL of sterile saline solution. The solution was serial diluted up to 10^{-6} mL and plated on Petri dishes containing BHI agar to count the colonies and to calculate (Equation 3) the colony-forming units per milliliter (CFU/mL).

Equation (3):

$$\textbf{CFU/mL} = \left(\frac{Avarage\ number\ of\ colonies \times Dilution\ factor}{Volume\ of\ culture\ plate}\right) \quad (3)$$

To evaluate the antibacterial activity against planktonic bacteria, the BHI broth that was in contact with the polymerized samples along the 24 h mentioned above was used. From each well of the 48-well plate, 100 µL was collected after the 24 h of bacteria–sample contact and was inserted into Eppendorf tubes with 900 µL of saline solution to be vortexed, diluted until 10^{-6}, and plated on BHI agar Petri dishes. For this test, an additional group was added as a negative control. The negative control was composed of BHI broth and *S. mutans* at the same proportion (10% of initial inoculum of bacteria in each well) without samples' contact. Colonies were visually counted, and the results were also expressed in CFU/mL.

4.5. Cytotoxicity

To evaluate the possible cytotoxic effect of the experimental dental resin, human keratinocytes (HaCaT, CLS Cell Lines Service GmbH, Eppelheim, Germany) were used in this test [40]. All reagents used to evaluate the possible cytotoxicity of the dental resins were purchased from Aldrich Chemical Company (St. Louis, MO, USA). The keratinocytes were kept in contact (5×10^3 cells/per well) with 100 µL of Dulbecco's modified Eagle's medium (DMEM) in 96-well plates for 24 h at 37 °C. On the same day, five samples per group (1 mm thickness and 4 mm diameter) were placed separately in Eppendorf tubes containing 1 mL of DMEM and kept for 24 h at 37 °C. Thus, eluates from possible leaching from the samples were formed. These eluates (100 µL) were kept in contact with the keratinocytes in the

96-well plates for 72 h at 37 °C. There was one group that was maintained without eluates with 100 µL of pure DMEM that was used as a control for the test. In addition to using five samples per group, the eluates were applied in five replications, totalizing 100 wells containing eluates from the samples. After 72 h, 50 µL of trichloroacetic acid/distilled water solution (50:50) was added in each well and kept at 4 °C for 1 hour to fix the cells on the bottom. Running water (30 s) was used to wash the 96-well plates six times. The 96-well plates were kept at room temperature until drying.

Sulforhodamine B at 0.4% (50 µL) was added in each well, and the 96-well plates were kept at room temperature for 30 min. The 96-well plates were washed four times with acetic acid at 1% and kept at room temperature until drying. Trizma solution at 10 mM (100 µL) was added in each well, and the 96-well plates were incubated for 1 hour at room temperature. The absorbance of plates' wells was analyzed at 560 nm, and the cell viability of the wells that contained eluates was compared to the wells without eluates. The viability of the keratinocytes used was normalized against the viability of cells without contact with eluates (negative control). The results were expressed in percentage using the viability of negative control as 100%.

4.6. Statistical Analysis

Data normality was evaluated by the Shapiro–Wilk test. One-way ANOVA and Tukey's post-hoc test was used to compare groups for all tests at a level of 0.05 of significance.

5. Conclusions

The present study incorporated a not yet investigated quaternary ammonium MYTAB into a dental resin, which achieved potent antibacterial effects against dental plaque microcosm biofilms for the first time. The results showed that (1) MYTAB at 2% greatly decreased *S. mutans* in planktonic and biofilm stages compared to G_{Ctrl}, (2) higher MYTAB mass fraction rendered the adhesive more strongly antibacterial, (3) MYTAB at 1% mass fraction greatly reduced the cell viability, and (4) the incorporation of MYTAB at a maximum concentration of 2% had no detrimental effect on the degree of conversion and ultimate tensile strength. MYTAB-containing dental resin is a promising strategy for dental applications.

Author Contributions: P.A.M.S. contributed to design, data acquisition, and analysis, and drafted the manuscript; I.M.G. contributed to design, data acquisition, and analysis, and drafted the manuscript; J.N. contributed to data acquisition and analysis; F.V. contributed to data acquisition and analysis; V.C.B.L. contributed to design, data analysis, and data interpretation; M.A.M. contributed to data analysis and data interpretation, and drafted the manuscript; F.M.C. contributed to conception, design, data acquisition, analysis, and interpretation, and drafted and critically revised the manuscript. All authors have read and agreed to the published version of the manuscript.

Funding: This research received no external funding.

Acknowledgments: This study was financed in part by the Coordenação de Aperfeiçoamento de Pessoal de Nível Superior—Brasil (CAPES)—Finance Code 001.

Conflicts of Interest: The authors declare no conflict of interest.

References

1. Lin, N.J. Biofilm over teeth and restorations: What do we need to know? *Dent. Mater.* **2017**, *33*, 667–680. [CrossRef] [PubMed]
2. Pandey, V.K.; Srivastava, K.R.; Ajmal, G.; Thakur, V.K.; Gupta, V.K.; Upadhyay, S.N.; Mishra, P.K. Differential Susceptibility of Catheter Biomaterials to Biofilm-Associated Infections and Their Remedy by Drug-Encapsulated Eudragit RL100 Nanoparticles. *Int. J. Mol. Sci.* **2019**, *20*, 5110. [CrossRef] [PubMed]
3. Alves, D.; Vaz, A.T.; Grainha, T.; Rodrigues, C.F.; Pereira, M.O. Design of an Antifungal Surface Embedding Liposomal Amphotericin B Through a Mussel Adhesive-Inspired Coating Strategy. *Front. Chem.* **2019**, *7*, 431. [CrossRef] [PubMed]
4. Donlan, R.M. Biofilms and device-associated infections. *Emerg. Infect. Dis.* **2001**, *7*, 277–281. [CrossRef]

5. He, Y.; Wan, X.; Xiao, K.; Lin, W.; Li, J.; Li, Z.; Luo, F.; Tan, H.; Li, J.; Fu, Q. Anti-biofilm surfaces from mixed dopamine-modified polymer brushes: Synergistic role of cationic and zwitterionic chains to resist staphyloccocus aureus. *Biomater. Sci.* **2019**, *7*, 5369–5382. [CrossRef]
6. Jung, J.; Li, L.; Yeh, C.K.; Ren, X.; Sun, Y. Amphiphilic quaternary ammonium chitosan/sodium alginate multilayer coatings kill fungal cells and inhibit fungal biofilm on dental biomaterials. *Mater. Sci. Eng. C* **2019**, *104*, 109961. [CrossRef]
7. Melo, M.A.; Orrego, S.; Weir, M.D.; Xu, H.H.; Arola, D.D. Designing Multiagent Dental Materials for Enhanced Resistance to Biofilm Damage at the Bonded Interface. *ACS Appl. Mater. Interfaces* **2016**, *8*, 11779–11787. [CrossRef]
8. Degrazia, F.W.; Genari, B.; Leitune, V.C.B.; Arthur, R.A.; Luxan, S.A.; Samuel, S.M.W.; Collares, F.M.; Sauro, S. Polymerisation, antibacterial and bioactivity properties of experimental orthodontic adhesives containing triclosan-loaded halloysite nanotubes. *J. Dent.* **2018**, *69*, 77–82. [CrossRef]
9. Makvandi, P.; Jamaledin, R.; Jabbari, M.; Nikfarjam, N.; Borzacchiello, A. Antibacterial quaternary ammonium compounds in dental materials: A systematic review. *Dent. Mater.* **2018**, *34*, 851–867. [CrossRef]
10. Braga, R.R.; Fronza, B.M. The use of bioactive particles and biomimetic analogues for increasing the longevity of resin-dentin interfaces: A literature review. *Dent. Mater. J.* **2019**. [CrossRef]
11. Andre, C.B.; Chan, D.C.; Giannini, M. Antibacterial-containing dental adhesives' effects on oral pathogens and on Streptococcus mutans biofilm: Current perspectives. *Am. J. Dent.* **2018**, *31*, 37B–41B. [PubMed]
12. Chi, M.; Qi, M.; A, L.; Wang, P.; Weir, M.D.; Melo, M.A.; Sun, X.; Dong, B.; Li, C.; Wu, J.; et al. Novel Bioactive and Therapeutic Dental Polymeric Materials to Inhibit Periodontal Pathogens and Biofilms. *Int. J. Mol. Sci.* **2019**, *20*, 278. [CrossRef] [PubMed]
13. Garcia, I.M.; Souza, V.S.; Hellriegel, C.; Scholten, J.D.; Collares, F.M. Ionic Liquid-Stabilized Titania Quantum Dots Applied in Adhesive Resin. *J. Dent. Res.* **2019**, *98*, 682–688. [CrossRef] [PubMed]
14. Schwendicke, F.; Frencken, J.E.; Bjorndal, L.; Maltz, M.; Manton, D.J.; Ricketts, D.; Van Landuyt, K.; Banerjee, A.; Campus, G.; Domejean, S.; et al. Managing Carious Lesions: Consensus Recommendations on Carious Tissue Removal. *Adv. Dent. Res.* **2016**, *28*, 58–67. [CrossRef] [PubMed]
15. Lynch, C.D.; Opdam, N.J.; Hickel, R.; Brunton, P.A.; Gurgan, S.; Kakaboura, A.; Shearer, A.C.; Vanherle, G.; Wilson, N.H. Guidance on posterior resin composites: Academy of Operative Dentistry—European Section. *J. Dent.* **2014**, *42*, 377–383. [CrossRef]
16. Tyas, M.J.; Anusavice, K.J.; Frencken, J.E.; Mount, G.J. Minimal intervention dentistry—A review. FDI Commission Project 1–97. *Int. Dent. J.* **2000**, *50*, 1–12. [CrossRef]
17. Ferracane, J.L. Resin composite—State of the art. *Dent. Mater.* **2011**, *27*, 29–38. [CrossRef]
18. Ahovuo-Saloranta, A.; Forss, H.; Walsh, T.; Nordblad, A.; Makela, M.; Worthington, H.V. Pit and fissure sealants for preventing dental decay in permanent teeth. *Cochrane Database Syst. Rev.* **2017**, *7*, CD001830. [CrossRef]
19. De Munck, J.; Van Meerbeek, B.; Yoshida, Y.; Inoue, S.; Vargas, M.; Suzuki, K.; Lambrechts, P.; Vanherle, G. Four-year water degradation of total-etch adhesives bonded to dentin. *J. Dent. Res.* **2003**, *82*, 136–140. [CrossRef]
20. Frassetto, A.; Breschi, L.; Turco, G.; Marchesi, G.; Di Lenarda, R.; Tay, F.R.; Pashley, D.H.; Cadenaro, M. Mechanisms of degradation of the hybrid layer in adhesive dentistry and therapeutic agents to improve bond durability—A literature review. *Dent. Mater.* **2016**, *32*, e41–e53. [CrossRef]
21. Kusuma Yulianto, H.D.; Rinastiti, M.; Cune, M.S.; de Haan-Visser, W.; Atema-Smit, J.; Busscher, H.J.; van der Mei, H.C. Biofilm composition and composite degradation during intra-oral wear. *Dent. Mater.* **2019**, *35*, 740–750. [CrossRef] [PubMed]
22. Opdam, N.J.; van de Sande, F.H.; Bronkhorst, E.; Cenci, M.S.; Bottenberg, P.; Pallesen, U.; Gaengler, P.; Lindberg, A.; Huysmans, M.C.; van Dijken, J.W. Longevity of posterior composite restorations: A systematic review and meta-analysis. *J. Dent. Res.* **2014**, *93*, 943–949. [CrossRef] [PubMed]
23. Imazato, S.; Ma, S.; Chen, J.H.; Xu, H.H. Therapeutic polymers for dental adhesives: Loading resins with bio-active components. *Dent. Mater.* **2014**, *30*, 97–104. [CrossRef] [PubMed]
24. Kenawy, E.-R.; Abdel-Hay, F.I.; El-Shanshoury, A.E.-R.R.; El-Newehy, M.H. Biologically active polymers. V. Synthesis and antimicrobial activity of modified poly(glycidyl methacrylate-co-2-hydroxyethyl methacrylate) derivatives with quaternary ammonium and phosphonium salts. *J. Polym. Sci.* **2002**, *40*, 2384–2393. [CrossRef]

25. Lu, G.; Wu, D.; Fu, R. Studies on the synthesis and antibacterial activities of polymeric quaternary ammonium salts from dimethylaminoethyl methacrylate. *React. Funct. Polymer* **2007**, *67*, 355–366. [CrossRef]
26. Quan, A.; McGeachie, A.B.; Keating, D.J.; van Dam, E.M.; Rusak, J.; Chau, N.; Malladi, C.S.; Chen, C.; McCluskey, A.; Cousin, M.A.; et al. Myristyl trimethyl ammonium bromide and octadecyl trimethyl ammonium bromide are surface-active small molecule dynamin inhibitors that block endocytosis mediated by dynamin I or dynamin II. *Mol. Pharmacol.* **2007**, *72*, 1425–1439. [CrossRef]
27. Monteiro, J.C.; Garcia, I.M.; Leitune, V.C.B.; Visioli, F.; de Souza Balbinot, G.; Samuel, S.M.W.; Makeeva, I.; Collares, F.M.; Sauro, S. Halloysite nanotubes loaded with alkyl trimethyl ammonium bromide as antibacterial agent for root canal sealers. *Dent. Mater.* **2019**, *35*, 789–796. [CrossRef]
28. Pallesen, U.; van Dijken, J.W. A randomized controlled 27 years follow up of three resin composites in Class II restorations. *J. Dent.* **2015**, *43*, 1547–1558. [CrossRef]
29. Munoz-Sandoval, C.; Gambetta-Tessini, K.; Giacaman, R.A. Microcavitated (ICDAS 3) carious lesion arrest with resin or glass ionomer sealants in first permanent molars: A randomized controlled trial. *J. Dent.* **2019**, *88*, 103163. [CrossRef]
30. Chisini, L.A.; Collares, K.; Cademartori, M.G.; de Oliveira, L.J.C.; Conde, M.C.M.; Demarco, F.F. Restorations in primary teeth: A systematic review on survival and reasons for failures. *Int. J. Paediatr. Dent.* **2018**, *28*, 123–139. [CrossRef]
31. Collares, F.M.; Ogliari, F.A.; Zanchi, C.H.; Petzhold, C.L.; Piva, E.; Samuel, S.M. Influence of 2-hydroxyethyl methacrylate concentration on polymer network of adhesive resin. *J. Adhes. Dent.* **2011**, *13*, 125–129. [PubMed]
32. Barszczewska-Rybarek, I.M. Characterization of urethane-dimethacrylate derivatives as alternative monomers for the restorative composite matrix. *Dent. Mater.* **2014**, *30*, 1336–1344. [CrossRef] [PubMed]
33. Martini Garcia, I.; Jung Ferreira, C.; de Souza, V.S.; Castelo Branco Leitune, V.; Samuel, S.M.W.; de Souza Balbinot, G.; de Souza da Motta, A.; Visioli, F.; Damiani Scholten, J.; Mezzomo Collares, F. Ionic liquid as antibacterial agent for an experimental orthodontic adhesive. *Dent. Mater.* **2019**, *35*, 1155–1165. [CrossRef] [PubMed]
34. Borges, B.C.; Bezerra, G.V.; Mesquita Jde, A.; Pereira, M.R.; Aguiar, F.H.; Santos, A.J.; Pinheiro, I.V. Effect of irradiation times on the polymerization depth of contemporary fissure sealants with different opacities. *Braz. Oral Res.* **2011**, *25*, 135–142. [CrossRef] [PubMed]
35. Rodrigues, S.B.; Collares, F.M.; Leitune, V.C.; Schneider, L.F.; Ogliari, F.A.; Petzhold, C.L.; Samuel, S.M. Influence of hydroxyethyl acrylamide addition to dental adhesive resin. *Dent. Mater.* **2015**, *31*, 1579–1586. [CrossRef]
36. Vidal, M.L.; Rego, G.F.; Viana, G.M.; Cabral, L.M.; Souza, J.P.B.; Silikas, N.; Schneider, L.F.; Cavalcante, L.M. Physical and chemical properties of model composites containing quaternary ammonium methacrylates. *Dent. Mater.* **2018**, *34*, 143–151. [CrossRef]
37. Liang, Y.; Deng, Z.; Dai, X.; Tian, J.; Zhao, W. Micro-invasive interventions for managing non-cavitated proximal caries of different depths: A systematic review and meta-analysis. *Clin. Oral Investig.* **2018**, *22*, 2675–2684. [CrossRef]
38. Yoshimura, T.; Chiba, N.; Matsuoka, K. Supra-long chain surfactants with double or triple quaternary ammonium headgroups. *J. Colloid Interface Sci.* **2012**, *374*, 157–163. [CrossRef]
39. Wang, Y.; Costin, S.; Zhang, J.F.; Liao, S.; Wen, Z.T.; Lallier, T.; Yu, Q.; Xu, X. Synthesis, antibacterial activity, and biocompatibility of new antibacterial dental monomers. *Am. J. Dent.* **2018**, *31*, 17b–23b.
40. Garcia, I.M.; Rodrigues, S.B.; de Souza Balbinot, G.; Visioli, F.; Leitune, V.C.B.; Collares, F.M. Quaternary ammonium compound as antimicrobial agent in resin-based sealants. *Clin. Oral Investig.* **2019**. [CrossRef]
41. Garcia, I.M.; Rodrigues, S.B.; Leitune, V.C.B.; Collares, F.M. Antibacterial, chemical and physical properties of sealants with polyhexamethylene guanidine hydrochloride. *Braz. Oral Res.* **2019**, *33*, e019. [CrossRef] [PubMed]
42. Machado, A.H.S.; Garcia, I.M.; Motta, A.S.D.; Leitune, V.C.B.; Collares, F.M. Triclosan-loaded chitosan as antibacterial agent for adhesive resin. *J. Dent.* **2019**, *83*, 33–39. [CrossRef] [PubMed]
43. Metwalli, K.H.; Khan, S.A.; Krom, B.P.; Jabra-Rizk, M.A. Streptococcus mutans, Candida albicans, and the human mouth: A sticky situation. *PLoS Pathog.* **2013**, *9*, e1003616. [CrossRef] [PubMed]

44. Kirstila, V.; Hakkinen, P.; Jentsch, H.; Vilja, P.; Tenovuo, J. Longitudinal analysis of the association of human salivary antimicrobial agents with caries increment and cariogenic micro-organisms: A two-year cohort study. *J. Dent. Res.* **1998**, *77*, 73–80. [CrossRef]
45. van Tonder, A.; Joubert, A.M.; Cromarty, A.D. Limitations of the 3-(4,5-dimethylthiazol-2-yl)-2,5-diphenyl-2H-tetrazolium bromide (MTT) assay when compared to three commonly used cell enumeration assays. *BMC Res. Notes* **2015**, *8*, 47. [CrossRef]
46. Gindri, I.M.; Siddiqui, D.A.; Bhardwaj, P.; Rodriguez, L.C.; Palmer, K.L.; Frizzo, C.P.; Martins, M.A.P.; Rodrigues, D.C. Dicationic imidazolium-based ionic liquids: A new strategy for non-toxic and antimicrobial materials. *RSC Adv.* **2014**, *4*, 62594–62602. [CrossRef]
47. International Organization for Standardization (ISO). *Biological Evaluation of Medical Devices—Part 5: Tests for in vitro Cytotoxicity*; ISO 10993-5:2009(E); ISO: Vernier, Geneva, Switzerland, 2009; Volume 1, p. 34.
48. Stuart, C.H.; Schwartz, S.A.; Beeson, T.J.; Owatz, C.B. Enterococcus faecalis: Its role in root canal treatment failure and current concepts in retreatment. *J. Endod.* **2006**, *32*, 93–98. [CrossRef]
49. Genari, B.; Leitune, V.C.B.; Jornada, D.S.; Camassola, M.; Arthur, R.A.; Pohlmann, A.R.; Guterres, S.S.; Collares, F.M.; Samuel, S.M.W. Antimicrobial effect and physicochemical properties of an adhesive system containing nanocapsules. *Dent. Mater.* **2017**, *33*, 735–742. [CrossRef]

© 2020 by the authors. Licensee MDPI, Basel, Switzerland. This article is an open access article distributed under the terms and conditions of the Creative Commons Attribution (CC BY) license (http://creativecommons.org/licenses/by/4.0/).

Article

Novel CaF₂ Nanocomposites with Antibacterial Function and Fluoride and Calcium Ion Release to Inhibit Oral Biofilm and Protect Teeth

Heba Mitwalli [1,2], Abdulrahman A. Balhaddad [1,3], Rashed AlSahafi [1,4], Thomas W. Oates [5], Mary Anne S. Melo [5,6], Hockin H. K. Xu [5,7,8] and Michael D. Weir [5,*]

1. Program in Dental Biomedical Sciences, University of Maryland School of Dentistry, Baltimore, MD 21201, USA; hmitwalli@umaryland.edu (H.M.); aabalhaddad@umaryland.edu (A.A.B.); rashed.alsahafi@umaryland.edu (R.A.)
2. Department of Restorative Dental Science, College of Dentistry, King Saud University, Riyadh 11451, Saudi Arabia
3. Department of Restorative Dental Sciences, College of Dentistry, Imam Abdulrahman bin Faisal University, Dammam 31441, Saudi Arabia
4. Department of Restorative Dental Sciences, College of Dentistry, Umm Al-Qura University, Makkah 24211, Saudi Arabia
5. Department of Advanced Oral Sciences and Therapeutics, School of Dentistry, University of Maryland, Baltimore, MD 21201, USA; TOates@umaryland.edu (T.W.O.); MMelo@umaryland.edu (M.A.S.M.); HXu2@umaryland.edu (H.H.K.X.)
6. Division of Operative Dentistry, Department of General Dentistry, University of Maryland School of Dentistry, Baltimore, MD 21201, USA
7. Center for Stem Cell Biology & Regenerative Medicine, University of Maryland School of Medicine, Baltimore, MD 21201, USA
8. Marlene and Stewart Greenebaum Cancer Center, University of Maryland School of Medicine, Baltimore, MD 21201, USA
* Correspondence: michael.weir@umaryland.edu

Received: 18 May 2020; Accepted: 28 July 2020; Published: 1 August 2020

Abstract: (1) Background: The objective of this study was to develop a novel dental nanocomposite containing dimethylaminohexadecyl methacrylate (DMAHDM), 2-methacryloyloxyethyl phosphorylcholine (MPC), and nanoparticles of calcium fluoride (nCaF$_2$) for preventing recurrent caries via antibacterial, protein repellent and fluoride releasing capabilities. (2) Methods: Composites were made by adding 3% MPC, 3% DMAHDM and 15% nCaF$_2$ into bisphenol A glycidyl dimethacrylate (Bis-GMA) and triethylene glycol dimethacrylate (TEGDMA) (denoted BT). Calcium and fluoride ion releases were evaluated. Biofilms of human saliva were assessed. (3) Results: nCaF$_2$+DMAHDM+MPC composite had the lowest biofilm colony forming units (CFU) and the greatest ion release; however, its mechanical properties were lower than commercial control composite ($p < 0.05$). nCaF$_2$+DMAHDM composite had similarly potent biofilm reduction, with mechanical properties matching commercial control composite ($p > 0.05$). Fluoride and calcium ion releases from nCaF$_2$+DMAHDM were much more than commercial composite. Biofilm CFU on composite was reduced by 4 logs ($n = 9$, $p < 0.05$). Biofilm metabolic activity and lactic acid were also substantially reduced by nCaF$_2$+DMAHDM, compared to commercial control composite ($p < 0.05$). (4) Conclusions: The novel nanocomposite nCaF$_2$+DMAHDM achieved strong antibacterial and ion release capabilities, without compromising the mechanical properties. This bioactive nanocomposite is promising to reduce biofilm acid production, inhibit recurrent caries, and increase restoration longevity.

Keywords: dental nanocomposite; calcium fluoride nanoparticles; remineralization; antibacterial; protein repellent; oral biofilm

1. Introduction

Dental resin composites are an excellent material for direct restorations of anterior teeth and in many cases posterior teeth due to their esthetics and ease of placement [1]. Nevertheless, composites are known to accumulate more oral bacterial plaque and biofilm than other direct restorative materials, which could expose the restored tooth to a higher risk for future recurrent caries [2]. Indeed, most failed restorations due to secondary caries are restored with composites [3,4]. The formation of plaque starts with the salivary-acquired pellicle formation. The glycoprotein found in the acquired pellicle promotes bacterial cell adherence. The microbes in the biofilm then produce acids which lowers the pH and lead to mineral loss over time resulting in dissolution of the tooth structure, the formation of caries, and failure of the restoration [5–7]. Unfortunately, currently available commercial composites lack antibacterial properties. Accordingly, efforts were made to overcome the presence of cariogenic bacteria, in an effort to prevent recurrent caries [8].

The incorporation of calcium fluoride nanoparticles (nCaF$_2$) into composites has the potential to reduce demineralization [9]. Fluoride (F) ions work by stimulating the remineralization and suppressing the oral microorganisms [10,11]. The presence of F ions in the event of demineralization enhances the precipitation of calcium and phosphate ions and forms fluorapatite [Ca$_5$(PO$_4$)$_3$F] to protect the tooth surface [12–14]. Fluoride was also shown to have the advantage of reducing bacterial acid production to reduce recurrent caries.

Designing a composite containing calcium fluoride nanoparticles would enhance the fluorapatite deposition in the affected tooth structure. When the tooth structure is subjected to acidic attack by the cariogenic pathogens, calcium and phosphate ions are lost from enamel. Using remineralization approaches to restore the lost minerals is required to enforce and strengthen the tooth structure. Therefore, the composite with calcium fluoride nanoparticles would enhance remineralization and form fluorapatite that is able to resist future acidic challenges. Several studies have demonstrated the ability of forming fluorapatite using nanotechnology [15,16]. In one study, they manufactured fluorapatite nanoparticles and examined its doping with silver ion nanoparticles and evaluated its physical and antimicrobial effects. The results showed 30% inhibition of bacterial growth after 4 h of incubation while maintaining the natural morphology of fluorapatite [15]. In another study, fluorapatite was incorporated into chitosan scaffolds. Fluorapatite maintained its structure, granted antimicrobial effects, and showed osteoconductive capability [16].

Furthermore, the incorporation of antibacterial agents into composites have also been investigated. Imazato et al. integrated 12-methacryloyloxydodecylpyridinium bromide (MDPB) into composites and showed successful antibacterial effects [17–20]. The incorporation of quaternary ammonium polyethylenimine (QPEI) into composites also produced a potent and wide-spectrum antimicrobial effect against salivary microorganisms [21]. Antimicrobial peptides (AMPs) were also demonstrated to have antimicrobial properties by bacterial membrane permeabilization and intracellular targeting [22]. Other studies developed antibacterial agents such as dimethylaminohexadecyl methacrylate (DMAHDM) [23,24] and showed a strong antibiofilm activity without compromising the mechanical properties [25].

Previous studies indicated that the salivary protein accumulation on composite surface could lower the efficiency of "contact-killing" mechanisms [26,27]. Accordingly, efforts were made to improve protein-repellent strategies including the addition of protein-repellent agents such as (2-methacryloyloxyethyl phosphorylcholine, or MPC) into resins [28–30]. This method provided resistance to protein adsorption and bacterial adhesion due to the hydrophilic characteristic of MPC [29,30]. However, to date, there has been no report on the development of a novel bioactive dental composite that contains nCaF$_2$, DMAHDM, and MPC in combination.

The objectives of this study were to develop a new composite consisting nCaF$_2$, DMAHDM, and MPC, and to investigate the mechanical, ion release and oral biofilm properties for the first time. The following hypotheses were tested: (1) Adding DMAHDM and MPC into the nCaF$_2$ composite would have mechanical properties similar to a commercial control composite; (2) Adding DMAHDM and MPC into the nCaF$_2$ composite would not compromise the F and Ca ion release; and (3) The new bioactive composite would have much less microorganisms, produce less biofilm acid, and have better remineralizing properties than the commercial control composite.

2. Materials and Methods

2.1. Fabrication of Composites

The experimental resin consisted of bisphenol A glycidyl dimethacrylate (BisGMA, Esstech, Essington, PA, USA), and triethylene glycol dimethacrylate (TEGDMA, Esstech) at 50:50 mass ratio. Camphorquinone at 0.2% (Millipore Sigma, Burlington, MA, USA) and 0.8% ethyl 4-N, N-diethylaminobenzoate (Millipore Sigma) were incorporated for photoactivation. The resin is referred to as BT resin. MPC (Millipore Sigma) was added at a mass fraction of 3% and incorporated into the BT resin with magnetic stirring bar at 150 rpm to be dissolved completely into the resin.

The synthesis of DMAHDM was performed using a modified Menschutkin reaction [31]. Briefly, 10 mmol of 2-(dimethylamino) ethyl methacrylate (Millipore Sigma), 10 mmol of 1-bromohexadecane (TCI America, Portland, OR, USA), and 3 g of ethanol were combined in a reaction vessel and then stirred for 24 h at 70 °C. After the evaporation of the solvent and removal of impurities, the DMADHM was collected. DMAHDM was added into the BT resin at a mass fraction of 3% and was stirred using a magnetic stirring bar at 150 rpm until it was completely dissolved into the resin.

The nCaF$_2$ was manufactured using a spray-dry method as described in previous studies, yielding a mean particle size of 32 nm [9,32–34]. The mass fraction of nCaF$_2$ incorporated into BT resin was 15%, based on our preliminary study. A previous study tested different concentrations of nCaF$_2$ in composite and, after long-term water-aging, the composite with 20% nCaF$_2$ had a flexural strength of 60 MPa [33]. In the present study, 15% nCaF$_2$ was integrated into the resin to achieve good mechanical strength. Silanized barium boroaluminosilicate glass particles with a mean size of 1.4 μm (Dentsply Sirona, Milford, DE, USA) were incorporated into the BT resin for mechanical enhancement. As a commercial control composite, Heliomolar (Ivoclar Vivadent, Mississauga, ON, Canada) was also tested. Heliomolar contains 66.7% filler mass fraction of ytterbium-trifluoride and nanofillers of 40–200 nm of silica. The following groups were tested (Table 1 summarizes the materials used in the study):

1. Heliomolar (referred to as commercial control (CC));
2. BT Resin + 70% glass (referred to as experimental control (EC));
3. Remineralizing composite: BT + 15% nCaF$_2$ + 55% glass (referred to as nCaF$_2$);
4. Antibacterial and remineralizing composite: BT + 15% nCaF$_2$ + 3% DMAHDM + 55% glass (referred to as nCaF$_2$+DMAHDM);
5. Protein-repellent and remineralizing composite: BT + 15% nCaF$_2$ + 3% MPC + 55% glass (referred to as nCaF$_2$+MPC);
6. Antibacterial, protein-repellent, and remineralizing composite: BT + 15% nCaF$_2$ + 3% DMAHDM + 3% MPC + 55% glass (referred to as nCaF$_2$+DMAHDM+MPC).

Table 1. Materials used in the study.

Formulation/Manufacturer	Abbreviation	Fluoride	DMAHDM	MPC
Heliomolar, Ivoclar Vivadent, Mississauga, ON, Canada (Commercial control)	CC	+	-	-
30% BT+70% Glass (Experimental Control)	EC	-	-	-
30% BT+15% nCaF$_2$+55% Glass	nCaF$_2$	+	-	-
27% BT+15% nCaF$_2$+3%DMAHDM +55% Glass	nCaF$_2$+ DMAHDM	+	+	-
27% BT+15% nCaF$_2$+3% MPC +55% Glass	nCaF$_2$+MPC	+	-	+
24% BT+15% nCaF$_2$+3% DMAHDM +3% MPC+55% Glass	nCaF$_2$+DMAHDM+MPC	+	+	+

2.2. Characterization of nCaF$_2$

Transmission electron microscopy (TEM, Tecnai T12, FEI, Hillsboro, OR, USA) was used to assess the nanoparticles. Samples were prepared through placing nanoparticles on a perforated copper grid coated by a carbon film. To avoid particle agglomeration, the sample was ultrasonicated for 5 min in acetone prior to deposition. Particle size distribution was measured using a laser diffraction particle size analyzer (SALD-2300, Shimadzu North America, Columbia, MD, USA).

2.3. Mechanical Properties Testing

Each composite paste was mixed in a disposable plastic container using a speed mixer (DAC 150.1 FVZ-K SpeedMixer™, FlackTec Inc., Landrum, SC, USA) at a speed of 2800 rpm for 1 min, and then thoroughly mixed by hand on a plastic slab for 5 min. The paste was then placed in a rectangular mold of 2 × 2 × 25 mm^3. Mylar strips were placed on both sides, followed by two glass slides. The specimen was light-cured using a curing unit at 1200 mW/cm^2 (Labolight DUO, GC America, Alsip, IL, USA) on each side for 1 min [35]. After demolding, the samples were stored in a 100% humidity chamber for 24 h at 37 °C. Flexural strength and elastic modulus were tested at a crosshead-speed of 1 mm/min with a 10 mm span with a three-point flexural test using a computer-controlled universal testing system (Insight 1, MTS, Eden Prairie, MN, USA) [36,37]. Flexural strength and elastic modulus were measured after 24 h of specimen immersion in distilled water at 37 °C. Flexural strength: $S = 3P_{max}/L(2bh^2)$, where P_{max} is the fracture load, L is span, b is sample width and h is thickness. Elastic modulus: $E = (P/d) (L^3/[4bh^3])$, where load P was divided by displacement d which is the slope in the linear elastic region. Six specimens were tested for each group ($n = 6$).

2.4. Ca and F Ion Release

The ion releases for all groups containing nCaF$_2$ were tested. A solution of sodium chloride (NaCl) (133 mmol/L) was buffered with 50 mmol/L HEPES to pH 7 [36,38]. Three specimens of 2 × 2 × 12 mm^3 were placed into 50 mL of solution, accommodating a specimen volume/solution ratio of 3.0 mm^3/mL, similar to those in previous studies [36,38,39]. The specimen's F and Ca ions release were measured at 1, 2, 4, 7, 14, 21, 28, 35, 42, 49, 56, 63, and 70 days. At every time point, aliquots of 2 mL were collected and substituted by a fresh 2 mL solution of NaCl. The aliquots were investigated for Ca ions by a colorimetric assay using a microplate reader (SpectraMax M5, Molecular Probes, San Jose, CA, USA) as previously described, using known standard and calibration curves [36,38,39]. The F ion release was tested with a F ion selective electrode (Orion, Cambridge, MA, USA). Fluoride standard solutions were measured to form a standard curve. The standard curve was used to establish the F concentration. The F ion concentration measurement was performed by combining 0.5 mL of sample and 0.5 mL of undiluted TISAB solution (Fisher Scientific, Pittsburgh, PA, USA).

2.5. Sample Preparation for Biofilm Tests

The cover of a 96 well plate was used to fabricate composite discs for microbiological experiments yielding samples 0.5 mm in thickness and 8 mm in diameter [31]. Composite paste was placed at each indent in the in 96-well plate cover then covered with Mylar strips and glass slides to form a smooth surface. It was then light cured as described previously and then stored for 24 h at 37 °C. The following day discs were magnetically stirred for 1 h at 100 rpm in distilled water to remove uncured monomers [18,40,41]. The specimens were sterilized using ethylene oxide (Anprolene AN 74i, Andersen Products, Haw River, NC, USA) for 24 and allowed to de-gas for 7 days, following the instructions of the manufacturer.

2.6. Saliva Collection and Dental Plaque Microcosm Biofilm Model

Saliva collection was conducted in accordance with the Declaration of Helsinki, and the protocol was approved by the Institutional Review Board at the University of Maryland Baltimore (IRB #: HP-00050407). The advantage of the dental plaque microcosm biofilm model is the use of an inoculum of human saliva to mimic the heterogeneity and complexity of the bacteria that are present in human dental plaque [18]. An equal amount of saliva was simultaneously gathered from ten healthy contributors with normal dentition, free of active caries, and no antibiotic use within the prior 3 months. Contributors were instructed not to brush their teeth 24 h preceding collection and not to eat or drink 2 h preceding the collection. Subsequently, the collected saliva from all participants was mixed and diluted to 70% in sterile glycerol. Then the saliva–glycerol solution was stored at −80 °C until use [42].

For all biofilm experiments, McBain artificial saliva growth medium was used. McBain medium contained 2.5 g/L Type II mucin (porcine, gastric, Millipore Sigma), 2.0 g/L bacteriological peptone (Becton Dickinson, Sparks, MD, USA), 2.0 g/L tryptone (Becton Dickinson), 0.35 g/L NaCl, 1.0 g/L yeast extract (Fisher Scientific), 0.2 g/L potassium chloride (Millipore Sigma), 0.1 g/L cysteine hydrochloride (Millipore Sigma), 0.2 g/L calcium chloride (Millipore Sigma). The pH of the medium was adjusted to 7 and autoclaved. After cooling the medium, 0.0002 g/L vitamin K_1, 0.001 g/L hemin were added. During biofilm experiments, 2% sucrose solution and the saliva–glycerol solution were used as an inoculum at a ratio of 1:50. The sucrose and inoculum were added to the medium and 1.5 mL of the medium was placed in each well of a 24-well plate containing a composite specimen from each groups. Specimen were incubated in 5% CO_2 at 37 °C for 8 h to permit biofilm growth on the samples. The same procedure was repeated after 8 h without the addition of saliva and incubation occurred again for 16 h. After 16 h the samples were moved to a new 24-well plate which contained fresh medium and sucrose, and was further incubated for 24 h. Composites were exposed to bacterial culture for a total of 48 h, which resulted in reasonably mature dental plaque microcosm biofilms on composites [25,43].

2.7. Biofilm Colony Forming Units (CFU) Counts

Nine discs were prepared for each group. Following the 48 h incubation, the disc samples containing biofilm were transported into a vial filled with 1 mL of cysteine peptone water (CPW). This was vortexed for 5 s then sonicated for 5 min and vortexed again to harvest the biofilm [29]. Serial dilutions of the suspensions of bacteria were prepared and transported to agar plates to grow. The CFU were counted on three different agar plates. To determine total streptococci count, mitis salivarius agar (MSA, Becton Dickinson, Sparks, MD, USA) were used. To determine the growth of mutans streptococci, 0.2 units per mL bacitracin (Millipore Sigma) was added to the mitis salivarius agar (MSB). To evaluate the growth of the total microorganisms, tryptic soy blood agar (TSBA) agar plates were used by adding defibrinated sheep blood to tryptic soy agar (TSA, Becton Dickinson). The agar plates were kept at 37 °C in a 5% CO_2 incubator for 48 h. CFU calculation was based on the colony number and multiplied by the dilution factor [29].

2.8. Biofilms Metabolic Activity Evaluation (MTT)

The MTT (3-[4,5-dimethylthiazol-2-yl]-2,5- diphenyltetrazolium bromide) assay was performed to investigate the biofilm metabolic activity. Following 48 h of incubation, the discs ($n = 9$) were transferred into a clean 24-well plate, then 1 mL of tetrazolium dye was placed to every disc. The discs were then incubated in an incubator of 5% CO_2 at 37 °C. Discs were then transported to another 24-well plate, and 1 mL of dimethyl sulfoxide (DMSO) was added to every disc and incubated for 20 min in a dark room [29,44]. After incubation, 200 µL of the DMSO solution was collected and the absorbance at 540 nm was measured [29,44] using a microplate reader (SpectraMax® M5, Molecular Devices, San Jose, CA, USA).

2.9. Biofilms Lactic Acid Production

Following 48 h incubation, discs ($n = 9$) were moved to a different 24-well plate comprising 1.5 mL buffered-peptone water (BPW) with 0.2% sucrose then incubated for 3 h in a 5% CO_2 incubator at 37 °C to release acids. After 3 h, the BPW solution lactic acid concentrations were measured by recording the absorbance at 340 nm [29,44] using a microplate reader (SpectraMax® M5, Molecular Devices). Standard curves were produced by means of lactic acid standards.

2.10. Scanning Electron Microscopy (SEM) of Biofilms

For biofilm visualization and confirmation of bacterial attachments on composite discs, biofilms formed at 24 h, 48 h, and 96 h were sputter-coated with platinum. Scanning electron microscopy (SEM, Quanta 200, FEI Company, Hillsboro, OR, USA) was used to examine the bacterial accumulation (Figure 7).

3. Statistical Analysis

All data were evaluated with one-way analysis of variance (ANOVA), and post hoc multiple comparison using Tukey's honestly significant difference test was performed. All statistical analysis was completed using the GraphPad Prism 8 software package (GraphPad Software, San Diego, CA, USA) at 0.05 level of significance.

4. Results

A representative TEM image of calcium fluoride ($nCaF_2$) nanoparticles is shown in Figure 1A. The nanoparticle size distribution ranged from 22 nm to 57 nm, with a mean particle size of 32 nm and is illustrated in Figure 1B.

Flexural strength and elastic modulus of the six composite groups (mean ± sd; $n = 6$) are shown in Figure 2A,B, respectively. The flexural strength was measured after one day of immersion in water at 37 °C. The flexural strength was significantly higher in EC and $nCaF_2$ composites, when compared to the commercial Heliomolar control ($p < 0.05$). Flexural strength in groups with $nCaF_2$+DMAHDM and $nCaF_2$+MPC matched those of Heliomolar control composite ($p > 0.05$). However, flexural strength in $nCaF_2$+DMAHDM+MPC was significantly lower than commercial Heliomolar control ($p < 0.05$).

The elastic modulus values of EC and $nCaF_2$ were significantly greater than all other groups ($p < 0.05$). Other groups had comparable elastic modulus values to Heliomolar control ($p < 0.05$).

The accumulative F ion release is shown in Figure 3. At 70 days, the composite $nCaF_2$+MPC had the highest F release of (0.40 ± 0.02) mmol/L ($p < 0.05$). Meanwhile, $nCaF_2$+DMAHDM+MPC had F release of (0.25 ± 0.03) mmol/L, $nCaF_2$+DMAHDM had (0.20 ± 0.03) mmol/L, and $nCaF_2$ had (0.04 ± 0.01) mmol/L of fluoride ion release. Heliomolar control had the lowest F release of (0.004 ± 0.0003) mmol/L.

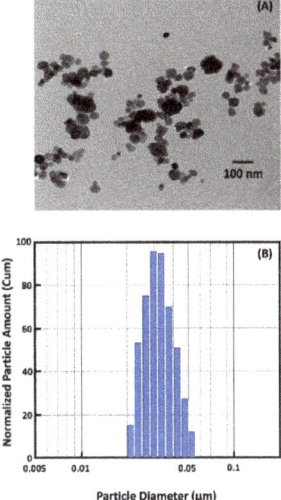

Figure 1. (**A**) TEM of nanoparticles of CaF_2 (nCaF_2) synthesized in this study. (**B**) Particle size distribution of nCaF_2. The nCaF_2 were synthesized via a spray-drying technique and collected using an electrostatic precipitator.

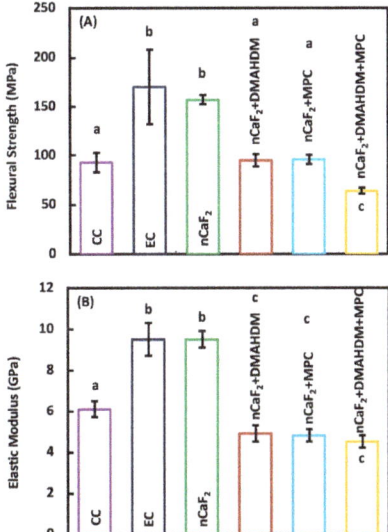

Figure 2. Mechanical properties of composites: (**A**) Flexural strength, and (**B**) Elastic modulus (mean ± sd; $n = 6$). The flexural strength was higher in experimental control (EC) and nCaF_2 than commercial Heliomolar control ($p < 0.05$). Flexural strength in nCaF_2+DMAHDM and nCaF_2+MPC matched Heliomolar control ($p > 0.05$). However, the flexural strength in nCaF_2+DMAHDM+MPC was reduced ($p < 0.05$). The elastic modulus of the EC and nCaF_2 were higher than Heliomolar control ($p < 0.05$). All other groups had comparable elastic moduli to Heliomolar control after 1 day of immersion ($p > 0.05$). In each plot, different letters (a, b, c) indicate values that are significantly different from each other ($p < 0.05$).

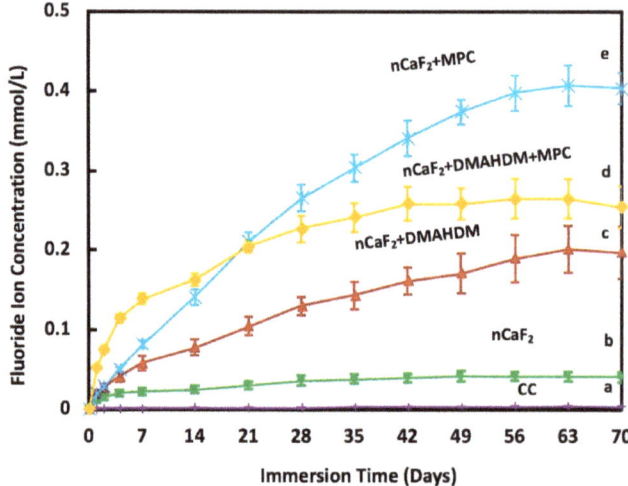

Figure 3. Fluoride (F) ion release from composites (mean ± sd; $n = 6$) at pH 7.0. The incorporation of nCaF$_2$, DMAHDM, and MPC increased the release of F ions with time ($p < 0.05$). Different letters (a, b, c, d, e) indicate significant differences between groups at day 70 ($p < 0.05$).

The calcium ion release is plotted in Figure 4. At 70 days, nCaF$_2$+MPC had a Ca ion release of (0.32 ± 0.005) mmol/L, and nCaF$_2$+DMAHDM+MPC had a similar ion release at (0.35 ± 0.006) mmol/L. Groups containing MPC had ion releases that were significantly higher when compared to other groups ($p < 0.05$). nCaF$_2$ had a Ca ion release of (0.11 ± 0.004) mmol/L. nCaF$_2$+DMAHDM had (0.18 ± 0.005) mmol/L, and Heliomolar control had close to zero Ca ion release.

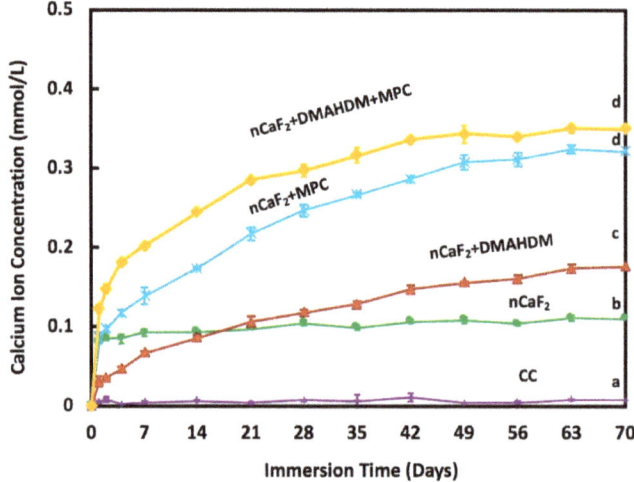

Figure 4. Calcium (Ca) ion release from composite resins (mean ± sd; $n = 6$) at pH 7.0. The incorporation of nCaF$_2$, DMAHDM, and MPC increased the release of Ca ions with time ($p < 0.05$). Different letters (a, b, c, d) indicate significant differences between groups at day 70 ($p < 0.05$).

Two-day biofilm colony forming units CFU on composites are shown in Figure 5: (A) Total microorganisms, (B) total streptococci and (C) mutans streptococci (mean ± SD; $n = 9$). CFU was reduced by 6 logs from a mean of 2.51×10^8 counts for Heliomolar control to 1.00×10^2 counts for the new nCaF$_2$+DMAHDM+MPC composite ($p < 0.05$). nCaF$_2$+DMAHDM reduced the CFU by 4 logs ($p < 0.05$). The combination of nCaF$_2$+DMAHDM+MPC yielded the smallest CFU counts.

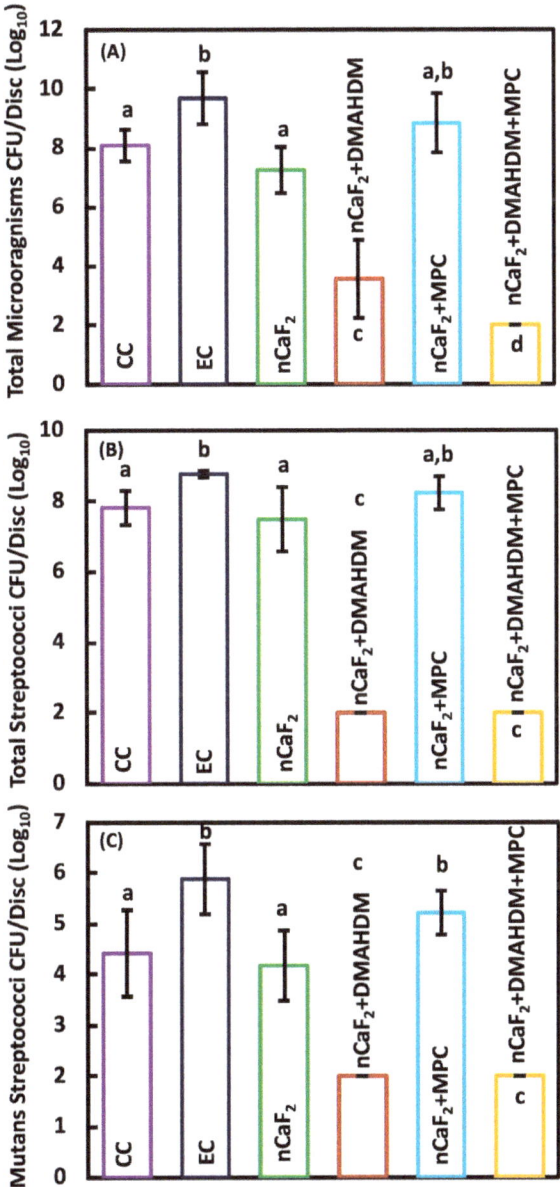

Figure 5. Colony-forming unit (CFU) counts of 2-day biofilm on composite discs (mean ± sd; $n = 9$): (**A**) Total microorganisms, (**B**) total streptococci, and (**C**) mutans streptococci (mean ± sd; $n = 9$). The incorporation of nCaF$_2$+DMAHDM+MPC had the lowest CFU, followed by nCaF$_2$+DMAHDM in the total microorganisms. The reduction of CFU in nCaF$_2$+DMAHDM+MPC and nCaF$_2$+DMAHDM was similar in total streptococci and mutans streptococci ($p < 0.05$). In each plot, different letters (a, b, c, d) indicate significant differences between groups ($p < 0.05$).

The bacterial metabolic activity of 2 days biofilm on composites is shown in Figure 6A. Metabolic activity was decreased from 0.18 (OD_{540}/cm^2) for commercial Heliomolar composite control, to 0.02 for the $nCaF_2$+DMAHDM+MPC composite ($p < 0.05$). Biofilm lactic acid production results can be seen in Figure 6B. The lactic acid production was reduced from 0.72 mmol/L on commercial Heliomolar control composite to 0.29 mmol/L in the $nCaF_2$+DMAHDM+MPC composite.

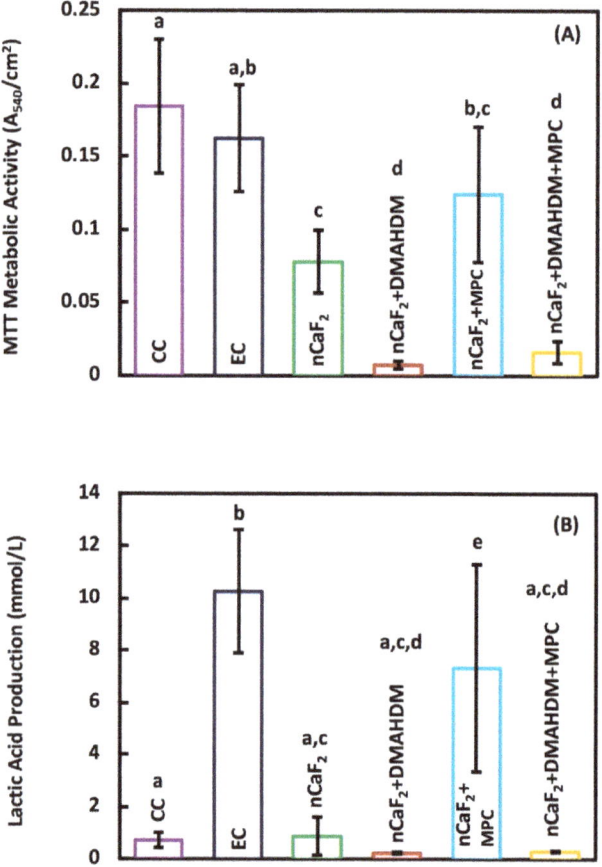

Figure 6. (**A**) The total metabolic activity, and (**B**) Lactic acid production (mean ± sd; $n = 9$) of composite specimens exposed to salivary biofilm. The incorporation of $nCaF_2$+ DMAHDM had the best reduction in the metabolic activity and lactic acid production ($p < 0.05$). In each plot, different letters (a, b, c, d) indicate significant differences between groups ($p < 0.05$).

SEM results in Figure 7 indicate an extensive biofilm formation at 24 h, 48 h, and 96 h in all groups except those containing the antimicrobial DMAHDM. Heliomolar (CC), experimental control (EC), and $nCaF_2$+MPC groups had the most biofilm formation at all time points, followed by the $nCaF_2$ group. Biofilm formation increased with time. However, in DMAHDM-containing groups there was minimal attachment of biofilm observed. At 96 h, the DMAHDM-containing groups showed a reduction in biofilm attachment compared with earlier time points.

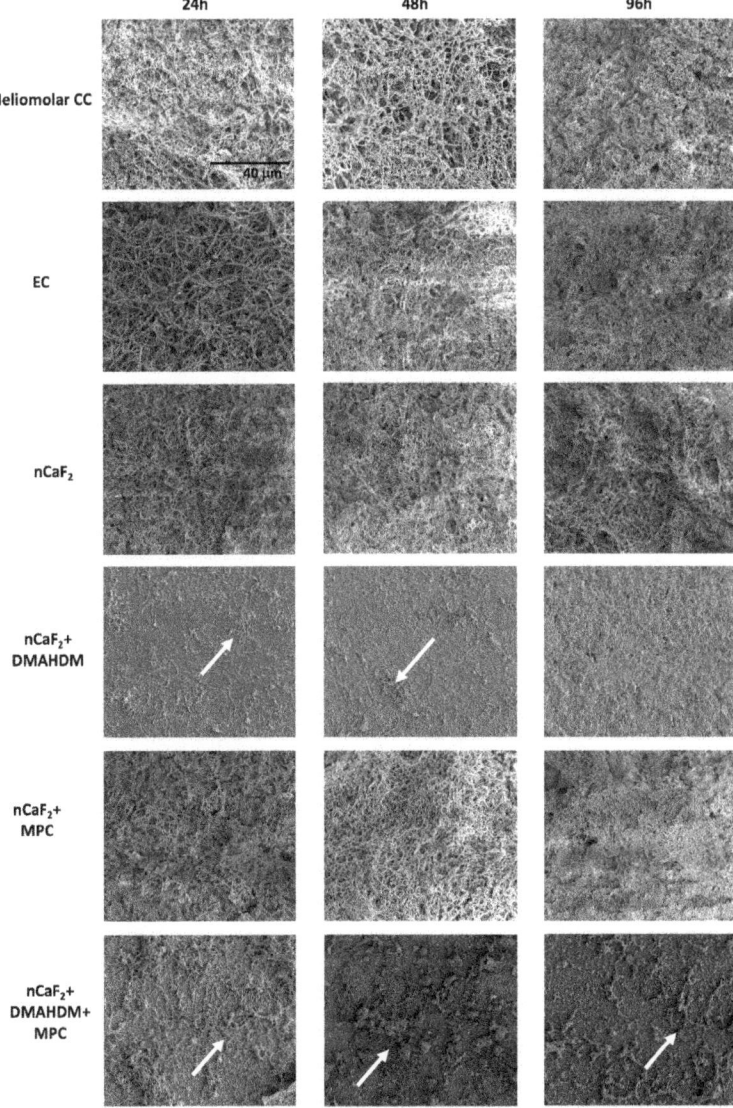

Figure 7. Scanning Electron Microscopy (SEM) images of biofilm formed on composite disc surfaces at 24 h, 48 h, and 96 h. All images are at 3000× magnification. Arrows indicate presence of minimal biofilm on DMAHDM-containing specimens. All other groups had full coverage of a substantial and mature biofilm.

5. Discussion

The present study developed a new composite consisting of $nCaF_2$, DMAHDM and MPC, and investigated the mechanical, ion release and oral biofilm properties. The protein-repellant $nCaF_2$+DMAHDM+MPC had slightly lower mechanical properties when compared to all other groups. However, $nCaF_2$+DMAHDM had good mechanical properties matching those of a commercial composite, while possessing high levels of F and Ca ion release and a strong antibacterial effect.

The new bioactive composite nCaF$_2$+DMAHDM decreased biofilm CFU by four orders of magnitude over that of commercial control composite, and substantially reduced acid production to inhibit caries.

Critical to the development of any new composite restorative material are its mechanical properties. The flexural strength of EC was around (170 ± 38) MPa. With the addition of 15% nCaF$_2$, the strength was reduced slightly to (157 ± 5) MPa. Those values are consistent with the results in a previous study [32]. The composite in that study had 65% total fillers and 20% CaF$_2$, similar to the 70% total fillers and 15% nCaF$_2$ composition in the present study. The flexural strength for the control group containing no nCaF$_2$ in the previous study was (145 ± 9) MPa, while the group which had 20% nCaF$_2$ had a flexural strength of 121 MPa. The nCaF$_2$+DMAHDM and nCaF$_2$+MPC had similar strengths of (95 ± 6), and (96 ± 5) MPa, respectively. Meanwhile, for nCaF$_2$+DMAHDM+MPC, the flexural strength was around (64 ±3) MPa. Those are similar to the results of a previous study using the same filler mass fraction where the flexural strength was around 90 MPa when 3% MPC and 3% DMAHDM were added to the BT resin [43]. The decreased strength in nCaF$_2$+DMAHDM+MPC could be due to two reasons. First, the BT base resin is composed of BisGMA and TEGDMA, which contain two reactive groups each. However, both DMAHDM and MPC are monomethacrylates and contain one reactive group. The addition of these monomethacrylates may substantially change the reaction kinetics and resulting crosslinked network, which can result in inferior mechanical properties. Another possible explanation for the decrease in mechanical properties is due to the hydrophilic nature of MPC. The specimens for flexural strength were stored for 24 h in water at 37 °C. The hydrophilicity of MPC makes it more likely that the composite would absorb water during the 24 h immersion prior to testing. This would likely lead to the reduction in flexural strength compared with the non-MPC groups.

Fluoride releasing materials have been known to render teeth more resistant to decay [45]. However, many available materials with fluoride-releasing properties, such as glass ionomer and resin modified glass ionomers, have inferior mechanical properties that do not serve the high load requirements for restorations in stress-bearing areas. Therefore, the incorporation of F ions into composite restorations would have the dual benefit of fluoride release and improved mechanical properties [46–48]. The current study synthesized nCaF$_2$ through a spray-drying method, which produced nanoparticles of CaF$_2$ with a median particle size of about 32 nm. The use of small amounts of nCaF$_2$ fillers prevented the mechanical properties to be compromised. As seen in previous studies, the use of fluoride fillers with larger particle sizes require higher mass fractions of fillers to achieve significant fluoride ion release. This would lead to a decrease in the composite mechanical properties [49].

The incorporation of nanoparticles with a high surface area resulted in the release of high levels of F and Ca ions using a low nCaF$_2$ filler level [33]. F ions foster remineralization by forming fluoroapatite [Ca$_5$(PO$_4$)$_3$F] [50]. Furthermore, it was shown that the lower the pH, the higher the release of ions. In this study, the release of F ions was significantly increased with the integration of 15% nCaF$_2$ at a pH of 7, measuring (0.04 ± 0.01) mmol/L. This was comparable to the 0.03 mmol/L F release level in a study that had a pH of 6, yet with an increased percentage of CaF$_2$ (23%) [51]. Groups containing MPC had the highest F ion release over time. The levels of release in nCaF$_2$+MPC and nCaF$_2$+DMAHDM+MPC were (0.40 ± 0.02) mmol/L and (0.25 ± 0.03) mmol/L, respectively. This could be due to the hydrophilic nature of MPC, resulting in a higher level of water uptake. As more water was absorbed by the composite, the Ca and F ions were solubilized, leading to an increase in the initial ion release. However, nCaF$_2$+DMAHDM F ion release was (0.20 ± 0.03) mmol/L, which was superior to the control. While the amount of nCaF$_2$ was the same in the two groups, the lower level of initial release in nCaF$_2$+DMAHDM, when compared to MPC containing groups, could result in a more sustained level of F ion release over a longer period. The releasing trend of F and Ca ions were the same for all groups. The groups containing nCaF$_2$ had a high F release initially, and subsequently were followed by a steady state release. The higher release of Ca and F ions in all nCaF$_2$ composites revealed that CaF$_2$ nanoparticles were fast released from the nCaF$_2$ composites because of the hydrolytic breakdown of the interface between the resin matrix and the CaF$_2$ nanoparticles. A likely explanation for such a trend is that the ions were released from the near-surface Ca and F reservoir of the nCaF$_2$ composite, and the reservoir near the surface diminished with the increase in immersion time. However, in clinical situations, the superficial layer of nCaF$_2$ would experience repeated chewing forces and tooth-brushing, hence the CaF$_2$-depleted surface would be removed by wear. As a result, a fresh surface would be exposed with the ability to further continue the calcium and fluoride release. Further studies are needed to investigate the outcome of wear on Ca and F ion release, as well as whether these materials can be made rechargeable to extend the lifetime of ion release.

Recurrent caries is considered a major drawback of composite restorations [1,52,53]. Studies proposed that the greater vulnerability to recurrent caries may be associated with the absence of composites' antibacterial capability when compared to other commonly used restorative materials such as glass ionomers and amalgams [2,54]. Therefore, the improvement of composites by incorporating antibacterial, protein repellent, and remineralizing properties is essential to lengthen the service life of composite restorations. DMAHDM has demonstrated strong antibacterial effects on an extensive antimicrobial spectrum through its contact-killing properties [26,55]. Since DMAHDM contains a reactive methacrylate group, it is copolymerized within the resin matrix, rendering it immobilized and preventing its release or loss over time [26,56]. Hence, its antibacterial capability is long-lasting. However, the antibacterial ability of DMAHDM is limited to biofilms that contact the composite surface [26], as biofilms that are not in contact with DMAHDM cannot be killed via DMAHDM's mechanism of action. Even with this limitation, the composite containing nCaF$_2$+DMAHDM substantially lowered the biofilm development and production of lactic acid, reducing the CFU of biofilm by 4 logs.

It can be seen in Figure 6B that the experimental control group and the group containing nCaF$_2$ and MPC exhibited a greater lactic acid production when compared with the other groups. DMAHDM has been shown in previous work to have a strong antibacterial effect, as illustrated by the CFU data in Figure 5. This antibacterial activity is reflected indirectly by the lactic acid production. When an *S mutans* biofilm accumulates, the secretion of lactic acid occurs. In the absence of a viable biofilm, the concentration of lactic acid is expected to be very low, as is the case in the DMAHDM-containing groups. Interestingly, the nCaF$_2$ and nCaF$_2$+MPC groups did not have a significant reduction in CFU (Figure 5), but the nCaF$_2$ group showed a significant reduction in lactic acid production in contrast to the nCaF$_2$+MPC group. Previously, it has been shown that the inclusion of CaF$_2$ in a composite formulation has a moderating effect on the lactic acid production [57]. It has been speculated that the F ion release helped reduce the acid production of the bacteria via the inhibition of metabolic pathways such as the fermentation pathway for lactic acid production, biofilm plaque, and hydrodynamic effects on mass transfer, fluoride delivery, and caries [58]. However, when MPC was added to the composite containing nCaF$_2$, there was no decrease in lactic acid production. This response may be related to the interactions between the calcium and fluoride ions and the phosphorylcholine fragment of MPC. The protein-repellent nature of MPC from the hydrogen bond network of water molecules surrounding the phosphorylcholine fragment limited the ability of the protein to adsorb on the surface. It was found, however, that the presence of halide ions (Cl$^-$, Br$^-$, I$^-$) influenced the hydrogen bond network and the diffusion through it [59]. It is possible that the presence of F$^-$ ions would have a similar result, which may affect F ion diffusion and disrupt the creation of a protein-repellant coating, leading to biofilm formation.

SEM images in Figure 7 confirm the presence of substantial biofilm on all groups except those containing DMAHDM. At 24 h and 48 h, Heliomolar (CC), experimental control (EC), nCaF$_2$, and nCaF$_2$+MPC groups exhibited the formation of microcolonies and early biofilm multilayers. The biofilms on these groups at 96 h were fully mature, substantial, and dense. However, in DMAHDM-containing groups there was minimal attachment of biofilm observed. At 96 h the DMAHDM-containing groups showed a reduction in biofilm attachment compared to earlier time points. It is possible that the antimicrobial effect of DMAHDM as a contact-killing agent had eliminated the bacteria left on the samples over time. These results correlate very well with the CFU results.

It is possible that the salivary proteins adsorption on composite resin surfaces might reduce the effectiveness of this "contact-killing" mechanism. Therefore, a protein repellent agent was added. MPC is a methacrylate that contains phospholipid polar groups. MPC is highly hydrophilic and has been shown to decrease the adsorption of protein and reduce bacterial attachments [60]. Previous work indicated that integrating MPC into the resin decreased the adsorption of proteins by approximately one order of magnitude [43,61]. Additionally, it was shown that surfaces containing MPC were resistant to brushing mechanical stresses [62]. In this study, MPC was incorporated into the resin, and copolymerized into the composite resin. This is analogous to the process of incorporating DMAHDM, with a similar result of a stable, covalently bonded, and non-releasing functionality.

In the current study, the antibacterial capability of the experimental composite was improved with $nCaF_2$ and was further substantially improved with the use of both $nCaF_2$ and DMAHDM. However, the $nCaF_2$+MPC group showed no reduction in the CFU counts. The composite that had $nCaF_2$+DMAHDM+MPC displayed the most potent antibacterial effect with CFU 6 log biofilm reduction. Therefore, these results confirmed that the addition of DMAHDM is vital to improving the antibacterial effect. It is thought that DMAHDM serves to reduce the accumulation of biofilm and, as a result, improves the efficacy of MPC in resisting protein adsorption. However, in the absence of DMAHDM, the bacterial biofilm will accumulate and render MPC less effective.

The combination of $nCaF_2$+DMAHDM promoted remineralization and had a significant antibacterial effect. Another potential benefit of using $nCaF_2$ is that the release of F ions could act as a transmembrane proton carrier and an inhibitor to the glycolytic enzyme, thus preventing oral microorganisms by stimulating acidification of the cytoplasm and providing antibacterial effect at a long-distance to inhibit caries [10,63]. These properties would be highly valuable to inhibit recurrent caries around the margins of the restoration, since this is the location that most dental plaque tends to accumulate. In addition to its superior antibacterial and remineralization properties, the $nCaF_2$+DMAHDM composite exhibits excellent mechanical properties which make it suitable in a variety of restoration applications. In comparison, the composite consisting of $nCaF_2$+DMAHDM+MPC exhibited lower mechanical properties. However, the strength and modulus achieved by this group was still high enough to be used for restorations in low load areas such as class V restorations [37]. This is a critical area of utilization, since older patients tend to have exposed root surfaces that are more prone to caries. Further studies are needed to investigate the $nCaF_2$+DMAHDM combination in dental cements, bonding agents, fissure sealants, and composites to remineralize tooth lesions and suppress biofilm and plaque buildup, especially in patients with a high caries risk.

6. Conclusions

A novel composite with remineralization and antibacterial properties was established by combining $nCaF_2$+DMAHDM for the first time. Release of F and Ca ions from the composite was significantly achieved through the incorporation of $nCaF_2$. The new composite exhibited strong antibacterial and ion release capabilities, prevented biofilm production of lactic acid, and decreased biofilm CFU by 4 log. This bioactive nanocomposite is promising to protect tooth structures, inhibit demineralization, and provide a reservoir for the release of calcium and fluoride ions. The combined effect of $nCaF_2$+DMAHDM shows promise in a variety of dental applications where remineralization and prevention of recurrent caries is a priority.

Author Contributions: H.M. contributed to design, data acquisition, and analysis, and drafted the manuscript; A.A.B. contributed to data acquisition, and analysis; R.A. contributed to data acquisition and analysis; T.W.O. contributed to data interpretation and funding acquisition; M.A.S.M. contributed to data acquisition, analysis, and interpretation; H.H.K.X. contributed to conception, design, data interpretation and critically revised the manuscript; M.D.W. contributed to design, data acquisition, analysis, and interpretation, and critically revised the manuscript. All authors have read and agreed to the published version of the manuscript.

Funding: This research received no external funding.

Acknowledgments: We thank Quan Dai, Jianghong Gao, Maria Ibrahim, Bashayer Baras, Pei Feng, Nancy J. Lin, Jirun Sun, and Laurence C. Chow for discussions and assistance. We thank the technical support of the Core Imaging Facility of the University of Maryland Baltimore. We are grateful to Dentsply Sirona (Milford, DE) for the donation of the glass fillers. This work was supported by The University of Maryland School of Dentistry bridge fund (HX) and University of Maryland seed grant (HX).

Conflicts of Interest: The authors declare no conflict of interest.

References

1. Ferracane, J.L. Resin composite—State of the art. *Dent. Mater.* **2011**, *27*, 29–38. [CrossRef]
2. Beyth, N.; Domb, A.J.; Weiss, E.I. An in vitro quantitative antibacterial analysis of amalgam and composite resins. *J. Dent.* **2007**, *35*, 201–206. [CrossRef] [PubMed]
3. Deligeorgi, V.; Mjör, I.A.; Wilson, N.H. An overview of reasons for the placement and replacement of restorations. *Prim. Dent. Care* **2001**, *8*, 5–11. [CrossRef] [PubMed]
4. Jokstad, A.; Bayne, S.; Blunck, U.; Tyas, M.; Wilson, N. Quality of dental restorations. FDI Commission Project 2-95. *Int. Dent. J.* **2001**, *51*, 117–158. [CrossRef]
5. Takahashi, N.; Nyvad, B. Ecological Hypothesis of Dentin and Root Caries. *Caries Res.* **2016**, *50*, 422–431. [CrossRef]
6. Kidd, E.A.M.; Fejerskov, O. What constitutes dental caries? Histopathology of carious enamel and dentin related to the action of cariogenic biofilms. *J. Dent. Res.* **2004**, *83*, C35–C38. [CrossRef]
7. Kidd, E. *Essentials of Dental Caries: The Disease and Its Management*, 3rd ed.; OUP Oxford: Oxford, UK, 2005; ISBN 978-0-19-852978-1.
8. Xu, X.; Ling, L.; Wang, R.; Burgess, J.O. Formulation and characterization of a novel fluoride-releasing dental composite. *Dent. Mater.* **2006**, *22*, 1014–1023. [CrossRef]
9. Xu, H.H.K.; Moreau, J.L.; Sun, L.; Chow, L.C. Novel CaF_2 Nanocomposite with High Strength and Fluoride Ion Release. *J. Dent. Res.* **2010**, *89*, 739–745. [CrossRef]
10. Zheng, X.; Cheng, X.; Wang, L.; Qiu, W.; Wang, S.; Zhou, Y.; Li, M.; Li, Y.; Cheng, L.; Li, J.; et al. Combinatorial effects of arginine and fluoride on oral bacteria. *J. Dent. Res.* **2015**, *94*, 344–353. [CrossRef]
11. Koo, H. Strategies to enhance the biological effects of fluoride on dental biofilms. *Adv. Dent. Res.* **2008**, *20*, 17–21. [CrossRef]
12. Rošin-Grget, K. The cariostatic mechanisms of fluoride. *Acta Med. Acad.* **2013**, *42*, 179–188. [CrossRef]
13. ten Cate, J.M. Current concepts on the theories of the mechanism of action of fluoride. *Acta Odontol. Scand.* **1999**, *57*, 325–329. [CrossRef]
14. ten Cate, J.M. Contemporary perspective on the use of fluoride products in caries prevention. *Br. Dent. J.* **2013**, *214*, 161–167. [CrossRef]
15. Wojnarowska-Nowak, R.; Rzeszutko, J.; Barylyak, A.; Nechyporenko, G.; Zinchenko, V.; Leszczyńska, D.; Bobitski, Y.; Kus-Liśkiewicz, M. Structural, physical and antibacterial properties of pristine and Ag+ doped fluoroapatite nanomaterials. *Adv. Appl. Ceram.* **2017**, *116*, 108–117. [CrossRef]
16. Anastasiou, A.D.; Nerantzaki, M.; Gounari, E.; Duggal, M.S.; Giannoudis, P.V.; Jha, A.; Bikiaris, D. Antibacterial properties and regenerative potential of Sr^{2+} and Ce^{3+} doped fluorapatites; a potential solution for peri-implantitis. *Sci. Rep.* **2019**, *9*. [CrossRef]
17. Imazato, S.; Russell, R.R.; McCabe, J.F. Antibacterial activity of MDPB polymer incorporated in dental resin. *J. Dent.* **1995**, *23*, 177–181. [CrossRef]
18. Imazato, S.; Ehara, A.; Torii, M.; Ebisu, S. Antibacterial activity of dentine primer containing MDPB after curing. *J. Dent.* **1998**, *26*, 267–271. [CrossRef]
19. Imazato, S. Bio-active restorative materials with antibacterial effects: New dimension of innovation in restorative dentistry. *Dent. Mater. J.* **2009**, *28*, 11–19. [CrossRef]
20. Imazato, S.; Torii, M.; Tsuchitani, Y.; McCabe, J.F.; Russell, R.R. Incorporation of bacterial inhibitor into resin composite. *J. Dent. Res.* **1994**, *73*, 1437–1443. [CrossRef]
21. Beyth, N.; Yudovin-Farber, I.; Perez-Davidi, M.; Domb, A.J.; Weiss, E.I. Polyethyleneimine nanoparticles incorporated into resin composite cause cell death and trigger biofilm stress in vivo. *Proc. Natl. Acad. Sci. USA* **2010**, *107*, 22038–22043. [CrossRef]
22. Jiao, Y.; Tay, F.R.; Niu, L.-N.; Chen, J.-H. Advancing antimicrobial strategies for managing oral biofilm infections. *Int. J. Oral Sci.* **2019**, *11*, 28. [CrossRef] [PubMed]

23. Imazato, S. Antibacterial properties of resin composites and dentin bonding systems. *Dent. Mater.* **2003**, *19*, 449–457. [CrossRef]
24. Faust, D.; Dolado, I.; Cuadrado, A.; Oesch, F.; Weiss, C.; Nebreda, A.R.; Dietrich, C. p38alpha MAPK is required for contact inhibition. *Oncogene* **2005**, *24*, 7941–7945. [CrossRef] [PubMed]
25. Li, F.; Weir, M.D.; Xu, H.H.K. Effects of quaternary ammonium chain length on antibacterial bonding agents. *J. Dent. Res.* **2013**, *92*, 932–938. [CrossRef]
26. Beyth, N.; Yudovin-Farber, I.; Bahir, R.; Domb, A.J.; Weiss, E.I. Antibacterial activity of dental composites containing quaternary ammonium polyethylenimine nanoparticles against Streptococcus mutans. *Biomaterials* **2006**, *27*, 3995–4002. [CrossRef]
27. Namba, N.; Yoshida, Y.; Nagaoka, N.; Takashima, S.; Matsuura-Yoshimoto, K.; Maeda, H.; Van Meerbeek, B.; Suzuki, K.; Takashiba, S. Antibacterial effect of bactericide immobilized in resin matrix. *Dent. Mater.* **2009**, *25*, 424–430. [CrossRef]
28. Müller, R.; Eidt, A.; Hiller, K.-A.; Katzur, V.; Subat, M.; Schweikl, H.; Imazato, S.; Ruhl, S.; Schmalz, G. Influences of protein films on antibacterial or bacteria-repellent surface coatings in a model system using silicon wafers. *Biomaterials* **2009**, *30*, 4921–4929. [CrossRef]
29. Xie, X.; Wang, L.; Xing, D.; Zhang, K.; Weir, M.D.; Liu, H.; Bai, Y.; Xu, H.H.K. Novel dental adhesive with triple benefits of calcium phosphate recharge, protein-repellent and antibacterial functions. *Dent. Mater.* **2017**, *33*, 553–563. [CrossRef]
30. Zhang, N.; Weir, M.D.; Romberg, E.; Bai, Y.; Xu, H.H.K. Development of novel dental adhesive with double benefits of protein-repellent and antibacterial capabilities. *Dent. Mater.* **2015**, *31*, 845–854. [CrossRef]
31. Zhou, H.; Li, F.; Weir, M.D.; Xu, H.H.K. Dental plaque microcosm response to bonding agents containing quaternary ammonium methacrylates with different chain lengths and charge densities. *J. Dent.* **2013**, *41*, 1122–1131. [CrossRef]
32. Xu, H.H.K.; Moreau, J.L.; Sun, L.; Chow, L.C. Strength and fluoride release characteristics of a calcium fluoride based dental nanocomposite. *Biomaterials* **2008**, *29*, 4261–4267. [CrossRef] [PubMed]
33. Weir, M.D.; Moreau, J.L.; Levine, E.D.; Strassler, H.D.; Chow, L.C.; Xu, H.H.K. Nanocomposite containing CaF2 nanoparticles: Thermal cycling, wear and long-term water-aging. *Dent. Mater.* **2012**, *28*, 642–652. [CrossRef] [PubMed]
34. Sun, L.; Chow, L.C. Preparation and properties of nano-sized calcium fluoride for dental applications. *Dent. Mater.* **2008**, *24*, 111–116. [CrossRef]
35. Trujillo-Lemon, M.; Ge, J.; Lu, H.; Tanaka, J.; Stansbury, J.W. Dimethacrylate derivatives of dimer acid. *J. Polym. Sci. Part A Polym. Chem.* **2006**, *44*, 3921–3929. [CrossRef]
36. Xu, H.H.K.; Moreau, J.L.; Sun, L.; Chow, L.C. Nanocomposite containing amorphous calcium phosphate nanoparticles for caries inhibition. *Dent. Mater.* **2011**, *27*, 762–769. [CrossRef]
37. ISO 4049:200.9. *Dentistry—Polymer-Based Restorative Materials*; International Organization for Standardization: Geneva, Switzerland, 2009.
38. Xu, H.H.K.; Weir, M.D.; Sun, L.; Takagi, S.; Chow, L.C. Effects of calcium phosphate nanoparticles on Ca-PO4 composite. *J. Dent. Res.* **2007**, *86*, 378–383. [CrossRef]
39. Skrtic, D.; Antonucci, J.M.; Eanes, E.D. Improved properties of amorphous calcium phosphate fillers in remineralizing resin composites. *Dent. Mater.* **1996**, *12*, 295–301. [CrossRef]
40. Wang, L.; Xie, X.; Li, C.; Liu, H.; Zhang, K.; Zhou, Y.; Chang, X.; Xu, H.H.K. Novel bioactive root canal sealer to inhibit endodontic multispecies biofilms with remineralizing calcium phosphate ions. *J. Dent.* **2017**, *60*, 25–35. [CrossRef]
41. Xie, X.; Wang, L.; Xing, D.; Arola, D.D.; Weir, M.D.; Bai, Y.; Xu, H.H.K. Protein-repellent and antibacterial functions of a calcium phosphate rechargeable nanocomposite. *J. Dent.* **2016**, *52*, 15–22. [CrossRef]
42. McBain, A.J. In Vitro Biofilm Models: An Overview. In *Advances in Applied Microbiology*; Elsevier: Cambridge, MA, USA, 2009; Volume 69, pp. 99–132.
43. Zhang, N.; Ma, J.; Melo, M.A.S.; Weir, M.D.; Bai, Y.; Xu, H.H.K. Protein-repellent and antibacterial dental composite to inhibit biofilms and caries. *J. Dent.* **2015**, *43*, 225–234. [CrossRef]
44. Al-Dulaijan, Y.A.; Cheng, L.; Weir, M.D.; Melo, M.A.S.; Liu, H.; Oates, T.W.; Wang, L.; Xu, H.H.K. Novel rechargeable calcium phosphate nanocomposite with antibacterial activity to suppress biofilm acids and dental caries. *J. Dent.* **2018**, *72*, 44–52. [CrossRef] [PubMed]

45. Deng, D.M.; van Loveren, C.; ten Cate, J.M. Caries-preventive agents induce remineralization of dentin in a biofilm model. *Caries Res.* **2005**, *39*, 216–223. [CrossRef] [PubMed]
46. Kus-Liśkiewicz, M.; Rzeszutko, J.; Bobitski, Y.; Barylyak, A.; Nechyporenko, G.; Zinchenko, V.; Zebrowski, J. Alternative Approach for Fighting Bacteria and Fungi: Use of Modified Fluorapatite. *J. Biomed. Nanotechnol.* **2019**, *15*, 848–855. [CrossRef] [PubMed]
47. Montazeri, N.; Jahandideh, R.; Biazar, E. Synthesis of fluorapatite–hydroxyapatite nanoparticles and toxicity investigations. *Int. J. Nanomed.* **2011**, *6*, 197–201. [CrossRef]
48. Wiegand, A.; Buchalla, W.; Attin, T. Review on fluoride-releasing restorative materials—Fluoride release and uptake characteristics, antibacterial activity and influence on caries formation. *Dent. Mater.* **2007**, *23*, 343–362. [CrossRef]
49. Balhaddad, A.A.; Kansara, A.A.; Hidan, D.; Weir, M.D.; Xu, H.H.K.; Melo, M.A.S. Toward dental caries: Exploring nanoparticle-based platforms and calcium phosphate compounds for dental restorative materials. *Bioact. Mater.* **2019**, *4*, 43–55. [CrossRef]
50. Lata, S.; Varghese, N.O.; Varughese, J.M. Remineralization potential of fluoride and amorphous calcium phosphate-casein phospho peptide on enamel lesions: An in vitro comparative evaluation. *J. Conserv. Dent.* **2010**, *13*, 42–46. [CrossRef]
51. Anusavice, K.J.; Zhang, N.-Z.; Shen, C. Effect of CaF2 content on rate of fluoride release from filled resins. *J. Dent. Res.* **2005**, *84*, 440–444. [CrossRef]
52. Sakaguchi, R.L. Review of the current status and challenges for dental posterior restorative composites: Clinical, chemistry, and physical behavior considerations. Summary of discussion from the Portland Composites Symposium (POCOS) June 17–19, 2004, Oregon Health and Science University, Portland, Oregon. *Dent. Mater.* **2005**, *21*, 3–6. [CrossRef]
53. Demarco, F.F.; Corrêa, M.B.; Cenci, M.S.; Moraes, R.R.; Opdam, N.J.M. Longevity of posterior composite restorations: Not only a matter of materials. *Dent. Mater.* **2012**, *28*, 87–101. [CrossRef]
54. Nedeljkovic, I.; Teughels, W.; De Munck, J.; Van Meerbeek, B.; Van Landuyt, K.L. Is secondary caries with composites a material-based problem? *Dent. Mater.* **2015**, *31*, e247–e277. [CrossRef] [PubMed]
55. Imazato, S.; Ebi, N.; Takahashi, Y.; Kaneko, T.; Ebisu, S.; Russell, R.R.B. Antibacterial activity of bactericide-immobilized filler for resin-based restoratives. *Biomaterials* **2003**, *24*, 3605–3609. [CrossRef]
56. Xu, X.; Wang, Y.; Liao, S.; Wen, Z.T.; Fan, Y. Synthesis and characterization of antibacterial dental monomers and composites. *J. Biomed. Mater. Res. Part B Appl. Biomater.* **2012**, *100B*, 1151–1162. [CrossRef]
57. Cheng, L.; Weir, M.D.; Xu, H.H.K.; Kraigsley, A.M.; Lin, N.J.; Lin-Gibson, S.; Zhou, X. Antibacterial and physical properties of calcium-phosphate and calcium-fluoride nanocomposites with chlorhexidine. *Dent. Mater.* **2012**, *28*, 573–583. [CrossRef]
58. Stoodley, P.; Wefel, J.; Gieseke, A.; Debeer, D.; von Ohle, C. Biofilm plaque and hydrodynamic effects on mass transfer, fluoride delivery and caries. *J. Am. Dent. Assoc.* **2008**, *139*, 1182–1190. [CrossRef]
59. Zhang, Z.J.; Madsen, J.; Warren, N.J.; Mears, M.; Leggett, G.J.; Lewis, A.L.; Geoghegan, M. Influence of salt on the solution dynamics of a phosphorylcholine-based polyzwitterion. *Eur. Polym. J.* **2017**, *87*, 449–457. [CrossRef]
60. Sibarani, J.; Takai, M.; Ishihara, K. Surface modification on microfluidic devices with 2-methacryloyloxyethyl phosphorylcholine polymers for reducing unfavorable protein adsorption. *Colloids Surf. B Biointerfaces* **2007**, *54*, 88–93. [CrossRef]
61. Zhang, N.; Chen, C.; Melo, M.A.; Bai, Y.-X.; Cheng, L.; Xu, H.H. A novel protein-repellent dental composite containing 2-methacryloyloxyethyl phosphorylcholine. *Int. J. Oral Sci.* **2015**, *7*, 103–109. [CrossRef]
62. Tateishi, T.; Kyomoto, M.; Kakinoki, S.; Yamaoka, T.; Ishihara, K. Reduced platelets and bacteria adhesion on poly(ether ether ketone) by photoinduced and self-initiated graft polymerization of 2-methacryloyloxyethyl phosphorylcholine. *J. Biomed. Mater. Res. A* **2014**, *102*, 1342–1349. [CrossRef]
63. Van Loveren, C. Antimicrobial activity of fluoride and its in vivo importance: Identification of research questions. *Caries Res.* **2001**, *35* (Suppl. S1), 65–70. [CrossRef]

© 2020 by the authors. Licensee MDPI, Basel, Switzerland. This article is an open access article distributed under the terms and conditions of the Creative Commons Attribution (CC BY) license (http://creativecommons.org/licenses/by/4.0/).

Article

Early Polylysine Release from Dental Composites and Its Effects on Planktonic *Streptococcus mutans* Growth

Nikos N. Lygidakis [1], Elaine Allan [2], Wendy Xia [3], Paul F. Ashley [1] and Anne M. Young [3,*]

1. Unit of Paediatric Dentistry, Department of Craniofacial Growth and Development, UCL Eastman Dental Institute, London WC1X 8LD, UK; nikos.lygidakis@gmail.com (N.N.L.); p.ashley@ucl.ac.uk (P.F.A.)
2. Division of Microbial Diseases, UCL Eastman Dental Institute, London WC1X 8LD, UK; e.allan@ucl.ac.uk
3. Department of Biomaterials and Tissue Engineering, UCL Eastman Dental Institute, London WC1X 8LD, UK; wendy.xia2011@yahoo.co.uk
* Correspondence: anne.young@ucl.ac.uk

Received: 19 June 2020; Accepted: 24 July 2020; Published: 27 July 2020

Abstract: The study aim was to assess the effect of incorporating polylysine (PLS) filler at different mass fractions (0.5, 1 and 2 wt%) on PLS release and *Streptococcus mutans* planktonic growth. Composite containing PLS mass and volume change and PLS release upon water immersion were assessed gravimetrically and via high-performance liquid chromatography (HPLC), respectively. Disc effects on bacterial counts in broth initially containing 8×10^5 versus 8×10^6 CFU/mL *Streptococcus mutans* UA159 were determined after 24 h. Survival of sedimented bacteria after 72 h was determined following LIVE/DEAD staining of composite surfaces using confocal microscopy. Water sorption-induced mass change at two months increased from 0.7 to 1.7% with increasing PLS concentration. Average volume increases were 2.3% at two months whilst polylysine release levelled at 4% at 3 weeks irrespective of composite PLS level. Early percentage PLS release, however, was faster with higher composite content. With 0.5, 1 and 2% polylysine initially in the composite filler phase, 24-h PLS release into 1 mL of water yielded 8, 25 and 93 ppm respectively. With initial bacterial counts of 8×10^5 CFU/mL, this PLS release reduced 24-h bacterial counts from 10^9 down to 10^8, 10^7 and 10^2 CFU/mL respectively. With a high initial inoculum, 24-h bacterial counts were 10^9 with 0, 0.5 or 1% PLS and 10^7 with 2% PLS. As the PLS composite content was raised, the ratio of dead to live sedimented bacteria increased. The antibacterial action of the experimental composites could reduce residual bacteria remaining following minimally invasive tooth restorations.

Keywords: polylysine; antibacterial; dental composites

1. Introduction

Dental caries is one of the most common diseases affecting humans. Traditionally, treatment involved the removal of all the affected tooth structure followed by replacement with amalgam. Silver mercury amalgam is a highly effective restorative material due to both its good mechanical and antibacterial properties. Following the Minamata agreement in 2013 on global mercury reduction signed by 140 counties, amalgam is being phased out. In Europe, amalgam use in primary teeth was banned in 2018 [1].

Modern dentistry advocates for the removal of highly infected surface dentine but not underlying demineralized and less infected dentine that may be close to the pulp [2]. Effective cavity sealing with Glass Ionomer Cement (GIC) restorations can reduce any underlying residual bacterial contamination. It can also encourage remineralization but have insufficient strength for larger cavities.

Conversely, dental composites have greater strength, but their placement is complex and their technique sensitive. Multiple adhesion-promoting steps are required, in addition to material placement and light curing in several increments to reduce polymerization shrinkage consequences [3]. Composite failure is usually due to secondary caries at the restoration margins. This is attributed to a combination of residual bacteria, shrinkage and enzyme-activated degradation of demineralized dentine, which may compromise the bonding interface, enabling further bacterial penetration and growth [4]. A composite material that could provide surface release of antibacterial and remineralizing agents upon damage of the tooth/restoration interface might therefore be beneficial.

Antibacterial components such as chlorhexidine, fluoride, quaternary ammonium methacrylate, silver and triclosan have previously been added to dental composites [5–9]. Polylysine (PLS) is an alternative antibacterial agent that has been included in composites more recently. Polylysine-greater eukaryotic cell-compatibility may address the limited biocompatibility of some other antibacterial agents [10,11]. Studies suggest electrostatic adsorption of polylysine on bacterial surfaces, and its abnormal distribution within their cytoplasm, leads to their cell damage [12,13]. Polylysine, however, is generally recognized as safe (GRAS). It is a natural polypeptide that is biodegradable, water-soluble, nontoxic and edible. Additionally, polylysine is extensively used in eukaryotic cell culture to promote cell attachment and as a food preservative [12,14]. In the human body, polylysine degrades harmlessly to give the essential amino acid, lysine [12,14]. Furthermore, polylysine addition into calcium phosphate-containing composites can enhance apatite precipitation on their surfaces from simulated body fluid. Polylysine may therefore also enhance composite remineralizing potential [10].

Whilst many properties of polylysine-containing dental composites have now been investigated [10,11], publications on their actions against bacteria are limited. In this study, therefore, the ability of composites with increasing levels of polylysine to affect bacterial growth was investigated. Effects on planktonic bacteria are assessed to model the feasibility of surface polylysine release reducing bacterial microleakage in gaps at the tooth restoration interface. Polylysine release kinetics are provided to help explain the varying levels of antibacterial action. Potential polylysine addition and release effects on set material stability are monitored through mass and volume changes in water.

2. Materials and Methods

2.1. Components

The resin matrix was prepared using Urethane Dimethacrylate (UDMA) (DMG, Hamburg, Germany) as a base monomer and PolyPropylene Glycol Dimethacrylate (PPGDMA) (Polysciences, Inc., Warrington, FL, USA) as a diluent monomer. 4-methacryloxyethyl trimellitate anhydride (4-META) (Polysciences, Inc., Warrington, FL, USA) and Camphorquinone (CQ) (DMG, Hamburg, Germany) were added at low level. The filler consisted of two different sizes of aluminosilicate glass of 7 μm and 0.7 μm (DMG, Hamburg, Germany), fumed silica (Aerosil OX 50, Evonik, Essen, Germany), monocalcium phosphate monohydrate (MCPM) (Himed, Old Bethpage, NY, USA) and polylysine (PLS) (Handary, Brussels, Belgium).

UDMA is a common base monomer, that has been used as an alternative to bisphenol A-glycidyl methacrylate (Bis-GMA) [15]. In previous studies, PPGDMA, a diluent monomer that can be used instead of triethylene glycol dimethacrylate (TEGDMA), enhanced paste stability, increased light activated polymerization and reduced associated shrinkage [11]. 4-META is an adhesion-promoting monomer [16]. Monocalcium phosphate monohydrate (MCPM) particles added to composites can promote water sorption and react to produce brushite crystals of greater volume thereby giving expansion to compensate polymerisation shrinkage. Furthermore, release of its phosphate ions can promote apatite precipitation from simulated dentinal fluid to remineralize dentine [17].

2.2. Paste Preparation

Four formulations with 1 variable (PLS) were prepared. The resin (liquid) phase consisted of UDMA (72 wt%), PPGDMA (24 wt%), 4-META (3 wt%), and CQ (1 wt%) [11]. This was prepared by mixing the components and stirring for 24 h at room temperature on a magnetic stirrer hot plate (Jeo Tech) until a clear liquid was achieved. The filler phase contained glass of 7 μm, 0.7 μm and fumed silica in the ratio 6:3:1. 10 wt% MCPM was added to the filler as in F2 in previous work [11] but the tricalcium phosphate was removed. Additionally, the filler PLS level was varied (0, 0.5, 1 or 2 wt%) instead of being fixed at 2 wt%. Filler: resin were mixed in the weight ratio 5:1 for 45 s at 3500 rpm using a centrifugal mixer (Speedmixer, Hauschild Engineering, Hamm, Germany). This high ratio produced pastes with a consistency comparable with the commercial packable composite Filtek Z250 (3M, Bracknell, UK) used as an additional control in antibacterial studies.

2.3. Composite Discs Preparation

Disc-shaped specimens were formed by applying the composite pastes within metal circlips with internal diameter 10 mm and thickness of 1 mm and pressing them between two sheets of acetate (Figure 1). The specimens were then photopolymerized using a blue light-emitting diode curing unit with a wavelength of 450–470 nm and power of 1100–1300 mW/cm^2 (Demi Plus, Kerr Dental, Bioggio, Switzerland) with the tip in contact with the acetate sheet. The curing duration was 40 s on each side of each disc. This method gives 72% monomer conversion for the formulations [11]. The discs were then removed from the circlip, any excesses were trimmed with a no.11 blade, and stored at room temperature in the dark until required.

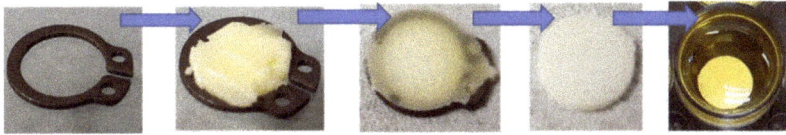

Figure 1. Preparation of composite disc and disc placed in well.

2.4. Mass and Volume Change

The mass and volume change of set discs versus time in deionised water (DW) were determined using a density kit and four-figure digital balance (AG204, Mettler, Toledo, OH, USA) according to the ISO 17304:2013. 1% sodium dodecyl sulphate (Sigma-Aldrich, Gillingham, UK) in DW was used as the buoyancy medium.

Discs (n = 3), prepared as above, were immersed in 1 mL of DW in individual sterilin tubes at 23 °C. At 1, 3, 6 h, 1, 2, 5, 7 days, 2, 4 and 8 weeks, the discs were removed from the solution, their surfaces blot dried using absorbent paper, weighed in air and in the buoyancy medium and then placed into new tubes with fresh 1 mL of DW. Initial mass and volume were calculated by extrapolation of early data versus square root of time to zero. Mass and volume change were then calculated as percentages of original mass as shown in detail in previous publications [10,11].

2.5. Polylysine Release

High-performance liquid chromatography (HPLC) (Shimadzu corporation, Kyoto, Japan) was used to measure polylysine release from composite discs. The composite discs (n = 3) were prepared as above, weighed and then immersed in 1 mL deionized water (DW) in individual sterilin tubes. The discs were removed from the DW and placed in a new tube with 1 mL of fresh DW at 1, 3, 6 h, 1, 2, 5, 7 days and 3 weeks. The storage solutions were stored at 23 °C prior to analysis.

A normal phase column in hydrophilic interaction liquid chromatography (HILIC) mode was used. The mobile phase was 50 vol% acetonitrile in DW with 0.1 vol% of phosphoric acid (97%) added. Flow rate, run time, temperature, and UV detection wavelength were 1.0 mL/min, 43 min, 30 °C and 210 nm respectively. PLS solutions of 10 to 100 ppm were used to generate a calibration curve which was then employed to determine PLS concentrations in each storage solution. Cumulative PLS versus time, as a percentage of that calculated to be in the original specimen, was then determined [10].

2.6. Bacterial Growth Inhibition

A single colony of *S. mutans* UA159 was inoculated into brain heart infusion broth (BHI, Oxoid, Basingstoke, UK) and incubated statically at 37 °C for 16 h in air enriched with 5% CO_2. The culture was diluted in BHI broth to generate an inoculum of 8×10^5 and 8×10^6 CFU/mL which was confirmed by viable counting.

On the day of each experiment the discs ($n = 3$) were placed in a custom made ultraviolet light box and irradiated for 30 min on each side to ensure sterility. Decontamination was verified by plating on agar. The discs were then placed into a 24 well plate as seen in Figure 1. 1 mL of the *S. mutans* inoculum was added to each well and the plates placed in the incubator on a shaking tray at 200 rpm at 37 °C in air. At 24 h, the cultures were 10-fold serially diluted and plated out on BHI agar (Oxoid, Basingstoke, UK). Bacterial colonies were counted after 3 days of incubation at 37 °C in 5% CO_2.

2.7. Surface Bacteria Observations

To assess viability of attached bacteria on material surfaces, composite discs ($n = 2$) were immersed in an *S. mutans* suspension with concentration of 5×10^6 CFU/mL and incubated statically for 72 h in air at 37 °C. The discs were removed and gently immersed in fresh BHI broth to remove unattached bacteria before placing in a clean well plate for staining using LIVE/DEAD Viability kit (Thermofisher Scientific, Loughborough, UK). Confocal laser scanning microscopy (Radiance 2100, Biorad, Hercules, CA, USA) with an objective lens of 10×–20× magnification was used to visualize bacteria. Images were saved and processed using Lasersharp 2000 (Biorad, Hercules, CA, USA) and ImageJ (ImageJ Developers).

2.8. Statistical Analysis

Data analysis was undertaken for volumetric analysis, polylysine release and bacterial viable counts using Analysis of Variance (ANOVA) and SPSS (IBM, Armonk, NY, USA) with $p < 0.05$ considered significant. To perform multiple comparisons of multiple formulations, the Bonferroni adjustment was used.

3. Results

3.1. Mass and Volume

Mass and volume change both increased proportional to the square root of time. Final mass change at 2 months increased linearly with polylysine content from 0.7 to 1.7 wt% (RSQ = 0.96). Final volume change, however, was not significantly affected by PLS level ($p > 0.05$) and was between 2.2 and 2.5 vol% (Figure 2).

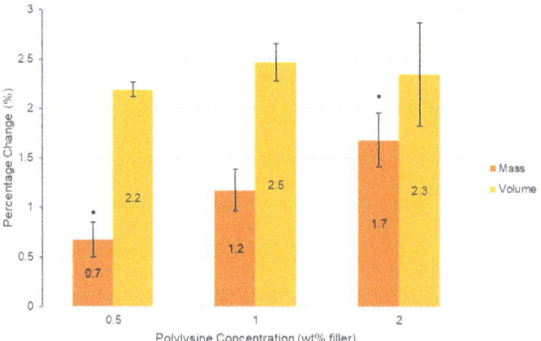

Figure 2. Graph of percentage change in mass and volume of the three formulations containing polylysine after 2 months. Error bars = st. dev. ($n = 3$). * indicates significance between results ($p < 0.05$).

3.2. Polylysine Release

Cumulative polylysine release was initially proportional to the square root of time but then levelled between 2 and 3 weeks (Figure 3). Cumulative percentage release was higher with 2% PLS at 1 and 2 days at 2.4 and 2.7% but within experimental error independent of concentration and 4% by 3 weeks. In Figure 4, the percentage release is converted to provide a calculated total concentration per disc that would have been released into 1 mL of storage solution by 6 h, 24 h and three weeks. At 3 weeks, the PLS release in grams was proportional to the level in the filler. At earlier time points, however, doubling the filler PLS content more than doubled the release in grams (see Figure 4).

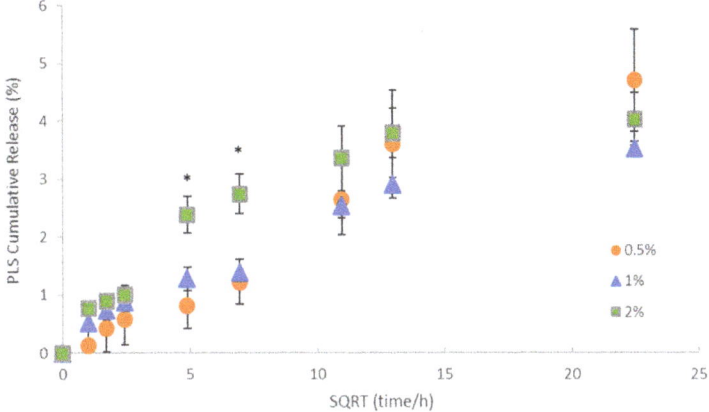

Figure 3. Cumulative percentage release of polylysine (PLS) versus the square root (SQRT) of time. Error bars = st. dev (n = 3). * indicates that formulation with 2% PLS has significantly higher percentage release at 24 and 48 h but not at early or later times.

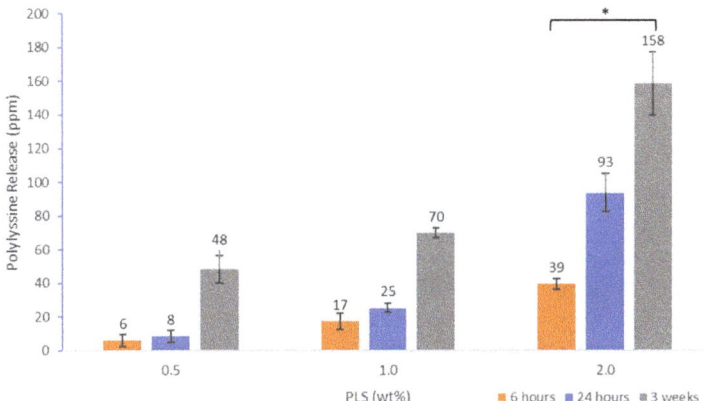

Figure 4. Cumulative polylysine release after 6 h, 24 h and 3 weeks at 23 °C for formulations with 0.5, 1 or 2% PLS. Error bars = st. dev (n = 3). * indicates 2% PLS containing formulation has significantly higher release at all time points compared with the other 2 formulations.

3.3. Determination of Antibacterial Activity

Colony forming units are shown in Figure 5, after 24 h incubation in air with different composite formulations for two different initial inoculum bacterial levels. Irrespective of initial inoculum level, CFU increased to 10^9 for all controls with no PLS. This was also observed with 0.5% and 1% PLS with the higher initial inoculum. Conversely, 0.5 and 1% PLS in the composite filler caused a 90 and

99% reduction in CFU with the lower initial inoculum concentration. With 2% PLS, the 24 h CFU count was down to 10^7 and 10^2 with high versus low initial inoculum levels, respectively. The 2% PLS formulations had statistically significantly less bacteria at 24 h when compared to all other composites at both inoculum concentrations.

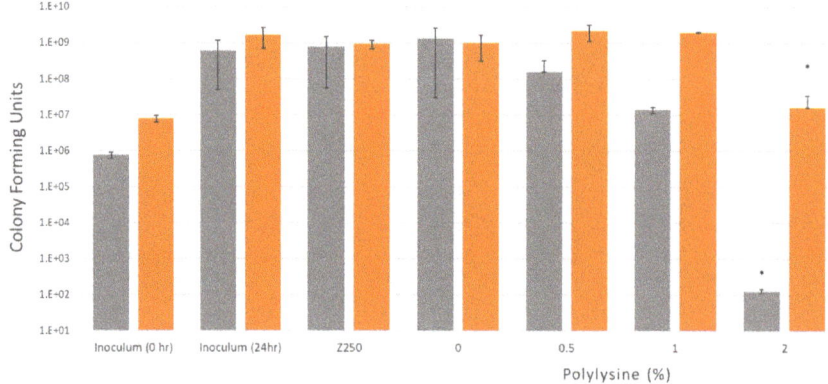

Figure 5. Bacterial growth after 24 h with two different inoculum concentrations of *S. mutans*, 8×10^5 and 8×10^6 CFU/mL. Four different formulations and a commercial composite were tested. Plates with discs and inoculum were incubated in air for 24 h at 37 °C while being shaken. Error bars = st. dev ($n = 3$), * indicates 2% is significantly different when compared to other formulations.

3.4. Confocal Laser Scanning Microscopy

Figure 6 shows confocal images of composite surfaces stained with live/dead stain. The bacteria do not form a biofilm due to the lack of sucrose but are seen as individual bacteria on the material surface. As the concentration of polylysine in the composite disc increases, an increase in the proportion of dead bacteria is seen.

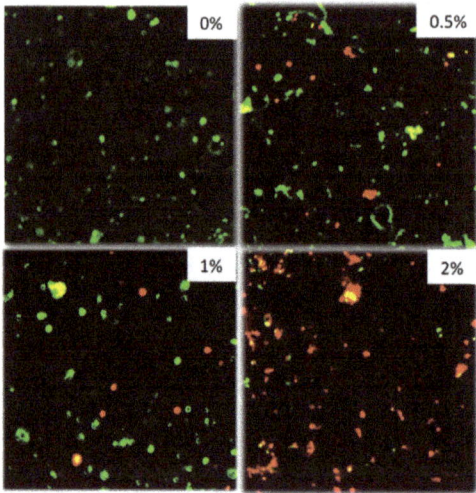

Figure 6. From top left to bottom right 0, 0.5, 1, 2% formulations. Inoculum: *S. mutans* 5×10^6 CFU/mL. Plates with discs and inoculum were incubated for 72 h in air at 37 °C before staining. Stained with Live/Dead staining. ($n = 2$).

4. Discussion

4.1. Mass and Volume Change

Dental restorations are exposed to fluids from the oral cavity continuously. Any dental composite will have micro voids after the setting reaction has taken place. When immersed in solution (i.e., deionized water), over time, the water will be absorbed by the voids in the composite as well as the composite matrix phase and the mass and volume will increase. Excessive expansion can result in a reduction in mechanical properties [18].

Mass and volume changes were undertaken for 2 months, as previous work has shown they tend to level off after this time [19]. With low PLS, the volume change, approximately double the mass change, suggests water is mostly expanding the matrix phase [19]. This is a consequence of the composite density being double that of water. With higher PLS, the mass increase, more comparable with the volume change, suggests there are more pores being filled by water which would cause mass increase but no expansion. The porosity may be caused by poor wetting and dispersion of the hydrophilic PLS particles by the monomer phase. The lack of change in the volume with increasing PLS level suggests it has minimal effect on the surrounding matrix expansion. The final levels of mass and volume change with 2% PLS (1.7 wt% and 2.3 vol%) are approximately two thirds of those seen with an earlier similar formulation (F2) (2.5 wt% and 3.5 vol%) that had an additional 10 wt% tricalcium phosphate [11]. This suggests the mass and volume changes are related to the total calcium phosphate levels in these composites. A previous study has shown that the commercial composite Z250 shows an increase in mass and volume after 7 weeks of 1.1 and 1.6% respectively [11]. The mass and volume changes of the experimental composites were comparable or above those observed for Z250.

4.2. Antibacterial Activity and Polylysine Release

Polylysine has been added to dental composites in previous research [10,11], but reports on the optimum concentration have not been carried out. This study attempts to replicate the conditions at the interface between the composite and the sealed affected dentine in which sucrose will be excluded. In this situation, the formation of biofilm on the discs was not observed. One recent study demonstrated that polylysine had satisfactory antibacterial efficacy against *S. mutans* in a liquid culture medium and as an application on biofilm–dentin surfaces. The study demonstrated that the susceptibility of microorganisms to polylysine was dependent on polylysine concentrations [20]. Additionally, it showed that polylysine can kill and inhibit the growth of the periodontal pathogen *Porphyromonas gingivalis* [20].

Increased polylysine release from the composite can have a negative effect on the mechanical properties of the material [11], so the minimum amount to cause bacterial kill would need to be used. This is a common issue for formulations with a released soluble antibacterial agent [5,6]. Other antibacterial agents, such as quaternary ammonium methacrylate, are non-released and their effect on physical properties is less evident [8,21]. Additionally, their surface antibacterial properties could be more stable over time. Surface antibacterial benefits, however, may be neutralized by a biofilm or adhesive covering. Conversely, the benefit of high early polylysine release is a greater chance of an effective reduction in residual bacteria deep within minimally excavated cavities. The above studies suggest, however, that higher levels of polylysine release may be required for highly infected cavities.

Previous experiments have shown greater early burst release and a higher release over time from a 2.5% PLS formulation with approximately 5.5% after 24 h and 9% at 3 weeks of being observed. As this was not previously affected by monocalcium and tricalcium phosphate addition, the differences may instead be due to the use of the more standard TEGDMA monomer. This can cause lower conversions [10] than seen with PPGDMA [11]. In this new study, the composite formulation with 2% PLS released 2.4% of polylysine in the first 24 h and after 3 weeks this had increased to just 4%. Initial polylysine release, proportional to the square root of time with no early burst release, is more in agreement with a recent report using the same monomer system as this study [22]. The work showed that increasing the polylysine concentration from 4 to 6 and then 8 wt% of the filler caused final release percentages to level at values of 13, 28 and 42% after 1, 2 and 3 months respectively [22]. This was attributed to surface release from layers of 65, 140 and 210 micron thick. The PLS release levelling between 1 and 3 weeks in this study fits with this trend. Given that the sample thickness is 100 microns, a final release of 4% suggests the PLS release may be from surface layers of just 20 microns thick.

Due to the initially higher concentration of PLS, the release in grams is higher as the concentration of polylysine in the formulations increased from 0.5 to 2%. The Minimum Inhibitory Concentration of polylysine on *S. mutans* in air with inoculum density of 5×10^5 CFU/mL is 20 μg/mL [23]. After 24 h, the 0.5%, 1% and 2% formulations had an average ($n = 3$) of 8, 25 and 93 μg/mL of polylysine released. This could explain why the 1% and 2% formulations inhibit growth in air conditions as they release more than 20 μg/mL of polylysine. Conversely, the 0.5% formulations which released less than 20 μg/mL did not inhibit growth. It is therefore expected that higher PLS concentrations will be required to inhibit growth of higher inoculum concentrations. This was also demonstrated in a recent study [20].

The aim of using Confocal was to visualize the live/dead state of bacteria sedimented on the composite disc surfaces. In many other studies, additional sucrose is added to promote biofilm formation [8,9]. The method chosen in this study with no additional sucrose, however, better replicates the environment beneath the restoration after effective sealing. With these conditions, largely only well-spaced individual bacteria were seen on the material surfaces and not biofilms. With the 0.5% and 1% PLS discs, some dead bacteria were present on disc surfaces at 72 h, whereas on the 2% PLS discs, the majority were dead. At the high level of initial inoculum chosen for the confocal study, the 2% PLS disc prevents any change in bacterial numbers in suspension at 24 h indicating bacterial increase is exactly balanced by killing. It is possible, therefore, that dead bacteria on the composite surfaces were killed when they made contact with remaining PLS on the composite surfaces or through more prolonged and increasing contact with PLS in the suspension.

5. Conclusions

Increasing polylysine concentration from 0.5% to 2% in composite discs leads to an increase in mass but not volume change over three weeks. Whilst the PLS-release percentage was faster in the first 24 h with 2% in the filler, the final percentage release was not affected by increasing composite PLS level. With 2% PLS, the amount released into 1 mL of broth in 24 h was sufficient to display significantly reduced bacterial levels when the initial inoculum level was at 8×10^5, but their growth in number was only inhibited when the initial bacterial counts were raised 10-fold. With an initial inoculum of 5×10^6 bacteria, those detected on the composite surfaces after 72 h were mostly live with 1% or less PLS in the composite but dead with 2% PLS. Higher PLS may enable the greater killing of residual bacteria when composites are used for minimally invasive tooth restorations. Considering the limitations of this study and taking into account work completed in previous studies [24–26], future research should include sucrose in the experiments to allow for biofilm growth on the discs to determine the effectiveness of PLS against biofilms, test multispecies biofilms and assess long term polylysine release and how this may affect the integrity of the restoration.

6. Patents

The work is covered by the following licensed patent families: Formulations and composites with reactive fillers (US8252851 B2, EP2066703B1, US20100069469, WO2008037991A1), and Formulations and materials with cationic polymers (PCT/GB2014/052349, WO2015015212 A1, EP3027164A1, US20160184190).

Author Contributions: Conceptualization, N.N.L., E.A., W.X., P.F.A., A.M.Y.; Methodology, N.N.L., E.A., W.X., P.F.A., A.M.Y.; Validation, N.N.L., E.A., P.F.A., A.M.Y.; Formal Analysis, N.N.L.; Investigation, N.N.L.; Resources, N.N.L., E.A., P.F.A., A.M.Y.; Writing—original Draft Preparation, N.N.L.; Writing—review & Editing, N.N.L., E.A.,

P.F.A., A.M.Y.; Supervision, E.A., W.X., P.F.A., A.M.Y.; Project Administration, A.M.Y.; Funding Acquisition, A.M.Y. All authors have read and agreed to the published version of the manuscript.

Funding: This research was funded by National Institute for Health Research (Invention for Innovation (i4i, http://www.nihr.ac.uk/funding/invention-for-innovation.htm)); Optimization and commercial manufacture of tooth-coloured composite dental-fillings with added poly-antimicrobial (PAM) and remineralising calcium phosphate (CaP), II-LB-0214-2002) (WX), UK EPSRC (Engineering and Physical Sciences Research Council, https://www.epsrc.ac.uk) (EPI/I02234/1)(WX); and Welcome Trust (Wtissf 3: Institutional Strategic Support Fund (Issf) Third Tranche, www.wellcome.ac.uk/Funding/WTP057769.htm) (WX).

Conflicts of Interest: In the future the corresponding author and inventor on the above patents (Anne Young) may receive royalties when a commercial product is produced. This author has had funding from NIHR and EPSRC and is working with Schottlander Dental Company to optimize a similar product to those in the publication.

References

1. Bourguignon, D. Briefing EU Legislation in Progress Mercury Aligning EU Legislation with Minamata. 2017. Available online: http://www.europarl.europa.eu/RegData/etudes/BRIE/2017/595887/EPRS_BRI(2017)595887_EN.pdf (accessed on 16 February 2020).
2. Ricketts, D.; Lamont, T.; Innes, N.; Kidd, E.; Clarkson, J.E. Operative caries management in adults and children. *Cochrane Database Syst. Rev.* **2013**, *28*. [CrossRef] [PubMed]
3. Ferracane, J.L. Resin composite—State of the art. *Dent. Mater.* **2011**, *27*, 29–38. [CrossRef]
4. Chisini, L.A.; Collares, K.; Cademartori, M.G.; de Oliveira, L.J.C.; Conde, M.C.M.; Demarco, F.F.; Corrêa, M.B. Restorations in primary teeth: A systematic review on survival and reasons for failures. *Int. J. Paediatr. Dent.* **2018**, *28*, 123–139. [CrossRef] [PubMed]
5. Leung, D.; Spratt, D.; Pratten, J.; Gulabivala, K.; Mordan, N.; Young, A.M. Chlorhexidine-releasing methacrylate dental composite materials. *Biomaterials* **2005**, *26*, 7145–7153. [CrossRef] [PubMed]
6. Zhang, J.F.; Wu, R.; Fan, Y.; Liao, S.; Wang, Y.; Wen, Z.T.; Xu, X. Antibacterial Dental Composites with Chlorhexidine and Mesoporous Silica. *J. Dent. Res.* **2014**, *93*, 1283–1289. [CrossRef] [PubMed]
7. Nedeljkovic, I.; Teughels, W.; De Munck, J.; Van Meerbeek, B.; Van Landuyt, K.L. Is secondary caries with composites a material-based problem? *Dent. Mater.* **2015**, *31*, e247–e277. [CrossRef] [PubMed]
8. Wang, H.; Wang, S.; Cheng, L.; Jiang, Y.; Melo, M.A.; Weir, M.; Oates, T.; Zhou, X.; Xu, H.K. Novel dental composite with capability to suppress cariogenic species and promote non-cariogenic species in oral biofilms. *Mater. Sci. Eng. C* **2019**, *94*, 587–596. [CrossRef]
9. Balhaddad, A.A.; Ibrahim, M.S.; Weir, M.D.; Xu, H.H.K.; Melo, M.A.S. Concentration dependence of quaternary ammonium monomer on the design of high-performance bioactive composite for root caries restorations. *Dent. Mater.* **2020**, *36*, e266–e278. [CrossRef]
10. Panpisut, P.; Liaqat, S.; Zacharaki, E.; Xia, W.; Petridis, H.; Young, A.M. Dental Composites with Calcium/Strontium Phosphates and Polylysine. *PLoS ONE* **2016**, *11*, e0164653. [CrossRef]
11. Kangwankai, K.; Sani, S.; Panpisut, P.; Xia, W.; Ashley, P.; Petridis, H.; Young, A.M. Monomer conversion, dimensional stability, strength, modulus, surface apatite precipitation and wear of novel, reactive calcium phosphate and polylysine-containing dental composites. *PLoS ONE* **2017**, *12*, e0187757. [CrossRef]
12. Ye, R.; Xu, H.; Wan, C.; Peng, S.; Wang, L.; Xu, H.; Aguilar, Z.; Xiong, Y.; Zeng, Z.; Wei, H. Antibacterial activity and mechanism of action of poly-l-lysine. *Biochem. Biophys. Res. Commun.* **2013**, *439*, 148–153. [CrossRef]
13. Bo, T.; Han, P.; Su, Q.; Fu, P.; Guo, F.; Zheng, Z.; Tan, Z.; Zhong, C.; Jia, S. Antimicrobial ε-poly-l-lysine induced changes in cell membrane compositions and properties of Saccharomyces cerevisiae. *Food Control* **2016**, *61*, 123–134. [CrossRef]
14. Sayed, S.; Jardine, M. Antimicrobial biopolymers. In *Advanced Functional Materials*; John Wiley & Sons Inc.: Hoboken, NJ, USA, 2015; pp. 493–533.
15. Walters, N.; Xia, W.; Salih, V.; Ashley, P.; Young, A.M. Poly(propylene glycol) and urethane dimethacrylates improve conversion of dental composites and reveal complexity of cytocompatibility testing. *Dent. Mater.* **2016**, *32*, 264–277. [CrossRef]
16. Van Landuyt, K.; Snauwaert, J.; De Munck, J.; Peumans, M.; Yoshida, Y.; Poitevin, A.; Coutinho, E.; Suzuki, K.; Lambrechts, P.; Van Meerbeek, B. Systematic review of the chemical composition of contemporary dental adhesives. *Biomaterials* **2007**, *28*, 3757–3785. [CrossRef] [PubMed]

17. Aljabo, A.; Abou Neel, E.A.; Knowles, J.C.; Young, A.M. Development of dental composites with reactive fillers that promote precipitation of antibacterial-hydroxyapatite layers. *Mater. Sci. Eng. C Mater. Biol. Appl.* **2016**, *60*, 285–292. [CrossRef] [PubMed]
18. Sideridou, I.D.; Karabela, M.M.; Bikiaris, D.N. Aging studies of light cured dimethacrylate-based dental resins and a resin composite in water or ethanol/water. *Dent. Mater.* **2007**, *23*, 1142–1149. [CrossRef] [PubMed]
19. Aljabo, A.; Xia, W.; Liaqat, S.; Khan, M.A.; Knowles, J.C.; Ashley, P.; Young, A.M. Conversion, shrinkage, water sorption, flexural strength and modulus of re-mineralizing dental composites. *Dent. Mater.* **2015**, *31*, 1279–1289. [CrossRef]
20. Dima, S.; Lee, Y.-Y.; Watanabe, I.; Chang, W.-J.; Pan, Y.-H.; Teng, N.-C. Antibacterial Effect of the Natural Polymer ε-Polylysine Against Oral Pathogens Associated with Periodontitis and Caries. *Polymers* **2020**, *12*, 1218. [CrossRef]
21. Chen, L.; Suh, B.I.; Yang, J. Antibacterial dental restorative materials: A review. *Am. J. Dent.* **2018**, *31*, 6B–12B.
22. Yaghmoor, R.B.; Xia, W.; Ashley, P.; Allan, E.; Young, A.M. Effect of Novel Antibacterial Composites on Bacterial Biofilms. *J. Funct. Biomater.* **2020**, *11*, accepted.
23. Badaoui Najjar, M.; Kashtanov, D.; Chikindas, M.L. Natural Antimicrobials ε-Poly-l-lysine and Nisin A for Control of Oral Microflora. *Probiotics Antimicrob. Proteins* **2009**, *1*, 143–147. [CrossRef] [PubMed]
24. Balhaddad, A.A.; Kansara, A.A.; Hidan, D.; Weir, M.D.; Xu, H.; Melo, M. Toward dental caries: Exploring nanoparticle-based platforms and calcium phosphate compounds for dental restorative materials. *Bioact. Mater.* **2018**, *4*, 43–55. [CrossRef] [PubMed]
25. Dahl, J.E.; Stenhagen, I. Optimizing quality and safety of dental materials. *Eur. J. Oral Sci.* **2018**, *126* (Suppl. S1), 102–105. [CrossRef] [PubMed]
26. Ibrahim, M.S.; Garcia, I.M.; Kensara, A.; Balhaddad, A.A.; Collares, F.M.; Williams, M.A.; Ibrahim, A.S.; Lin, N.J.; Weir, M.D.; Xu, H.H.K.; et al. How we are assessing the developing antibacterial resin-based dental materials? A scoping review. *J. Dent.* **2020**, e103369. [CrossRef] [PubMed]

© 2020 by the authors. Licensee MDPI, Basel, Switzerland. This article is an open access article distributed under the terms and conditions of the Creative Commons Attribution (CC BY) license (http://creativecommons.org/licenses/by/4.0/).

Article

Time-Transient Effects of Silver and Copper in the Porous Titanium Dioxide Layer on Antibacterial Properties

Masaya Shimabukuro [1,*], Akari Hiji [2], Tomoyo Manaka [2], Kosuke Nozaki [2], Peng Chen [3], Maki Ashida [3], Yusuke Tsutsumi [4], Akiko Nagai [5] and Takao Hanawa [3]

1. Department of Biomaterials, Faculty of Dental Science, Kyushu University, 3-1-1 Maidashi, Higashi-ku, Fukuoka 812-8582, Japan
2. Graduate School of Medical and Dental Sciences, Tokyo Medical and Dental University, 1-5-45 Yushima, Bunkyo-ku, Tokyo 113-8549, Japan; ma190081@tmd.ac.jp (A.H.); manaka.met@tmd.ac.jp (T.M.); k.nozaki.fpro@tmd.ac.jp (K.N.)
3. Institute of Biomaterials and Bioengineering, Tokyo Medical and Dental University, 2-3-10 Kanda-Surugadai, Chiyoda-ku, Tokyo 101-0062, Japan; chen.met@tmd.ac.jp (P.C.); ashida.met@tmd.ac.jp (M.A.); hanawa.met@tmd.ac.jp (T.H.)
4. Research Center for Structural Materials, National Institute for Materials Science (NIMS), 1-2-1 Sengen, Tsukuba, Ibaraki 305-0047, Japan; TSUTSUMI.Yusuke@nims.go.jp
5. Department of Anatomy, School of Dentistry, Aichi Gakuin University, 1-100 Kusumoto, Chikusa-ku, Nagoya 464-8650, Japan; aknagai@dpc.agu.ac.jp
* Correspondence: shimabukuro@dent.kyushu-u.ac.jp; Tel.: +81-92-642-6346

Received: 23 May 2020; Accepted: 19 June 2020; Published: 22 June 2020

Abstract: Recently, silver (Ag) and copper (Cu) have been incorporated into a titanium (Ti) surface to realize their antibacterial property. This study investigated both the durability of the antibacterial effect and the surface change of the Ag- and Cu-incorporated porous titanium dioxide (TiO_2) layer. Ag- and Cu-incorporated TiO_2 layers were formed by micro-arc oxidation (MAO) treatment using the electrolyte with Ag and Cu ions. Ag- and Cu-incorporated specimens were incubated in saline during a period of 0–28 days. The changes in both the concentrations and chemical states of the Ag and Cu were characterized using X-ray photoelectron spectroscopy (XPS). The durability of the antibacterial effects against *Escherichia coli* (*E. coli*) were evaluated by the international organization for standardization (ISO) method. As a result, the Ag- and Cu-incorporated porous TiO_2 layers were formed on a Ti surface by MAO. The chemical state of Ag changed from Ag_2O to metallic Ag, whilst that of Cu did not change by incubation in saline for up to 28 days. Cu existed as a stable Cu_2O compound in the TiO_2 layer during the 28 days of incubation in saline. The concentrations of Ag and Cu were dramatically decreased by incubation for up to 7 days, and remained a slight amount until 28 days. The antibacterial effect of Ag-incorporated specimens diminished, and that of Cu was maintained even after incubation in saline. Our study suggests the importance of the time-transient effects of Ag and Cu on develop their antibacterial effects.

Keywords: antibacterial; surface modification; coatings; implant; biofilm; silver; copper

1. Introduction

Prosthetic joint infection (PJI) is a devastating and threatening complication for patients and orthopedists [1,2]. Biofilm, which is a main cause of PJI, is the final state of infections. Invaded bacteria generally initiate the infection by adhering onto the implant surface, and grow through a specific mechanism such as extracellular polysaccharide (EPS) production [3–5]. The removal of matured biofilms from the implant is difficult because biofilms are resistant to antibiotics due to their bacterial diversity and the presence of the EPS [6–9]. Often, the only way to eradicate the infection and prevent sepsis is to remove the contaminated device from the patient. To avoid this, biofilm formation must be prevented by inhibiting the initial stage of biofilm formation, namely bacterial invasion, adhesion and growth.

Antibacterial property, which can kill the bacteria, is a necessary and required bio-function on implant surfaces. Silver (Ag) and copper (Cu) are major antibacterial elements, and their effects on various bacteria have been examined by many studies. Ag and Cu potently inhibit various pathogenic bacteria [10–14]. Various surface modification techniques have been used to incorporate these elements on implant surfaces, and their efficacies have been shown by in vitro [15,16] and in vivo [17,18] experiments.

Micro-arc oxidation (MAO), which is an anodic oxidation process with micro-discharges on the specimen surface under high voltage, form a connective porous oxide layer with the incorporation of elements from the electrolyte solution. MAO using the calcium (Ca)- and phosphorous (P)-containing electrolyte improves the hard-tissue compatibility of titanium (Ti), owing to calcium phosphate formation, the promotion of osteoblast adhesion and proliferation, as well as the acceleration of calcification [19–26]. The widespread use of Ti in metallic biomaterials reflects the good mechanical property and biocompatibility of these materials. In addition, some studies have focused on the incorporation of antibacterial elements on Ti surfaces by MAO. Ag- and Cu-incorporated TiO_2 coatings reportedly exhibit antibacterial activity [27–35].

PJI comprises an early infection (within three weeks after surgery) and a late-onset infection (approximately three to eight weeks after surgery), because the bacterial invasion can occur due to implant surgery or the hematogenous spread of bacteria [36–38]. Thus, the long-term inhibition of biofilm formation relies on the durability of antibacterial effects. The surface changes are key in the development of antibacterial effect. Therefore, biodegradation behavior must be precisely characterized to understand the antibacterial effect and its durability. However, little attention has been given to the biodegradation behavior of antibacterial coatings. Therefore, the relationship between the surface change and the durability of antibacterial property on Ag and Cu in an in vivo environment is still unknown.

We investigated the long-term behaviors of Ag and Cu in the porous TiO_2 layer formed by MAO treatment. The changes in both the concentrations and chemical states of the Ag and Cu in the oxide layer incubated for a prolonged period in physiological saline were characterized using X-ray photoelectron spectroscopy (XPS). Moreover, the change of antibacterial property was evaluated using the international organization for standardization (ISO) method with *Escherichia coli* (*E. coli*). In other words, the aim was to clarify the time-transient effects of Ag and Cu in porous TiO_2 layers on antibacterial property.

2. Materials and Methods

2.1. Specimen Preparation

Ag- and Cu-incorporated porous titanium dioxide layers were prepared on a commercially pure Ti (grade 2) surface. The ti disks were prepared from the rod of Ti, and each surface was mechanically polished using #320, #320, #600 and #800 silicon carbide abrasive papers. After polishing, all the specimens were washed by ultra-sonication in acetone and ethanol for 10 min.

The electrolyte compositions for MAO were 150 mM calcium acetate and 100 mM calcium glycerophosphate solution, containing 2.5 mM silver nitrate or 2.5 mM copper chloride. The electrochemical parameters were a voltage of 400 V and a current density of 251 Am^{-2}, and the treatment time was 10 min. The area in contact with the electrolyte was 39 mm^2 using the working electrode [39]. After the MAO treatment, all specimens were incubated in a saline during 0 to 28 days. The specimens were fixed onto a polyethylene container to allow the release of metal ions from the surface into the saline. Incubation was performed at 37 °C in a humidified chamber under constant shaking (100 rpm). Every seventh day, the pooled solution was changed with a fresh one. This process simulated a simple biodegradation of Ag- and Cu-incorporated specimens in the body. The specimens after incubation in saline for each period were used further for the surface characterization and evaluation of antibacterial activity. After the MAO and incubation, the surfaces were thoroughly washed in ultrapure water in order to remove any solution remaining in the porous oxide layer.

2.2. Surface Characterization

Surface morphologies on the specimens incubated during 0 and 28 days were observed by scanning electron microscopy with energy dispersive X-ray spectrometry (SEM/EDS; S-3400NX, Hitachi High-Technologies Corp., Tokyo, Japan). X-ray diffraction (D8 ADVANCE, Bruker, Billerica, MA, USA) was performed to characterize the crystal structure of each specimen. X-ray photoelectron spectroscopy (XPS; JPS-9010MC, JEOL, Tokyo, Japan) was used in this investigation. The detail of the measurement condition was 10 kV of acceleration voltage, 10 mA of current, MgKα (energy: 1253.6 eV) of X-ray source, and was described in our previous study [40]. The calibration of the binding energy was performed based on C 1s photoelectron energy region peak derived from contaminating carbon (285.0 eV). The integrated intensity of each peak was calculated using Shirley's method [41]. In addition, a modified Auger parameter (α') of Ag and Cu, calculated from the Ag $3d_{5/2}$, Ag M_4VV, Cu $2p_{3/2}$ and Cu L_3VV peaks, was used for the investigation of chemical state changes. According to a method described previously [42], the surface composition on each specimen was calculated using a photoionization cross-section of empirical data [43,44] and theoretically calculated data [45].

2.3. Evaluation of Antibacterial Activity

The antibacterial activity was evaluated using *E. coli* (NBRC3972). This evaluation was performed according to the ISO 22196: 2007 method. This experiment was approved by the Pathogenic Organisms Safety Management Committee in Tokyo Medical and Dental University (22012-025c). Luria–Bertani (LB) broth (LB-Medium, MP Biomedicals, Irvine, CA, USA) was used as culture medium, and *E. coli* was cultured in this at 37 °C for 24 h. After culturing, the bacterial density in the suspension of this bacterium strain was measured by ultraviolet–visible spectrometer (UV–vis; V-550, JASCO, Tokyo, Japan), and was adjusted by dilution to be 1.0×10^6 CFUs mL^{-1}. The specimens, which were used for this evaluation, were sterilized by immersion in 70% ethanol and washed with ultrapure water. The prepared bacterial suspension was dropped on the surface of each specimen and cultured at 37 °C for 24 h using a sterilized cover plastic film. After 24 h culturing, the suspension of *E. coli* was collected from the specimen surface, and transferred onto nutrient agar plates with dilution. The *E. coli* in collected suspension was cultured on the nutrient agar plates at 37 °C during 24 h. The number of viable bacteria was determined by counting the number of colonies formed on the plates. In this evaluation, the specimen without Ag and Cu was used as the negative control and the specimens with Ag or Cu before incubation were used as the positive control because our previous study revealed that these specimens exhibited antibacterial effects for *E. coli* [27,28].

2.4. Statistical Analysis

All values are shown as the means ± standard deviation, and commercial statistical software KaleidaGraph (Synergy Software, Reading, PA) was used for statistical analysis. One-way analysis of variance was used following the multiple comparisons with the Student–Newman–Keuls method to assess the data, and $p < 0.05$ was considered to indicate statistical significance.

3. Results

3.1. Surface Characterization

Typical connective porous morphology after MAO were observed from Ag- and Cu-incorporated TiO$_2$ layers. This morphology on each specimen was maintained after incubation in saline during 28 days (Figure 1).

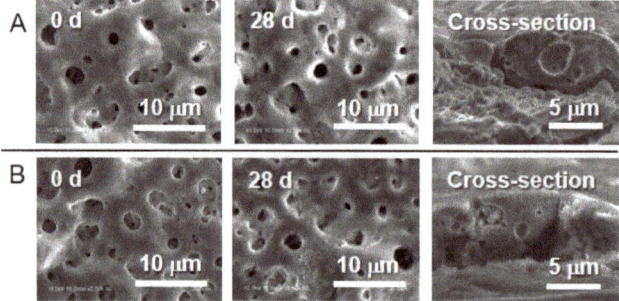

Figure 1. Scanning electron microscopy (SEM) images of (**A**) the Ag- and (**B**) the Cu-incorporated TiO$_2$ layers before and after incubation in saline during 28 days and the cross-sectional views of the specimens before incubation.

The XRD spectra obtained from the control, the specimens before incubation, and the specimens after incubation are presented in Figure 2, respectively, from the bottom to the top in each figure. Peaks corresponding to α-Ti and anatase TiO$_2$ were detected, and those of Ag were undetected in the Ag-incorporated specimens before and after incubation. Peaks corresponding to α-Ti, anatase TiO$_2$, and rutile TiO$_2$ were detected, while those of Cu were undetected. Furthermore, the chemical structures of Ag- and Cu-incorporated specimens did not change by the incubation in saline during 28 days.

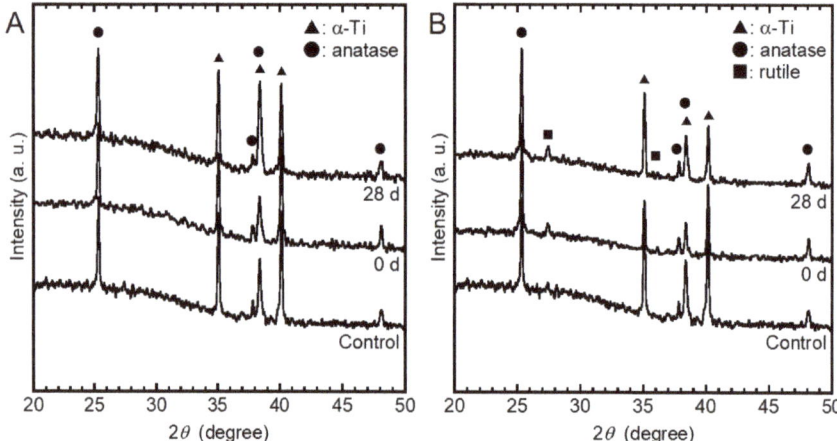

Figure 2. X-ray diffraction (XRD) spectra obtained from the (**A**) Ag- and (**B**) the Cu-incorporated specimens before and after incubation in saline during 28 days. The spectra presented at the bottom in each figure was obtained from the control.

Figure 3 shows the XPS survey scan spectra obtained from Ag- and Cu-incorporated specimens before and after incubation in saline during 28 days. The peaks originating from C, O, P, Ca, Ti, and Ag or Cu were detected from the XPS spectra of the Ag- and Cu-incorporated specimens. In addition to these elements, the peak originating from Na was detected in the specimens after incubation in saline from 7 to 28 days. P existed as a phosphate species and calcium existed as Ca^{2+}, because the binding energies of the corresponding peaks of P 2p and Ca $2p_{3/2}$ were 133.7–134.1 eV and 347.6–347.9 eV, respectively. The binding energy of the Ti $2p_{3/2}$ peak was 458.9–459.3 eV, indicating that Ti existed as TiO_2. The binding energies of the Na 1s peaks were 1071.7–1072.4 eV, indicating that Na exists as Na^+.

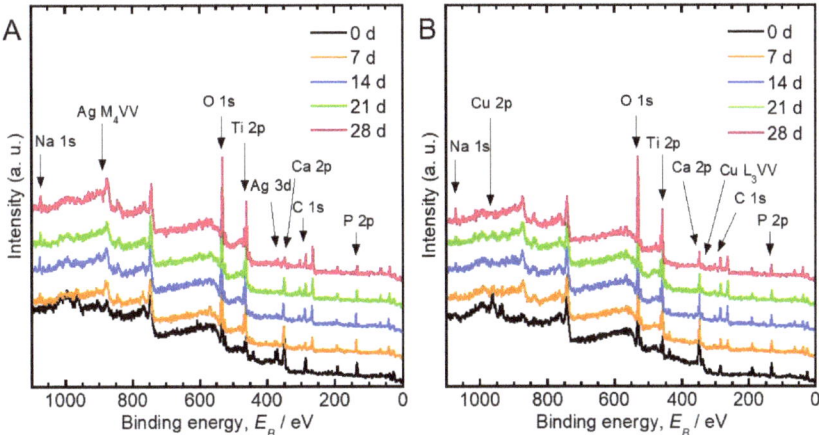

Figure 3. X-ray photoelectron spectroscopy (XPS) survey scan spectra obtained from the (**A**) Ag- and (**B**) the Cu-incorporated specimens before and after incubation in saline during 28 days.

Figure 4 depicts the Wagner plot of Ag and Cu obtained from this XPS characterization and previous studies [46–48]. The chemical state change of Ag and Cu is determined by the comparison of their binding energy with α′ values on the Wagner plot. According to the Wagner plot of Ag, the α′ value of the Ag-incorporated specimen before incubation (0 days) was 724.9 eV, indicating that Ag mainly exists as Ag_2O. The α′ values of Ag increased with incubation time. Since the α′ value finally converged to 726.1 eV, it became clear that the chemical state of Ag incorporated in TiO_2 by MAO approached that of metallic Ag with increasing incubation times. On the other hand, the α′ values of Cu from the specimens immersed in saline for 28 days were 1849.3–1849.7 eV, indicating that Cu exists as Cu_2O. The chemical state of Cu in the oxide layer did not change during the incubation in saline.

Figure 5 shows the changes in the concentrations of Ti, P, Ca and Ag or Cu detected from Figure 5A's Ag- and Figure 5B's Cu-incorporated specimens with the incubation time. The concentrations of Ag and Cu were relatively small (around 2.5 atom%), even before the incubation. Furthermore, from the results of the EDS analysis of the cross-section shown in Figure 1, 0.1 atom% of Ag and 0.1 atom% of Cu were detected from the inside oxide layers, respectively. Thus, the amounts of Ag and Cu incorporated during the MAO treatment were small. The amount of Ag and Cu dramatically decreased to approximately Ag 0.4 atom% and Cu 0.8 atom% upon incubation in saline from 0 to 7 days, respectively. These concentrations remained constant until 28 days. Moreover, the concentrations of Ca and P decreased, and that of Ti increased with the incubation time.

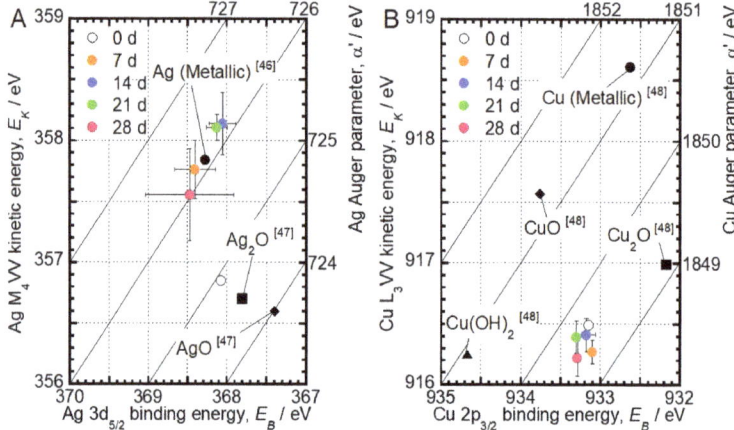

Figure 4. Wagner plot of (**A**) the Ag and (**B**) the Cu in the oxide layer incubated in saline for 28 days based on the photoelectron peaks and the Auger peaks. Each parameter of the Ag and the Cu compounds is plotted according to the previous studies [46–48].

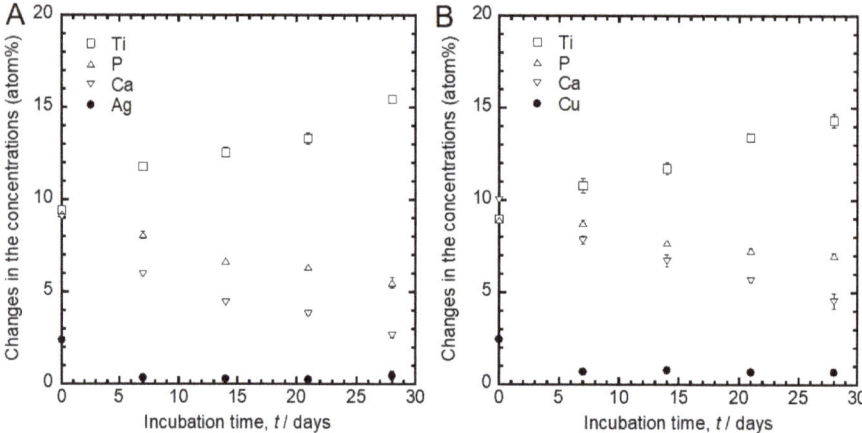

Figure 5. Changes in the atomic concentrations in (**A**) the Ag- and (**B**) the Cu-incorporated TiO_2 layers with the incubation time.

3.2. Evaluation of Antibacterial Activity

The normalized bacterial number of *E. coli* on each specimen is shown in Figure 6. The vertical axis represents the bacterial number normalized by the initial concentration of *E. coli*. The normalized bacterial number smaller than 1 (shown as a dashed line in the figure) indicates that the tested specimens exhibited an antibacterial effect. *E. coli* grew on the untreated Ti and control specimen, because the number of *E. coli* on those specimens significantly increased compared with the initial bacterial number. In contrast, Ag- and Cu-incorporated specimens developed antibacterial effects against *E. coli*.

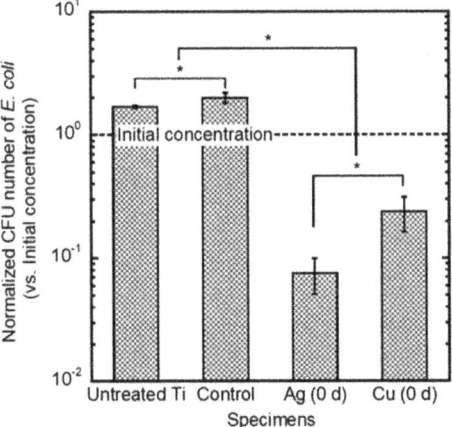

Figure 6. Comparison of the antibacterial effects of the untreated Ti control specimen (micro-arc oxidation (MAO)-treated Ti without antibacterial elements), the Ag- and the Cu-incorporated specimens. Data are shown as the mean ± standard deviation. * Significant difference between specimens ($p < 0.05$).

Changes in the antibacterial effects of Ag- and Cu-incorporated specimens before and after incubation in saline for 28 days are shown in Figure 7. The antibacterial effect of the Ag-incorporated specimen after incubation was significantly weakened compared to that before incubation. This effect was at the same level as that of Cu-incorporated specimens. On the other hand, the antibacterial effects of Cu-incorporated specimens did not change upon incubation, and were maintained even after the 28-day incubation in saline.

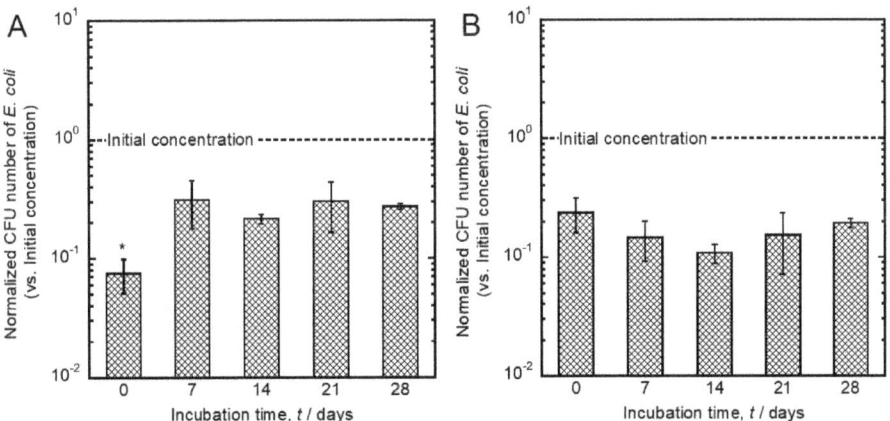

Figure 7. Changes of the antibacterial effects of (**A**) the Ag- and (**B**) the Cu-incorporated specimens before and after incubation in saline. Data are shown as the mean ± SD. * Significant difference between the specimens before and after incubation ($p < 0.05$).

4. Discussion

The porous oxide layers formed by MAO did not change their surface morphology and crystal structure during the incubation in saline (Figures 1 and 2). The lack of change is highly beneficial for the development of antibacterial property on implant surfaces.

The chemical states of P, Ca, and Ti in the Ag- and Cu-incorporated oxide layers were phosphate, Ca^{2+}, and TiO_2, respectively. These chemical states did not change upon incubation in saline for up to 28 days (Figure 3). The porous oxide layer consisted of TiO_2 as well as incorporated Ca, P, and antibacterial elements (Figures 3 and 5). The presence of Ca and P in the porous oxide layer makes the hard-tissue compatibility of Ti better [27]. The peak originating from Na^+ was detected from the specimens after incubation in saline for 7 to 28 days. It is conceivable that compounds related to Na^+ were generated on the specimen surfaces, owing to the interfacial reactions between oxide layer and saline.

Ag- and Cu-incorporated porous oxide layers were formed on Ti surfaces, and slight amounts of Ag and Cu were incorporated by MAO using the electrolyte containing Ag and Cu (Figures 1 and 5). These results indicate that the constituent elements in the electrolyte were incorporated into the porous oxide layer during the MAO treatment. The incorporations of Ag and Cu by MAO is beneficial for realizing the antibacterial property on the implant surface (Figure 6). In addition, our previous studies revealed that specimens with Cu and a suitable amount of Ag did not affect the cellular adhesion, proliferation, differentiation or the calcification of the osteoblast cells [27,28]. Therefore, MAO can be imparting dual-function to the Ti surface, namely antibacterial property and hard-tissue compatibility.

The concentration of Ag in the oxide layer was dramatically decreased, and the chemical state of Ag in the oxide layer was changed from Ag_2O to metallic Ag during the incubation in saline (Figures 4 and 5). These results indicate that Ag_2O was converted into chemically stable metallic Ag in saline, due to the release of Ag ions. A study that investigated the formation mechanism of Ag particles in sodium citrate solution described the reduction of Ag on Ag particles via the radical-to-particle electron transfer [49]. The α' values of Ag in the oxide layer increased with incubation time, indicating that the density of electrons increased. Therefore, like the reduction of Ag on Ag particles in the solution containing sodium citrate, the change of the chemical state of Ag in the oxide layer incubated in saline may be caused by radical-to-Ag electron transfer. Moreover, the antibacterial effects of Ag-incorporated specimens changed upon incubation in saline (Figure 7). This finding implicates changes of both the concentration and the chemical state of Ag in the oxide layer in the antibacterial effect. However, it can be considered that Ag changes its chemical state more drastically depending on components, such as sulfur, in the actual biological environment, since Ag changed its chemical state, even in the simple simulated body fluid. Therefore, changes in the chemical state of Ag in the actual biological environment could influence the antibacterial effect.

The concentration of Cu in the oxide layer was dramatically decreased, and the chemical state of Cu in the oxide layer did not change upon incubation in saline (Figures 4 and 5). The chemical state of Cu in the oxide layer was stabilized as Cu_2O despite the difference in incubation time. This result indicates that the Cu_2O is a stable chemical state in the TiO_2 layer. In addition, the antibacterial effect of the Cu-incorporated specimen was maintained even after the 28 days of incubation in saline. These results indicate that Cu_2O in a stable chemical state has a more important role in the development of an antibacterial effect, compared with the change of surface concentration of Cu. Previous studies revealed a substantial difference in antibacterial effects between CuO and Cu_2O. CuO inhibited the development of an antibacterial effect compared to metallic Cu. In contrast, thermally generated Cu_2O was as effective as metallic Cu [50,51]. In other words, the presence of Cu_2O and the slight amount of Cu in the TiO_2 layer are necessary to develop the antibacterial effect.

The results support the proposal that the concentrations of Ag and Cu in the oxide layer are easily and dramatically changed by the incubation in saline. In addition, the chemical state of Ag changed from Ag_2O to metallic Ag, while that of Cu did not change. The antibacterial effect of an Ag-incorporated specimen for Gram-negative bacteria changed by the 28-day incubation in saline, while the activity of Cu was maintained. The collective findings indicate the importance of the time-transient effects of Ag and Cu. This knowledge will be useful in the design of antibacterial implants based on the surface changes of Ag and Cu in vivo. Further study will be necessary to reveal the long-term effects of Ag and Cu for Gram-positive bacteria.

5. Conclusions

Ag- and Cu-incorporated porous titanium dioxide layers were formed on Ti surfaces by MAO treatment. The chemical states of P, Ca, and Ti in the Ag- and Cu-incorporated oxide layer were phosphate, Ca^{2+}, and TiO_2, respectively. These chemical states did not change upon incubation in saline for 28 days. Moreover, the chemical state of Ag changed from Ag_2O to metallic Ag, and that of Cu did not change by incubation in saline for up to 28 days. The concentrations of Ag and Cu were dramatically decreased by incubation for up to 7 days, and remained a slight amount until 28 days. The antibacterial effect of Ag-incorporated specimens changed, and the effect of Cu was maintained even after incubation in saline. The findings highlight the importance of the time-transient effects of antibacterial elements on their antibacterial properties.

Author Contributions: Conceptualization, M.S. and T.H.; formal analysis, M.S., Y.T., K.N., T.M. and P.C.; investigation, M.S., K.N., A.H. and M.A.; methodology, M.S., K.N., P.C. and A.H.; project administration, M.S., Y.T., A.N. and T.H.; supervision, A.N. and T.H.; validation, M.S., Y.T., K.N. and A.H.; writing—original draft, M.S.; writing—review and editing, T.H. All authors have read and agreed to the published version of the manuscript.

Funding: This research received no external funding.

Acknowledgments: This study was supported by the Research Center for Biomedical Engineering, Tokyo Medical and Dental University, Project "Creation of Life Innovation Materials for Interdisciplinary and International Researcher Development" and project "Cooperative project amount medicine, dentistry, and engineering for medical innovation-Construction of creative scientific research of the viable material via integration of biology and engineering" by the Ministry of Education, Culture, Sports, Science and Engineering, Japan.

Conflicts of Interest: The authors declare no conflict of interest.

References

1. Tande, J.; Patel, R. Prosthetic Joint Infection. *Clin. Microbiol. Rev.* **2014**, *27*, 302–345. [CrossRef] [PubMed]
2. Osmon, D.R.; Berbari, E.F.; Berendt, A.R.; Lew, D.; Zimmerli, W.; Steckelberg, J.M.; Rao, N.; Hanssen, A.; Wilson, W.R. Diagnosis and Management of Prosthetic Joint Infection: Clinical Practice Guidelines by the Infectious Diseases Society of America. *Clin. Infect. Dis.* **2013**, *56*, e1–e25. [CrossRef] [PubMed]
3. Hobley, L.; Harkins, C.; MacPhee, C.E.; Stanley-Wall, N.R. Giving structure to the biofilm matrix: An overview of individual strategies and emerging common themes. *FEMS Microbiol. Rev.* **2015**, *39*, 649–669. [CrossRef] [PubMed]
4. Hoiby, N.; Bjarnsholt, T.; Givskov, M.; Molin, S.; Ciofu, O. Antibiotic resistance of bacterial biofilms. *Int. J. Antimicrob. Agents* **2010**, *34*, 322–332. [CrossRef]
5. Koo, H.; Allan, R.N.; Howlin, R.P.; Stoodley, P.; Hall-Stoodley, L. Targeting microbial biofilms: Current and prospective therapeutic strategies. *Nat. Rev. Microbiol.* **2017**, *15*, 740–755. [CrossRef]
6. Lindsay, D.; von Holy, A. Bacterial biofilms within the clinical setting: What healthcare professionals should know. *J. Hosp. Infect.* **2006**, *64*, 313–325. [CrossRef]
7. Flemming, H.C.; Wingender, J.; Szewzyk, U.; Steinberg, P.; Rice, S.A.; Kjelleberg, S. Biofilms: An emergent form of bacterial life. *Nat. Rev. Microbiol.* **2016**, *14*, 563–575. [CrossRef]
8. Van Acker, H.; Van Dijck, P.; Coenye, T. Molecular mechanisms of antimicrobial tolerance and resistance in bacterial and fungal biofilms. *Trends Microbiol.* **2014**, *22*, 326–333. [CrossRef]
9. Lebeaux, D.; Ghigo, J.M.; Beloin, C. Biofilm-related infections: Bridging the gap between clinical management and fundamental aspects of recalcitrance toward antibiotics. *Microbiol. Mol. Biol. Rev.* **2014**, *78*, 510–543. [CrossRef]
10. Shimabukuro, M.; Ito, H.; Tsutsumi, Y.; Nozaki, K.; Chen, P.; Yamada, R.; Ashida, M.; Nagai, A.; Hanawa, T. The Effects of Various Metallic Surfaces on Cellular and Bacterial Adhesion. *Metals* **2019**, *9*, 1145. [CrossRef]
11. Rai, M.K.; Deshmukh, S.D.; Ingle, A.P.; Gade, A.K. Silver nanoparticles: The powerful nanoweapon against multidrug-resistant bacteria. *J. Appl. Microbiol.* **2012**, *112*, 841–852. [CrossRef] [PubMed]
12. Lara, H.H.; Ayala-Nunez, N.V.; Turrent, L.D.I.; Padilla, C.R. Bactericidal effect of silver nanoparticles against multidrug-resistant bacteria. *World J. Microbiol. Biotechnol.* **2010**, *26*, 615–621. [CrossRef]

13. Prakash, P.; Gnanaprakasam, P.; Emmanuel, R.; Arokiyaraj, S.; Saravanan, M. Green synthesis of silver nanoparticles from leaf extract of *Mimusops elengi*, Linn. for enhanced antibacterial activity against multi drug resistant clinical isolates. *Colloids Surf. B Biointerfaces* **2013**, *108*, 255–259. [CrossRef] [PubMed]
14. Pant, J.; Goudie, M.J.; Hopkins, S.P.; Brisbois, E.J.; Handa, H. Tunable Nitric Oxide Release from S-Nitroso-N-acetylpenicillamine via Catalytic Copper Nanoparticles for Biomedical Applications. *ACS Appl. Mater. Interfaces* **2017**, *9*, 15254–15264. [CrossRef] [PubMed]
15. Gasqueres, C.; Schneider, G.; Nusko, R.; Maier, G.; Dingeldein, E.; Eliezer, A. Innovative antibacterial coating by anodic spark deposition. *Surf. Coat. Technol.* **2012**, *206*, 3410–3414. [CrossRef]
16. Ferraris, S.; Spriano, S. Antibacterial titanium surfaces for medical implants. *Mater. Sci. Eng. C* **2016**, *61*, 965–978. [CrossRef]
17. Chai, H.; Guo, L.; Wang, X.; Fu, Y.; Guan, J.; Tan, L.; Ren, L.; Yang, K. Antibacterial effect of 317L stainless steel contained copper in prevention of implant-related infection in vitro and in vivo. *J. Mater. Sci. Mater. Med.* **2011**, *22*, 2525–2535. [CrossRef]
18. Tilmaciu, C.M.; Mathieu, M.; Lavigne, J.P.; Toupet, K.; Guerrero, G.; Ponche, A.; Amalric, J.; Noel, D.; Mutin, P.H. In vitro and in vivo characterization of antibacterial activity and biocompatibility: A study on silver-containing phosphonate monolayers on titanium. *Acta Biomater.* **2015**, *15*, 266–277. [CrossRef]
19. Ha, J.Y.; Tsutsumi, Y.; Doi, Y.; Nomura, N.; Kim, K.H.; Hanawa, T. Enhancement of calcium phosphate formation on zirconium by micro-arc oxidation and chemical treatments. *Surf. Coat. Technol.* **2011**, *205*, 4948–4955. [CrossRef]
20. Song, W.H.; Ryu, H.S.; Hong, S.H. Antibacterial properties of Ag (or Pt)-containing calcium phosphate coatings formed by micro-arc oxidation. *J. Biomed. Mater. Res. A* **2009**, *88*, 246–254. [CrossRef]
21. Li, L.H.; Kong, Y.M.; Kim, H.W.; Kim, Y.W.; Kim, H.E.; Heo, S.J.; Koak, J.Y. Improved biological performance of Ti implants due to surface modification by micro-arc oxidation. *Biomaterials* **2004**, *25*, 2867–2875. [CrossRef] [PubMed]
22. Li, Y.; Lee, I.S.; Cui, F.Z.; Choi, S.H. The biocompatibility of nanostructured calcium phosphate coated on micro-arc oxidized titanium. *Biomaterials* **2008**, *29*, 2025–2032. [CrossRef] [PubMed]
23. Suh, J.Y.; Janga, B.C.; Zhu, X.; Ong, J.L.; Kim, K.H. Effect of hydrothermally treated anodic oxide films on osteoblast attachment and proliferation. *Biomaterials* **2003**, *24*, 347–355. [CrossRef]
24. Son, W.W.; Zhu, X.; Shin, H.I.; Ong, J.L.; Kim, K.H. In vivo histological response to anodized and anodized/hydrothermally treated titanium implants. *J. Biomed. Mater. Res. B* **2003**, *66*, 520–525. [CrossRef]
25. Kim, D.Y.; Kim, M.; Kim, H.E.; Koh, Y.H.; Kim, H.W.; Jang, J.H. Formation of hydroxyapatite within porous TiO_2 layer by micro-arc oxidation coupled with electrophoretic deposition. *Acta Biomater.* **2009**, *5*, 2196–2205. [CrossRef]
26. Li, L.H.; Kim, H.W.; Lee, S.H.; Kong, Y.M.; Kim, H.E. Biocompatibility of titanium implants modified by microarc oxidation and hydroxyapatite coating. *J. Biomed. Mater. Res. A* **2005**, *73*, 48–54. [CrossRef]
27. Shimabukuro, M.; Tsutsumi, Y.; Yamada, R.; Ashida, M.; Chen, P.; Doi, H.; Nozaki, K.; Nagai, A.; Hanawa, T. Investigation of realizing both antibacterial property and osteogenic cell compatibility on titanium surface by simple electrochemical treatment. *ACS Biomater. Sci. Eng.* **2019**, *5*, 5623–5630. [CrossRef]
28. Shimabukuro, M.; Tsutsumi, Y.; Nozali, K.; Chen, P.; Yamada, R.; Aahida, M.; Doi, H.; Nagai, A.; Hanawa, T. Investigation of antibacterial effect of copper introduced titanium surface by electrochemical treatment against facultative anaerobic bacteria. *Dent. Mater. J.* **2020**. [CrossRef]
29. Cochis, A.; Azzimonti, B.; Della Valle, C.; De Giglio, E.; Bloise, N.; Visai, L.; Cometa, S.; Rimondini, L.; Chiesa, R. The effect of silver or gallium doped titanium against the multidrug resistant *Acinetobacter baumannii*. *Biomaterials* **2016**, *80*, 80–95. [CrossRef]
30. Tsutsumi, Y.; Niinomi, M.; Nakai, M.; Shimabukuro, M.; Ashida, M.; Chen, P.; Doi, H.; Hanawa, T. Electrochemical Surface Treatment of a b-titanium Alloy to Realize an Antibacterial Property and Bioactivity. *Metals* **2016**, *6*, 76. [CrossRef]
31. He, X.; Zhang, X.; Bai, L.; Hang, R.; Huang, X.; Qin, L.; Yao, X.; Tang, B. Antibacterial ability and osteogenic activity of porous Sr/Ag-containing TiO_2 coatings. *Biomed. Mater.* **2016**, *11*, 045008. [CrossRef] [PubMed]
32. Yao, X.H.; Zhang, X.Y.; Wu, H.B.; Tian, L.H.; Ma, Y.; Tang, B. Microstructure and antibacterial properties of Cu-doped TiO_2 coating on titanium by micro-arc oxidation. *Appl. Surf. Sci.* **2014**, *292*, 944–947. [CrossRef]
33. Wu, H.; Zhang, X.; Geng, Z.; Yin, Y.; Hang, R.; Huang, X.; Yao, X.; Tang, B. Preparation, antibacterial effects and corrosion resistant of porous Cu-TiO_2 coatings. *Appl. Surf. Sci.* **2014**, *308*, 43–49. [CrossRef]

34. Zhu, W.; Zhang, Z.; Gu, B.; Sun, J.; Zhu, L. Biological activity and antibacterial property of nano-structured TiO2 coating incorporated with Cu prepared by micro-arc oxidation. *J. Mater. Sci. Technol.* **2013**, *29*, 237–244. [CrossRef]
35. He, X.; Zhang, X.; Wang, X.; Qin, L. Review of antibacterial activity of titanium-based implants' surfaces fabricated by micro-arc oxidation. *Coatings* **2017**, *7*, 45. [CrossRef]
36. Tande, A.J.; Palraj, B.R.; Osmon, D.R.; Berbari, E.F.; Baddour, L.M.; Lohse, C.M.; Steckelberg, J.M.; Wilson, W.R.; Sohail, M.R. Clinical Presentation, Risk Factors, and Outcomes of Hematogenous Prosthetic Joint Infection in Patients with *Staphylococcus aureus* Bacteremia. *Am. J. Med.* **2016**, *129*, 221.e11–221.e20. [CrossRef]
37. Conventry, M.B. Treatment of infections occurring in total hip surgery. *Orthop. Clin. N. Am.* **1975**, *6*, 991–1003.
38. Rodríguez, D.; Pigrau, C.; Euba, G.; Cobo, J.; García-Lechuz, J.; Palomino, J.; Riera, M.; del Toro, M.D.; Granados, A.; Ariza, X. Acute haematogenous prosthetic joint infection: Prospective evaluation of medical and surgical management. *Clin. Microbiol. Infect.* **2010**, *16*, 1789–1795. [CrossRef]
39. Tanaka, Y.; Kobayashi, E.; Hiromoto, S.; Asami, K.; Imai, H.; Hanawa, T. Calcium phosphate formation on Ti by low-voltage electrolytic treatments. *J. Mater. Sci. Mater. Med.* **2007**, *18*, 797–806. [CrossRef]
40. Shimabukuro, M.; Tsutsumi, Y.; Nozaki, K.; Chen, P.; Yamada, R.; Ashida, M.; Doi, H.; Nagai, A.; Hanawa, T. Chemical and Biological Roles of Zinc in a Porous Titanium Dioxide Layer Formed by Micro-Arc Oxidation. *Coatings* **2019**, *9*, 705. [CrossRef]
41. Shirley, D.A. High-Resolution X-ray Photoemission Spectrum of the Valence Bands of Gold. *Phys. Rev. B* **1972**, *5*, 552–556. [CrossRef]
42. Asami, K.; Hashimoto, K.; Shimodaira, S. XPS determination of compositions of alloy surfaces and surface oxides on mechanically polished iron–chromium alloys. *Corros. Sci.* **1977**, *17*, 713–723. [CrossRef]
43. Asami, K.; Chen, S.C.; Habazaki, H.; Kawashima, A.; Hashimoto, K. A photoelectrochemical and ESCA study of passivity of amorphous nickel-valve metal alloys. *Corros. Sci.* **1990**, *31*, 727–732. [CrossRef]
44. Hashimoto, K.; Kasaya, M.; Asami, K.; Masumoto, T. Electrochemical and XPS studies on corrosion behavior of amorphous Ni–Cr–P–B alloys. *Corros. Eng.* **1977**, *26*, 445–452. [CrossRef]
45. Scofield, J.H. Hartree-Slater subshell photoionization cross-sections at 1254 and 1487 eV. *J. Electron. Spectrosc. Relat. Phenom.* **1976**, *8*, 129–137. [CrossRef]
46. Anthony, M.T.; Seah, M.P. XPS: Energy calibration of electron spectrometers. 1—An absolute, traceable energy calibration and the provision of atomic reference line energies. *Surf. Interface Anal.* **1984**, *6*, 95–106. [CrossRef]
47. Gaarenstroom, S.W.; Winograd, N. Initial and final state effects in the ESCA spectra of cadmium and silver oxides. *J. Chem. Phys.* **1977**, *67*, 3500–3506. [CrossRef]
48. Biesinger, M.C. Advanced analysis of copper X-ray photoelectron spectra. *Surf. Interface Anal.* **2017**, *49*, 1325–1334. [CrossRef]
49. Henglein, A.; Giersig, M. Formation of Colloidal Silver Nanoparticles: Capping Action of Citrate. *J. Phys. Chem. B* **1999**, *103*, 9533–9539. [CrossRef]
50. Hans, M.; Erbe, A.; Mathews, S.; Chen, Y.; Solioz, M.; Mucklich, F. Role of copper oxides in contact killing of bacteria. *Langmuir* **2013**, *29*, 16160–16166. [CrossRef]
51. Shimabukuro, M.; Manaka, T.; Tsutsumi, Y.; Nozaki, K.; Chen, P.; Ashida, M.; Nagai, A.; Hanawa, T. Corrosion Behavior and Bacterial Viability on Different Surface States of Copper. *Mater. Trans.* **2020**. [CrossRef]

© 2020 by the authors. Licensee MDPI, Basel, Switzerland. This article is an open access article distributed under the terms and conditions of the Creative Commons Attribution (CC BY) license (http://creativecommons.org/licenses/by/4.0/).

Article

CaSiO$_3$-HAp Structural Bioceramic by Sol-Gel and SPS-RS Techniques: Bacteria Test Assessment

Evgeniy Papynov [1,2,*], Oleg Shichalin [1,2], Igor Buravlev [1,2], Anton Belov [1,2], Arseniy Portnyagin [1], Vitaliy Mayorov [1], Evgeniy Merkulov [1], Taisiya Kaidalova [1], Yulia Skurikhina [3], Vyacheslav Turkutyukov [3], Alexander Fedorets [2] and Vladimir Apanasevich [3]

1. Institute of Chemistry, Far Eastern Branch of Russian Academy of Sciences, 159, Prosp. 100-letiya Vladivostoka, Vladivostok 690022, Russia; oleg_shich@mail.ru (O.S.); buravlev.i@gmail.com (I.B.); nefryty@gmail.com (A.B.); arsuha@gmail.com (A.P.); 024205@inbox.ru (V.M.); merkulov@ich.dvo.ru (E.M.); kaydalova@ich.dvo.ru (T.K.)
2. Far Eastern Federal University, 8, Sukhanova St., Vladivostok 690091, Russia; fedorets.alexander@gmail.com
3. Pacific State Medical University, 2, Ostryakov Aven., Vladivostok 690990, Russia; eesku@mail.ru (Y.S.); vyach.12593@mail.ru (V.T.); oncolog222@gmail.com (V.A.)
* Correspondence: papynov@mail.ru

Received: 2 April 2020; Accepted: 9 June 2020; Published: 12 June 2020

Abstract: The article presents an original way of getting porous and mechanically strong CaSiO$_3$-HAp ceramics, which is highly desirable for bone-ceramic implants in bone restoration surgery. The method combines wet and solid-phase approaches of inorganic synthesis: sol-gel (template) technology to produce the amorphous xonotlite (Ca$_6$Si$_6$O$_{17}$·2OH) as the raw material, followed by its spark plasma sintering–reactive synthesis (SPS-RS) into ceramics. Formation of both crystalline wollastonite (CaSiO$_3$) and hydroxyapatite (Ca$_{10}$(PO$_4$)$_6$(OH)$_2$) occurs "in situ" under SPS conditions, which is the main novelty of the method, due to combining the solid-phase transitions of the amorphous xonotlite with the chemical reaction within the powder mixture between CaO and CaHPO$_4$. Formation of pristine HAp and its composite derivative with wollastonite was studied by means of TGA and XRD with the temperatures of the "in situ" interactions also determined. A facile route to tailor a macroporous structure is suggested, with polymer (siloxane-acrylate latex) and carbon (fibers and powder) fillers being used as the pore-forming templates. Microbial tests were carried out to reveal the morphological features of the bacterial film *Pseudomonas aeruginosa* that formed on the surface of the ceramics, depending on the content of HAp (0, 20, and 50 wt%).

Keywords: porous bioceramics; wollastonite; hydroxyapatite; sol-gel technology; spark plasma sintering–reactive synthesis; bacterial test

---

1. Introduction

Sustainable development of modern biotechnology in the field of regenerative and reconstructive bone surgery is governed by the quality of the available biomaterials, a specific class of systems with a unique set of physico-chemical and mechanical characteristics (chemical inertness, microstructural diversity, mechanical strength, fracture resistance, and durability), as well as biocompatibility (non-toxicity, bio-inducivity, bio-conductivity, and bio-resistivity) [1,2]. The combination of such properties in one product is a challenging scientific and technological problem in the search for affordable raw materials and simple processing technologies to fabricate the final products.

Calcium monosilicate, β-wollastonite (CaSiO$_3$), is actively studied at the moment due to its applications in traumatology, orthopedics, dentistry, maxillofacial surgery, and other areas of medicine for the recovery, replacement, and reconstruction of the damaged tissue in a living organism [1,3–14]. Owing to β-wollastonite's ability to activate the growth of the apatite layer on its surface due to the pronounced osteoconduction and bio-resorption via exchange of Ca^{2+} and SiO_3^{2-} with the bioorganic medium, the material is promising as an artificial bone substitute [15,16]. In terms of chemical bonding, wollastonite ceramic is close to the inorganic bone matrix and has no toxic effect on the body. In addition, it is corrosion-resistant, thermally stable, and chemically inert or bioactive under long-term exposure in bioorganic environments [17]. In order to achieve maximum similarity to the functional and structural parameters of the bone tissue, β-wollastonite can be modified with hydroxyapatite (HAp), which is a complete analogue of a living bone [14,18,19].

Chemical composition is an important, but not the only characteristic, of bioceramics implants. Their uniqueness to a large extent is determined by a combination of structural and mechanical characteristics. Porous and mechanically stable ceramics represents a model of cellular "spongiose" material that serve as a matrix for ingrowth (implantation) of the bone tissue during osteointegration, a process of recovering lost tissular structures in the living organism in the presence of the implant. Osteointegration intensity depends on pores presented in the implant, particularly, on their size, quantity, and interconnectivity. The bio-integration process, which is based on the reproduction of osteogenic cells, requires large macropores, sized 100–135 μm, as well as on the thin submicron and nanosized pores that are commensurate with blood plasma proteins for their effective adsorption. Stability of such a matrix has to be optimal for the uniform distribution of mechanical load between natural and artificial bone that avoids the possibility of excess decomposition of the bone tissue [20,21]. Thus, it is obvious that load-resistant ceramics with a hierarchical pore size distribution is necessary for practical medicine.

From an implant manufacturing standpoint, fabrication of the β-wollastonite powder and its composites with the HAp-required characteristics and properties is not difficult. This can be done by sol-gel, hydrothermal, and precipitative synthesis technologies [6,22–30]. These approaches are easy to implement and allow to vary the size and shape of the crystallites and to tailor the surface morphology. On the contrary, to manufacture the bulk ceramics of the required geometry, characteristics, and properties is far more challenging due to the rigid thermal conditions of the powder treatment during consolidation into dense ceramics. During the fabrication of wollastonite ceramics, most of the conventional sintering methods do not provide the preservation of the porous structure of the wollastonite [31] due to the negative impact of the heat treatment on the composition and structure of the final product. The reason lies in the phase instability of the HAp above 1000 °C, as well as the destruction of the porous volume and the activation of grain growth, which leads to distortion or destruction of the porous frame in the solid body and negatively affects the final properties of the biocomposite [14,32,33].

The problems in the synthesis of porous ceramics described above are solved by using the technology of spark plasma sintering (SPS) [34–38]. The unique mechanism of powder consolidation in this technology is based on the spark plasma current flowing through the sample under pressure, which provides a number of advantages over traditional methods, because in this case ceramic wollastonite with a tailored microstructure and exceptional mechanical characteristics is rapidly formed [39–41]. The structural strength of the SPS ceramics is achieved without the need for additional reinforcement components that contaminate the final product. Our early studies have identified the above-described prospects of SPS application for the synthesis of nanostructured bioceramic wollastonite [42–45] with its bioactive properties being assessed "in vivo" [46]. Additionally, these studies showed several original methods for developing a porous structure of the ceramics that is similar to the texture of bone tissue by introducing various porous templates. However, the technology of spark plasma sintering–reactive synthesis (SPS-RS) should be considered even more promising for the production of innovative ceramics. SPS-RS is based on the chemical interaction between the starting components of the sintering mixture under the influence of spark plasma, resulting in a new type of the final product [47–49]. This approach allows one to directly obtain different materials with unique properties based on multi-component ceramic systems. The literature substantiates the advantages of SPS-RS compared to conventional SPS and hot pressing [41]. SPS-RS efficiency for bioceramic synthesis is based on the local character of the chemical interaction between the components of the reaction mixture (resulting in biocomponents) under the spark plasma heating, taking place on the interparticle contacts.

Such local heating favors the formation of the fine-crystal phases of the biocomponent, which will be more biologically compatible or resorbed depending on its chemical composition. Additionally, microlocal heating allows the reactions to proceed at lower temperatures; thus, not disrupting the metastability of the substances and increasing the limit of their thermodestruction. We have recently explored the feasibility of this approach for fabrication of ZrO_2 ceramics containing the bioactive phosphate compounds obtained "in situ" under SPS-RS conditions [50]. For the synthesis of ceramic HAp-containing wollastonite, similar studies have not been conducted before.

In this regard, the work intends to study the way to obtain a crystalline ceramic HAp–wollastonite composite via solid phase transformation of amorphous xonotlite and "in situ" interaction of the reaction mixture (CaO and $CaHPO_4$) under SPS conditions. Additionally, the way to tailor the porous structure of the ceramics using pore-forming templates has been investigated. Microbial tests were conducted to assess the possible risks of an infectious process caused by bacterial contamination of the ceramics.

The proposed non-standard SPS-RS approach can pave the way to fabrication of biocompatible ceramics for bone tissue engineering; thus, contributing another flexible strategy for the synthesis of biomaterials.

2. Materials and Methods

2.1. Materials

Sodium metasilicate ($Na_2SiO_3 \cdot 5H_2O$) and calcium chloride ($CaCl_2 \cdot 2H_2O$) have been used as the main precursors for calcium silicate synthesis. Calcium oxide (CaO) and calcium hydro-orthophosphate ($CaHPO_4$) were used to synthesize HAp. All reagents are of 99.98% purity (Sigma-Aldrich, St. Louis, MO, USA). Siloxane acrylate latex "KE 13–36" (LLC "Astrokhim", Electrostal, Russia, solid phase content 50%, average particle size 160 [51,52]), carbon fiber (CF) "AUT-M" (TS 1916-346-04838763-2009, ENPO "Inorganica", Electrostal, LTD Russia), and a graphite powder (GP) fraction 1–5 μm (GOST 7885-86, "Khazar", Turkmenistan) were used as the pore-forming templates.

2.2. Synthesis Technique

The composite $CaSiO_3$ (wollastonite)/HAp ceramics was prepared in three successive stages. Initially, the method of sol-gel (template) synthesis yielded a composite material (xerogel) based on hydrated calcium silica (xonotlite, $Ca_6Si_6O_{17} \cdot 2OH$) mixed with polymer latex (pore-forming template). Then, a sintering mixture (SM) was made by blending the obtained xerogel with the reaction mixture components (RM) for the formation of HAp. Finally, the consolidation of SM using SPS technology was carried out.

2.2.1. The Sol-Gel (Template) Synthesis of Amorphous Composite Powder $Ca_6Si_6O_{17} \cdot 2OH$ (Xonotlite)

A total of 50 mL of 1.0 M calcium chloride solution and 50 mL of 1.0 M sodium metasilicate solution were added batchwise to 150 mL of a siloxane–acrylate water solution (latex:water ratio 1:30) under intense stirring. Then, the solution was stirred for three hours at 100 °C until a dense gel was formed, which after boiling was cooled to room temperature (25 °C). After that, 16.6 mL of a 1.0 M calcium chloride solution and 10 mL of 1.0 M ammonium hydrophosphate were added to the obtained gel and stirred for 1 h at room temperature (25 °C). The resulting gel was filtered and washed with distilled water until a negative reaction to the chloride ions was obtained, and then dried for about 5 h at 90 °C.

Fabrication of wollastonite proceeded according to the chemical reaction below:

$$CaCl_2 + Na_2SiO_3 \rightarrow CaSiO_3 + 2NaCl \tag{1}$$

2.2.2. Preparation of the Sintering Mixture Added with Reaction Mixture Additive

The composite xerogel obtained according to Section 2.2.1 was mixed with the components of the reaction mixture (RM) $CaO+CaHPO_4$ at the planetary mill at a rotation rate of 870 rpm for three cycles for 15 min. Carbon fiber and powder graphite (Table 1) were added into this mixture as additional pore-forming agents during ball milling.

Reaction of HAp formation:

$$CaO + CaHPO_4 \rightarrow Ca_{10}(PO_4)_6(OH)_2 + 2H_2O \tag{2}$$

Table 1. Wollastonite/HAp and the amounts of added templates.

#	Sample Notation	Ratio, wt%		Quantity, wt%	
		Wollastonite	* HAp	** Carbon Fiber (CF)	** Graphite Powder (GP)
1	$CaSiO_3$-(50)Hap	50	50	-	-
2	$CaSiO_3$-(20)Hap	80	20	-	-
3	$CaSiO_3$-(20)Hap/5(CF)	80	20	5	-
4	$CaSiO_3$-(20)Hap/10(CF)	80	20	10	-
5	$CaSiO_3$-(20)Hap/5(CF)-10(GP)	80	20	5	10
6	$CaSiO_3$-(20)Hap/10(CF)-10(GP)	80	20	10	10

Note: * Estimated amount of HAp according to the reaction Equation (2); ** weight amount in terms of the total mass of the sintering mixture.

2.2.3. SPS-RS Fabrication of the Composite $CaSiO_3$ (Wollastonite)/HAp Ceramics

The SM powder (fraction 0.1–0.5 mm), obtained via ball milling (see Section 2.2.2), was put into a graphite die (outer diameter—30 mm, internal diameter—15.5 mm, and height—30 mm), prepressed (20.7 MPa), and then the green body was put into a vacuum chamber (pressure 6 Pa) and sintered on a LABOX-625 SPS machine "Sinter Land Incorporation, LTD" (Niigata, Japan). Sintering was executed at 900 °C at a heating rate of 100 °C/min and 5 min holding at maximal temperature. After sintering, the samples were cooled down to room temperature (25 °C) for 40 min. Pressure loaded onto the sample during sintering was 24.5 MPa and remained constant throughout the process. The frequency of the low-voltage pulse generated in the On/Off mode was 12/2, with a pulse packet duration of 39.6 ms and a 6.6 ms pause between the pulses. The maximum current and voltage during sintering were 500 A and 2 V, respectively. To prevent the consolidated powder from being sintered to the die walls and plungers, as well as to easily extract the resulting compound, 200 μm thick graphite paper was used. The die was wrapped in a thermal insulating fabric to reduce the heat loss when heated. The temperature of the process was controlled by an optical pyrometer focused on a hole (5.5 mm deep) located in the middle of the die's outer wall. In order to remove the templates, the samples were annealed in air at 800 °C for 1 h at a heating rate of 5 °C/min, in the "Nabertherm GmbH" furnace (Lilienthal, Germany).

2.3. Microbiological Test

Microbiological studies were done to assess the formation of *Pseudomonas aeruginosa* biofilms on the obtained samples of ceramic wollastonite. The samples were placed on the surface of the solid medium and pre-seeded with the reference strain *Pseudomonas aeruginosa*. Cultivation was carried out at 37 °C for 48 h. The biofilm was fixed on the sample by rinsing with 4% formaldehyde in a 1% solution of the phosphate buffer followed by exposure of 1% of the osmium tetroxide solution for 1 h. Dehydration was carried out with consistent treatment in ethanol of varying concentrations and at appropriate exposures (30%—10 min; 50%—10 min; 70%—10 min; and 96%—20 min), and then in acetone for 20 min. The morphology of the biofilms was studied using electron microscopy.

2.4. Characterization Methods

The identification of crystalline phases in the original powders and the ceramics derived from them was carried out using X-ray diffraction (XRD) with CuKα-radiation (Ni-filter, average wavelength (λ) 1.5418 Å, range of angles of 10–80°, scan step 0.02°, and spectrum registration speed—5°/min) on a multipurpose X-ray diffractometer D8 Advance "Bruker AXS" (Karlsruhe, Germany). XRD patterns were taken from the Powder Diffraction File™ (PDF, Soorya N Kabekkodu, 2007). The reference numbers from the PDF database are as follows: Wollastonite—01-084-0654 (C) Wollastonite 1A—$CaSiO_3$; hydroxyapatite: 01-074-0566 (C) Hydroxyapatite—$Ca_{10}(PO_4)_6(OH)_2$. Thermogravimetric analysis (TGA) was carried out on the derivativograph Q-1500 of the F. Paulik, J. Paulik, L. Erdey (Hungary) system in air in a platinum crucible at a heating rate of 10°/min to a maximum temperature of 1000 °C. Low-temperature nitrogen sorption was employed to determine the specific surface area ($S_{spec.}$) on an automated physisorption analyzer, Autosorb IQ "Quantachrome" (Boynton Beach, FL, USA), using the Brunauer–Emmett–Teller (BET) model. The pore size distribution was analyzed on a mercury porosimeter AutoPore IV "Micromeritics GmbH" (Norcross, GA, USA). The material's structure and biofilm's morphology were studied by the method of scanning electron microscopy (SEM) on an Ultra 55 "Carl Zeiss" (Oberkochen, Germany) with a field-emission cathode and Oxford X-Max detector for Energy-dispersive X-ray (EDX) spectroscopy. The composite ceramic samples were covered by 5 nm of platinum and were investigated at the fracture sites at accelerating voltages of 5 kV and at 20 kV for EDX analysis. To minimize irradiation and to exclude the samples charging, biofilms were investigated at an accelerating voltage of 1 kV after fixing on the samples (see Section 2.3). The beam current was I ≈ 100 pA. Vickers hardness (HV) was measured at a HV0.5 load on a hardness tester, the HMV-G-FA-D "Shimadzu" (Kyoto, Japan) micro-solid. Prior to microhardness tests, the ceramic surface was polished on an automated polishing machine MECATECH 234 (Grenoble, France). Compressive strength (σ_{cs}) of the cylindrically shaped samples (diameter 15.3 mm and 3–6 mm high) was determined by squeezing at a rate of 0.5 mm/min on the tensile machine Autograph AG-X plus 100 kN "Shimadzu" (Kyoto, Japan). The experimental density (ED) of the samples was determined by hydrostatic weighing on the balance Adventurer™ "OHAUS Corporation" (Parsippany, NJ, USA)

Young's modulus (E) was evaluated in the load range of 3000–5000 N according to the following formula:

$$E = \frac{F \times h}{S \times (l_2 - l_1)}$$

where F—load applied onto the sample (N); h—sample's height (mm); S—sample's surface area (mm²); and l_1 and l_2—starting and final height of the material (mm) in the range of applied loads.

Sample deformation along the height under applied pressure spanning 3000–5000 N was evaluated via the formula:

$$\text{Deformation}(\%) = 100 - \left(\frac{l_1}{l_2} \times 100\right)$$

where l_1 and l_2 is the starting and final sample height (mm) in the range of applied loads.

The elative density (RD) of the composite ceramic samples was carried out according to the formula:

$$RD(\%) = 100 / \frac{\omega_1}{\rho_1} + \frac{\omega_2}{\rho_2}$$

where ω—mass content of component; and ρ—theoretical density of the component.

To calculate the porosity of the samples the formula was used:

$$\text{Porosity}(\%) = 100 - (ED/RD \times 100)$$

ED—experimental density, RD—relative density.

The experimental density and Vickers microhardness were obtained as the mean values out of three measurements for each sample. Determination of the compressive strength is a destructive method of analysis, so only a single measurement was done on each sample. Young's modulus was calculated in the range of 3000–5000 N from the trends, obtained from the compressive strength measurements.

3. Results and Discussion

Primarily, the chemical interaction between the components of the reaction mixture (RM) was studied to reveal the HAp formation according to Reaction (2) in air as well as to simulate the conditions of SPS consolidation in the presence of the hydrated calcium silicate xerogel (xonotlite). The temperature when the RM components started to react (Reaction (2)) was determined by thermogravimetric analysis (Figure 1).

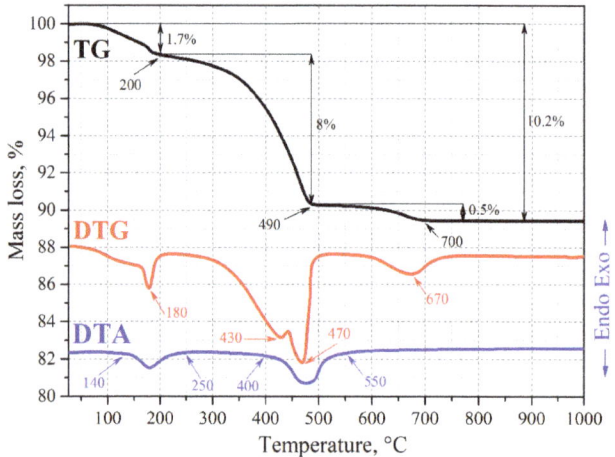

Figure 1. Thermogram of the reaction mixture (RM) sample heated in air.

According to the results of the TGA (Figure 1), a slight mass loss of 1.7 wt% is observed till 200 °C on the TG and DTG curves, while an endo-effect associated with it lasts till 250 °C on DTA. A further increase in temperature leads to the 8 wt% weight loss by 490 °C on TG due to the solid-phase interaction of the RM components yielding HAp and water by Reaction (2) with the onset of the endo-effect being observed from 400 to 550 °C on the DTA curve. Moreover, at these temperatures the non-reacted calcium hydrophosphate is likely to decompose according to Reaction (3).

$$2CaHPO_4 = Ca_2P_2O_7 + H_2O \qquad (3)$$

There is a 0.5% weight loss at 700 °C on the TG curve with a minimum at 670 °C on DTG, which can be caused by the partial decomposition of calcium carbonate according to Reaction (4).

$$CaCO_3 = CaO + CO_2\uparrow \qquad (4)$$

To confirm the results of the TGA, XRD was performed for the RM samples before and after annealing in air at different temperatures (Figure 2).

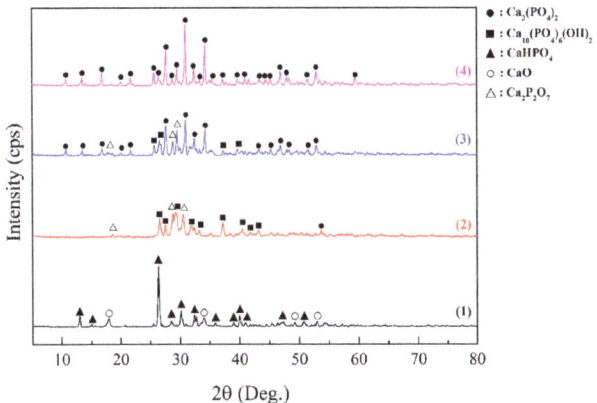

Figure 2. The XRD patterns of the RM powder in its original form (**1**) and after its heat treatment in air at the following temperatures: (**2**) 500–700 °C, (**3**) 900 °C, and (**4**) 1000 °C.

The RM composition consists of the CaO and $CaHPO_4$ crystalline phases (Figure 2, Pattern 1). Thermo-oxidative annealing initiates a solid phase reaction of the RM with the formation of HAp at the temperature ranging from 500 to 700 °C (Figure 2, Pattern 2) according to Reaction (2), which was also noted on TGA (Figure 1). An impurity of calcium pyrophosphate ($Ca_2P_2O_7$), which can be formed according to Reaction (3), has been identified in this sample.

An increased annealing temperature leads to partial (at 900 °C) and complete (at 1000 °C) decomposition of HAp with the formation of calcium orthophosphate ($Ca_3(PO_4)_2$) (Figure 2, Pattern 3) according to Reaction (5):

$$Ca_{10}(PO_4)_6(OH)_2 = 3Ca_3(PO_4)_2 + CaO + H_2O. \tag{5}$$

Then, we investigated formation of wollastonite combined with HAp from the sintering mixture (SM), which consists of a xerogel ($Ca_6Si_6O_{17} \cdot 2OH$ (xonotlite)) obtained via sol-gel and an RM along with the pore-forming templates (latex, carbon fiber, and graphite powder). To do this, the TGA was implemented and the corresponding thermograms (Figures 3 and 4) were obtained. According to them, the mass loss of 5.6 wt% in the samples by 250 °C on TG is mainly due to partial dehydration of xonotlite (Figure 3). When the temperature rises, the pore-forming agent, organo-silicon latex, gets oxidized and decomposed. This process results in a loss of sample's mass of 13.3 wt% and characterized by an exo-effect within 240–450 °C on DTA and a minimum at 360 °C on the DTG. Up to 450 °C, the dehydration of the xonotlite is complete and the latex becomes partially burnt out from the bulk. There is also a solid-phase interaction between the RM components in this temperature range, yielding HAp as it was studied above (Figure 1) according to Reaction (2). The remaining latex part is burnt out by 750 °C on the DTA along with the associated minima observed at 590 and 730 °C on DTG, resulting in a mass loss of 11.2 wt%. Partial HAp decomposition may also take place according to Reaction (4). The total weight loss was 30.1 wt%. The exo-effect with an inflection point at 790 °C on DTA is caused by the crystallization of the wollastonite.

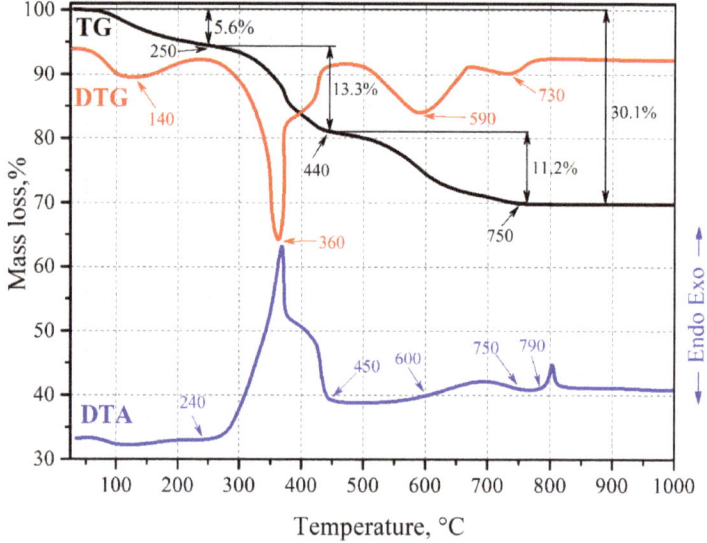

Figure 3. Thermograms of sintering mixture (SM) sample containing 20 wt% of the RM.

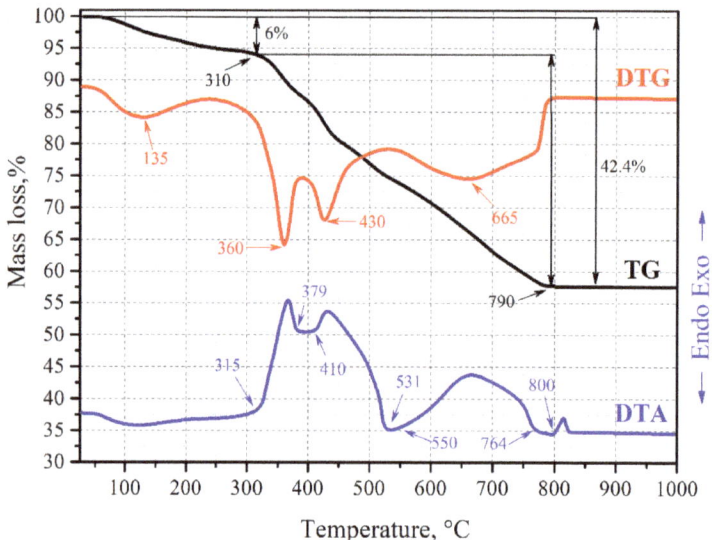

Figure 4. Thermograms of the SM sample containing 20 wt% of the RM as well as 10 wt% of carbon fiber and 10 wt% of graphite powder.

The initial stage of the SM sample annealing in the presence of carbon pore-formers manifests water removal with a mass loss of about 6 wt% by 310 °C on the TG curve (Figure 4), which is similar to the above sample. However, further heating of the sample has a significant difference associated with the imposed templates. Along with latex burnout from the xonotlite's bulk, the exo-effect at 315–450 °C on DTA merges with the exo-effect coming from the graphite powder burning within 410–531 °C. Decomposition of another template, carbon fiber, takes place in the range 550–764 °C on DTA. These effects correspond to the minima on the DTG at 360, 430, and 665 °C, related to the decomposition of latex, graphite powder, and carbon fibers, respectively. Besides, that region also encompasses latex burnout by 750 °C on TG, as was shown earlier on Figure 3. The decomposition processes of the carbon fiber and graphite powder occur simultaneously, causing a large area of exo-effect on the DTA curve in the range 531–764 °C. The formation of crystalline wollastonite is in this case similar to the above-described pattern without carbon additives, which occurs with the onset of heat at 800 °C on the DTA. It is worth noticing the continuous nature of the mass loss in this sample reaching 42.4 wt% by 790 °C on the TG curve, which is significantly higher in comparison with the sample without the carbon additives discussed above. This difference is caused by the thermo-oxidative destruction of the carbon pore-forming agents.

According to the XRD, the composition of the original SM sample includes the phases of the RM components (CaO and CaHPO$_4$), while the silicate phase cannot be found, evidencing its amorphous structure (Figure 5, Pattern 1). After annealing at 500 °C, the phase of calcium orthosilicate Ca$_2$SiO$_4$ is observed, which indicates the intermediate stage of crystalline wollastonite formation (Figure 5, Pattern 2). The presence of calcium silicate in this mixture slows down the interaction between the RM components at this temperature, as there is no HAp phase in the material. Annealing at 900 °C leads to the formation of crystalline wollastonite as well as HAp as a result of the interaction between the RM components (Figure 5, Pattern 3). In addition, there is an impurity of calcium orthophosphate Ca$_3$(PO$_4$)$_2$ caused by the HAp decomposition according to Reaction (5). Annealing of the SM sample containing carbon pore-forming agents at the same temperature (900 °C) also yields wollastonite and HAp (Figure 5, Pattern 4) without formation of calcium orthophosphate, which is likely caused by HAp phase stabilization during templates oxidation.

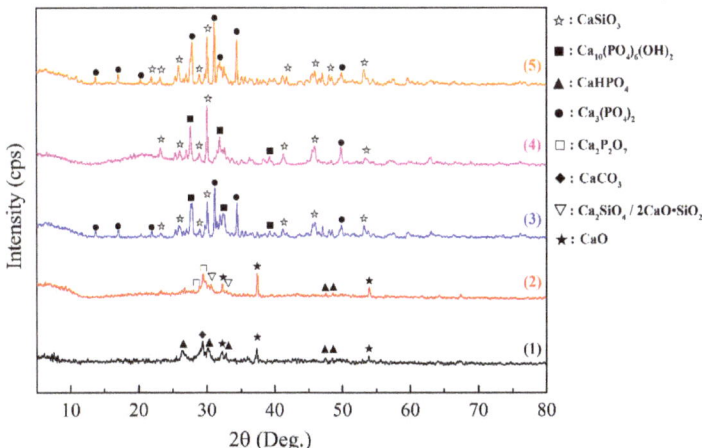

Figure 5. Thermograms of the original SM sample containing 20 wt% of the RM (1), as well as its derivatives annealed at different temperatures: (2)—500 °C; (3)—900 °C; (4)—900 °C, the sample containing pore-forming agents (10 wt% CF and 10 wt% GP); and (5)—1000 °C.

According to the described results, the optimal temperature to obtain crystalline wollastonite and to initiate an "in situ" reaction between the RM components for the formation of the HAp phase is found to be 900 °C. In addition, the temperature of SPS below 900 °C is not enough for sintering of wollastonite, because the required mechanical characteristics are not achieved as was shown in our previous study [44].

XRD patterns of the SPS ceramic samples obtained from the SM (Figure 6) shows the phase composition corresponding to a composite based on crystalline wollastonite and HAp. For all samples, the composition is identical regardless of the HAp content and the type of pore-forming template used.

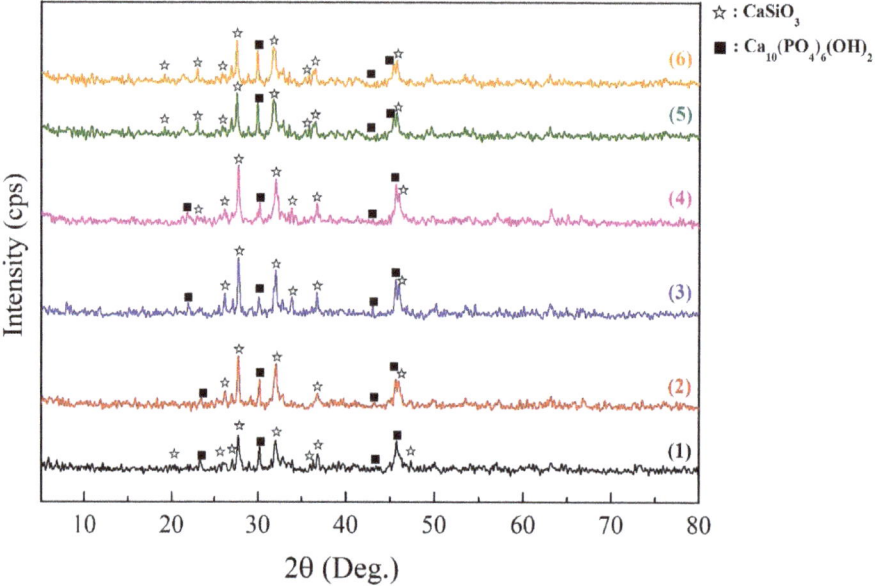

Figure 6. XRD patterns of the composite ceramics obtained by SPS-RS at 900 °C and their subsequent thermo-oxidative treatment at 800 °C: (**1**)—CaSiO$_3$-(50)HAp; (**2**)—CaSiO$_3$-(20)HAp; (**3**)—CaSiO$_3$-(20)Hap/5(Cw); (**4**)—CaSiO$_3$-(20)HAp/10(Cw); (**5**)—CaSiO$_3$-(20)HAp/5(Cw)-10(Cp); and (**6**)—CaSiO$_3$-(20)HAp/10(Cw)-10(Cp). The sample description is presented in Table 1.

The results of porosimetry and structural analysis revealed the effect of the templates on the structure of the obtained ceramics samples. Differential mercury intrusion analysis shows the average size of the macropores provided by the polymer latex is about 100 nm (Figure 7). Changing the amount of HAp in the ceramics from 20 to 50 wt% does not lead to significant structural changes, with the pore size distribution being maintained in the specified interval (samples CaSiO$_3$—(20)HAp and CaSiO$_3$—(50)HAp). The surface area for these samples, calculated by BET, lies in the same range of 2.6 and 2.8 m^2·g^{-1} (Table 2).

The introduction of carbon fiber (5 and 10 wt%) into the synthesis and its subsequent removal changes the range of the pore sizes in the resulting ceramics. When 5 wt% CF are added, the average pore size is approximately 100 nm but the base of the intrusion curve has extended, indicating the formation of larger pores sized above 100 nm (sample CaSiO$_3$-(20)HAp/5(CF)). The number of these pores increases with the amount of the additive. At 10 wt% of CF, there emerges a peak at about the 1–5 μm range on the differential intrusion curve (CaSiO$_3$-(20)HAp/10(CF) sample). At the same time, S$_{spec.}$ decreases from 2.9 to 1.9 m^2·g^{-1} (Table 2), which is probably due to the exclusion of the micro and mesopores (less than 50 nm) from the sample's structure at the moment of carbon template oxidation and macropores formation. There is an extended base of the intrusion curve observed for the sample containing 5 wt% of powder graphite (CaSiO$_3$-(20)HAp/5(CF)-10(GP) sample) with a separate maximum appearing in the range above 5 μm (CaSiO$_3$-(20)HAp/10(CF)-10(GP) sample). S$_{spec.}$ increases for the samples under consideration to 4.9 and 4 m^2·g^{-1}, respectively (Table 2). The general trend observed for all samples is the increase in the macropores' volume from 0.18 to 0.43 mL/g, depending on the type and quantity of the used template as well as the growth of the total porosity in the system from 8.3 to 27.7% (Table 2).

Changes in the structure of samples are confirmed by microscopic studies. SEM images (Figure 8) show that the samples synthesized with a different HAp content in the presence of a polymer latex

(CaSiO$_3$-(20)HAp and CaSiO$_3$-(50)HAp) samples) are characterized by the presence of pores sized below 200 nm, which is commensurate to the template's size. Additional carbon pore-forming agents produce larger pores of 1 µm and above, the shape and size of which are commensurate to that of carbon fiber as, e.g., for the CaSiO$_3$-(20)HAp/10(CF) sample. The introduction of graphite powder into the synthesis leads to a distorted arrangement of the pores derived from CF with the structure becoming looser (CaSiO$_3$-(20)HAp/10(CF)-10(GP)) (Figure 8). In both samples, when the large pores (larger than 1 µm) are forming, the presence of the small pores (less than 200 nm) is maintained.

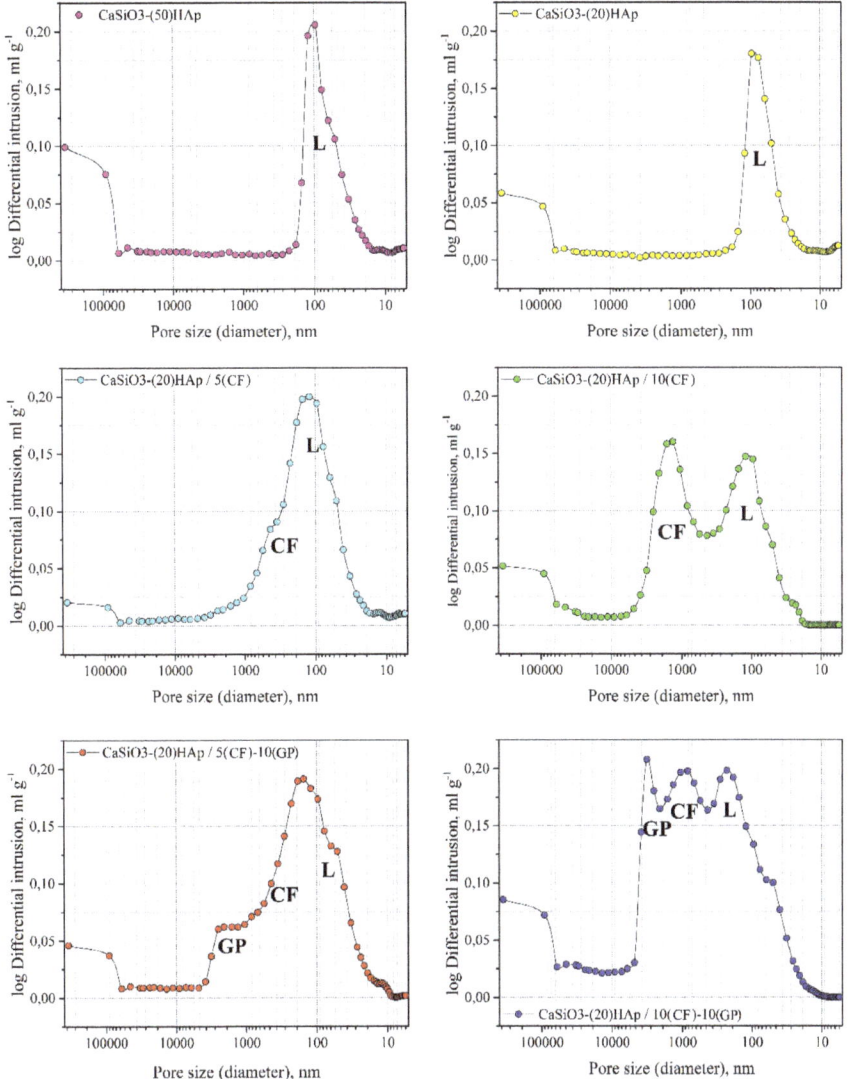

Figure 7. Differential mercury intrusion curves obtained for the porous SPS composite ceramics. Sample denotation is presented in Table 1, the structural characteristics are in Table 2. The indexes (GP, CF, and L) show the pore size ranges tailored by certain types of pore-forming agents (templates): L—siloxane-acrylate latex, CF—carbon fiber, and GP—graphite powder.

Table 2. Structural characteristics of the composite ceramics samples.

#	Sample	$S_{spec.}$, m^2·g^{-1}	V_{pore}, mL/g	Porosity, %
1	CaSiO$_3$-(50)HAp	2.8 ± 0.1	0.18 ± 0.01	8.3
2	CaSiO$_3$-(20)HAp	2.6 ± 0.1	0.13 ± 0.01	9.7
3	CaSiO$_3$-(20)HAp/5(CF)	2.9 ± 0.1	0.21 ± 0.01	18.5
4	CaSiO$_3$-(20)HAp/10(CF)	1.9 ± 0.1	0.26 ± 0.01	19.7
5	CaSiO$_3$-(20)HAp/5(CF)-10(GP)	4.9 ± 0.1	0.28 ± 0.01	25.4
6	CaSiO$_3$-(20)HAp/10(CF)-10(GP)	4 ± 0.1	0.43 ± 0.01	27.7

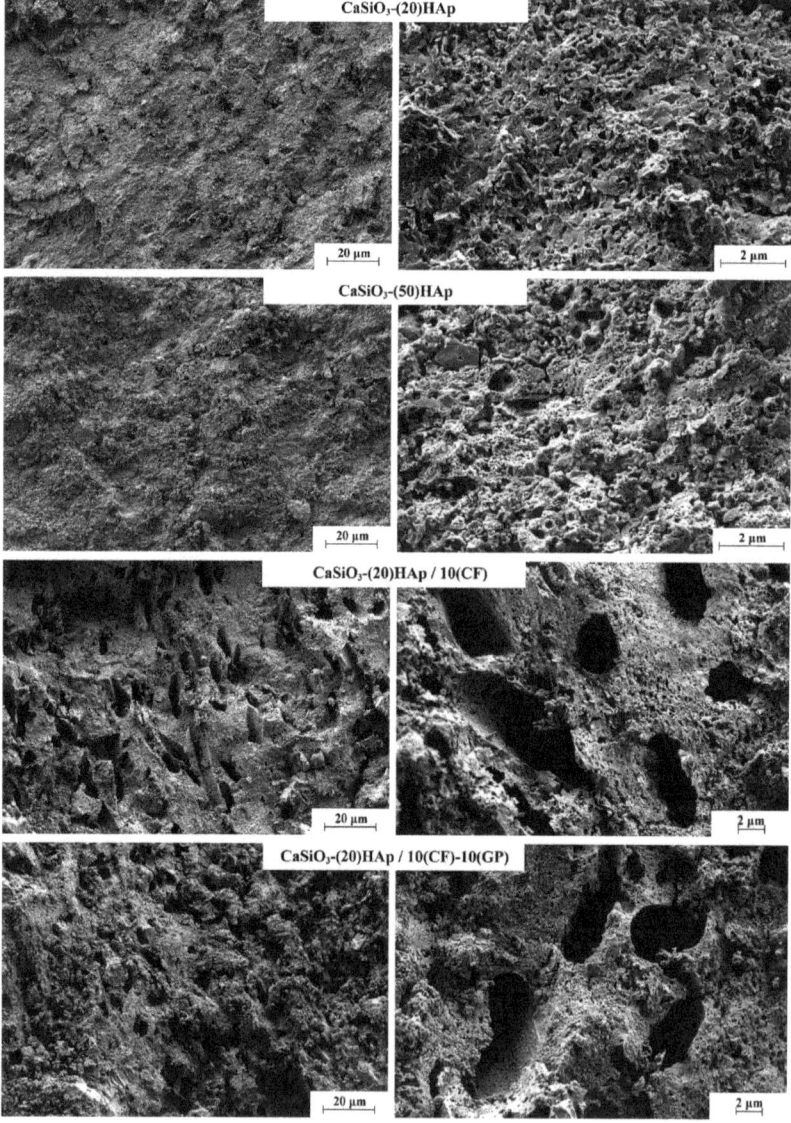

Figure 8. SEM images of the spark plasma sintering–reactive synthesis (SPS-RS) composite ceramics surface. The sample descriptions are presented in Table 1.

Based on the EDX data, the presence of phosphorus on the surface of the samples is found, which indicates the presence of HAp in the ceramic wollastonites. The uniform distribution of phosphorus in the samples can be observed on the map of the elements, which reflects the difference when the number of HAp increases from 20 to 50 wt% (Figure 9, Table 3). According to the quantitative EDX analysis, there is also a difference in the phosphorus content of the ceramics. Analysis of arbitrary surface areas of the samples with a different HAp content (Figure 9, Spectra 1 and 2) indicates a phosphorus content of 2.15 and 4.78 wt% (Table 3). A 2.2-fold difference in phosphorous content corresponds well to the amounts of introduced RM components into the sintering mixture within the error of the method.

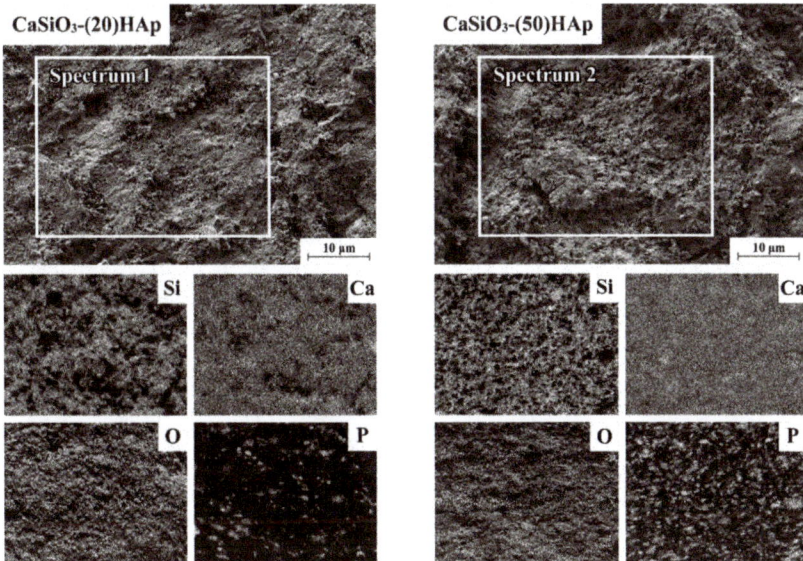

Figure 9. Elements mapping (EDX analysis) on the surface of the composite ceramic samples containing 20 wt% (the CaSiO3-(20)HAp sample) and 50 wt% HAp (the CaSiO3-(20)HAp sample).

Table 3. The quantitative surface elemental composition (EDX analysis) of the composite ceramic samples presented in Figure 9 (wt%).

Samples	Spectra	C	O	Si	P	Ca
CaSiO$_3$-(20)HAp	Spectrum 1	6.31 ± 0.03	67.49 ± 0.03	11.84 ± 0.03	2.15 ± 0.03	12.21 ± 0.03
CaSiO$_3$-(50)HAp	Spectrum 2	6.96 ± 0.03	68.81 ± 0.03	8.09 ± 0.03	4.78 ± 0.03	11.36 ± 0.03

The physical and mechanical characteristics of the biocomposite ceramics (Table 4) were determined. The relative density (RD) of the samples with a different HAp content ranges between 90.2 and 91.7% under certain conditions of SPS consolidation. The increase in the HAp content from 20 to 50 wt% in the ceramic composite leads to an increase in compressive strength (σ_{cs}) from 302 to 362 MPa and in microhardness (HV) from 134 to 146 (Table 4, Samples 1 and 2). The total porosity of the ceramics formed by polymer latex is reduced from 9.7 to 8.3% (Table 2) due to the fact that the amount of HAp in the sample is in direct relation to the original silicate raw material (xonotlite), which contains the latex template. When the ratio of xonotlite to HAp changes, the amount of latex responsible for porous structure formation thereby alters the strength characteristics of the final ceramics.

Table 4. Physical and mechanical characteristics of the composite ceramics samples.

#	Samples	ED, g·sm^{-3}	RD, %	HV	σ_{cs}, MPa	E, MPa (3000–5000 H)	Deformation, %
1	CaSiO$_3$-(50)HAp	2.723 ± 0.005	91.7	146	362	2564	5.3
2	CaSiO$_3$-(20)HAp	2.586 ± 0.005	90.2	134	302	2224	4.1
3	CaSiO$_3$-(20)HAp/5(CF)	2.334 ± 0.005	81.4	64	176	1997	5.2
4	CaSiO$_3$-(20)HAp/10(CF)	2.301 ± 0.005	80	34	111	1421	6.3
5	CaSiO$_3$-(20)HAp/5(CF)-10(GP)	2.137 ± 0.005	74.6	26	75	1379	8.7
6	CaSiO$_3$-(20)HAp/10(CF)-10(GP)	2.069 ± 0.005	72.2	1<	35	1358	9.2

The total porosity of the system was enhanced in this work by means of additional carbon pore-forming agents (Table 2), although it significantly reduces the density and affects the mechanical characteristics of the samples (Table 4). Adding 10 wt% carbon fiber into the synthesis is found to be optimal as it yields porous (19.7%) and dense (RD 80%) biocomposite ceramics, with the mechanical strength (σ_{cs}. 111 MPa) lying within the values of the normal strength of natural bone (110–120 MPa) [53–55]. Introducing powder carbon as an additional pore-forming agent is also useful for increasing the overall porosity of the ceramics, but its quantity should be clearly defined with the other characteristics achieved.

The work carried out a microbiological test to study the formation and morphological features of bacterial film (biofilm) based on *Pseudomonas aeruginosa* on the surface of composite ceramic samples with different HAp contents. In addition, a sample of pristine ceramic wollastonite without the HAp additive, which was obtained in our earlier studies, was used for comparison [42]. The type of bacteria implemented here lives in the external environment (water and soil) and is one of the leading pathogens that cause post-surgery infections requiring medical care [56]. According to the SEM images (Figure 10), all the samples clearly show that a mature biofilm had formed. On the sample of ceramic wollastonite without HAp, the biofilm is distributed in a denser, uniform layer (Figure 10a). On the contrary, in the case of composite ceramics, the monolithic agglomerates of the HAp phase are less intensely populated by bacteria with the latter occupying mostly places of porous and loose formations (Figure 10b,c). All samples between the cells clearly show cytoplasmic bridges, which is a sign of an actively developing biofilm. The main point to be noted is that the bacteria grown on the samples containing HAp are covered with a thick layer of alginate, which indicates the inclusion of a protective reaction of the bacteria under the negative impact of the environment caused by the presence of HAp in the ceramics and the completion of the biofilm formation (Figure 10b*,c*). The formation of an alginate layer indicates that the presence of HAp in the ceramic prevented the formation of a dense uniform (continuous) biofilm layer. In this regard, in terms of the risk of infection, the composite HAp-containing ceramics contained is more attractive for biomedical applications.

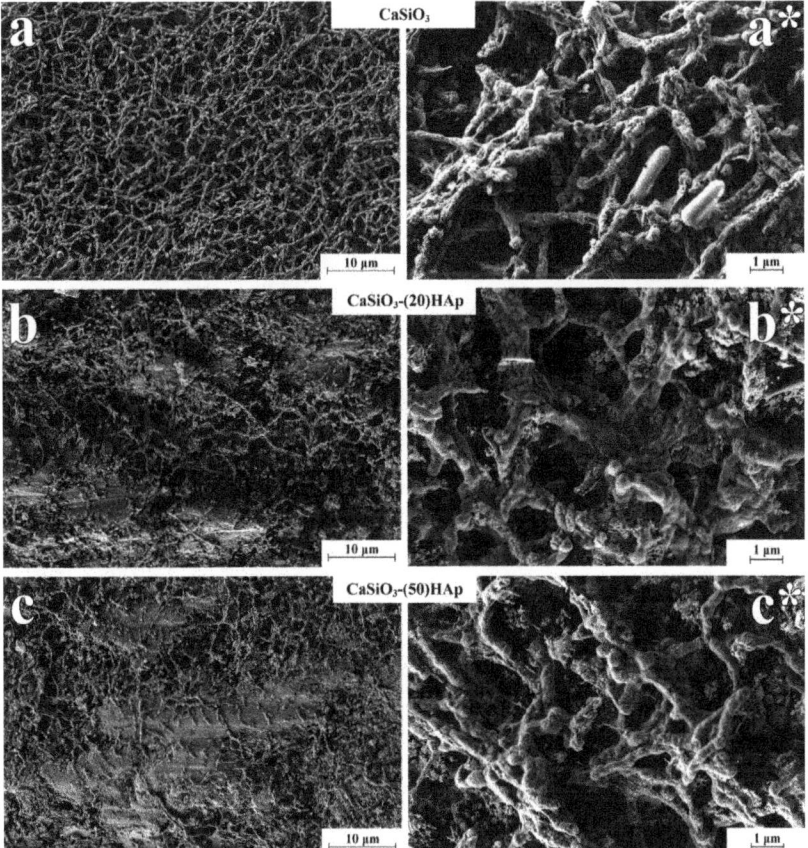

Figure 10. SEM images of bacterial film on ceramic samples with different HAp content obtained by SPS-RS: (**a,a***) 0 wt% HAp—continuous layer of bacterial cells without an alginate layer; (**b,b***) 20 wt% HAp—less dense layer of bacterial cells coated with an alginate layer; (**c,c***) 50 wt% HAp—rarefied layer of bacterial cells coated with a dense layer of alginate.

4. Conclusions

The work provides an original way to obtain $CaSiO_3$-HAp porous ceramics with the wet and solid-phase synthesis strategies being combined. The method involves sol-gel synthesis (template) of the starting raw material in the form of an amorphous composite material based on xonotlite and its subsequent spark plasma sintering yielding ceramic wollastonite. TGA, XRD, and EDX showed that the HAp phase formation in the resulting ceramics can be initiated by the solid-phase interaction of the reaction mixture (CaO and $CaHPO_4$) in the spark plasma heating by the "in situ" reaction directly at the moment of sintering of the amorphous xonotlite. It has been determined that the optimal temperature for the SPS-RS to obtain crystalline wollastonite from the amorphous xonotlite, as well as to initiate the reaction "in situ" with the formation of HAp, is 900 °C.

Nitrogen physisorption and mercury porosimetry allowed to reveal that the pore size and volume depend on the type and quantity of the template introduced at the different stages of the synthesis, leaving pores after the thermal-oxidative treatment of the ceramics. It is shown that latex introduced in the sol-gel synthesis allows to form pores of about 100 nm, while carbon pore-forming agents added prior to SPS-RS contribute to the formation of macropores of 1 µm and larger. We determined that the use of latex and 10 wt% carbon fiber is considered the optima, as porous (19.7%) and dense (RD 80%) biocomposite ceramics is formed that exhibit the required mechanical strength (σ_{cs} 111 MPa), which is within the values of the normal strength of natural bone (110–120 MPa). The use of graphite powder should be carefully adjusted due to its impact on the other important physical and mechanical characteristics of ceramics.

The results of the microbiological test revealed that the (20 and 50 wt%) HAp-containing ceramic samples show the bacteria *Pseudomonas aeruginosa* being covered with a thick layer of alginate, with the latter not being observed on pristine wollastonite. Alginate appearance indicates the protective response of the bacteria to a negative environmental impact. In this regard, in terms of the risk of infection, composite HAp-containing ceramics is more attractive for biomedical applications.

Author Contributions: Conceptualization, E.P.; data curation, V.T.; formal analysis, A.P.; investigation, V.M., E.M. and Y.S.; methodology, O.S. and A.B.; project administration, V.A.; resources, A.F.; software, Y.S.; validation, I.B.; visualization, T.K.; writing—review and editing, E.P. and A.P. All authors have read and agreed to the published version of the manuscript.

Funding: The investigation was financially supported by Russian Science Foundation (project No. 18-73-10107).

Acknowledgments: Equipment of CUC "Far Eastern center of structural investigation" (Institute of chemistry FEB RAS, Vladivostok, Russia), also interdisciplinary CUC in the field of nanotechnologies and new functional materials and CUC "Laboratory of mechanical tests and structural studies of materials" (FEFU, Vladivostok, Russia) used in the research is gratefully acknowledged. We thank Andrei Vladimirovich Shevel'kov, Head of Laboratory of Directed Inorganic Synthesis, Chair of Inorganic Chemistry, Chemical Faculty, Lomonosov Moscow State University, Moscow, Russia, and also Aleksandr Sergeevich Tyablikov for their help in performing experiments.

Conflicts of Interest: The authors declare no conflict of interest.

References

1. Dubok, V.A. Bioceramics—Yesterday, Today, Tomorrow. *Powder Metall. Met. Ceram.* **2001**, *39*, 381–394. [CrossRef]
2. Stevens, M.M. Biomaterials for bone materials that enhance bone regeneration have a wealth of potential. *Bone* **2008**, *11*, 18–25.
3. Sainitya, R.; Sriram, M.; Kalyanaraman, V.; Dhivya, S.; Saravanan, S.; Vairamani, M.; Sastry, T.P.; Selvamurugan, N. Scaffolds containing chitosan/carboxymethyl cellulose/mesoporous wollastonite for bone tissue engineering. *Int. J. Biol. Macromol.* **2015**, *80*, 481–488. [CrossRef] [PubMed]
4. Sautier, J.M.; Kokubo, T.; Ohtsuki, T.; Nefussi, J.R.; Boulekbache, H.; Oboeuf, M.; Loty, S.; Loty, C.; Forest, N. Bioactive glass-ceramic containing crystalline apatite and wollastonite initiates biomineralization in bone cell cultures. *Calcif. Tissue Int.* **1994**, *55*, 458–466. [CrossRef] [PubMed]
5. Saadaldin, S.A.; Rizkalla, A.S. Synthesis and characterization of wollastonite glass-ceramics for dental implant applications. *Dent. Mater.* **2014**, *30*, 364–371. [CrossRef] [PubMed]
6. Papynov, E.K.; Shichalin, O.O.; Apanasevich, V.I.; Portnyagin, A.S.; Yu, M.V.; Yu, B.I.; Merkulov, E.B.; Kaidalova, T.A.; Modin, E.B.; Afonin, I.S.; et al. Sol-gel (template) synthesis of osteoplastic CaSiO$_3$/HAp powder biocomposite: "In vitro" and "in vivo" biocompatibility assessment. *Powder Technol.* **2020**, *367*, 762–773. [CrossRef]
7. Bheemaneni, G.; Saravana, S.; Kandaswamy, R. Processing and characterization of poly(butylene adipate-co-terephthalate)/wollastonite biocomposites for medical applications. *Mater. Today Proc.* **2018**, *5*, 1807–1816. [CrossRef]
8. Maxim, L.D.; Niebo, R.; Utell, M.J.; McConnell, E.E.; Larosa, S.; Segrave, A.M. Wollastonite toxicity: An update. *Inhal. Toxicol.* **2014**, *26*, 95–112. [CrossRef]
9. Lin, K.; Lin, C.; Zeng, Y. High mechanical strength bioactive wollastonite bioceramics sintered from nanofibers. *RSC Adv.* **2016**, *6*, 13867–13872. [CrossRef]
10. Hu, Y.; Xiao, Z.; Wang, H.; Ye, C.; Wu, Y.; Xu, S. Fabrication and characterization of porous CaSiO$_3$ ceramics. *Ceram. Int.* **2018**, *45*, 3710–3714. [CrossRef]
11. Adams, L.A.; Essien, E.R.; Kaufmann, E.E. A new route to sol-gel crystalline wollastonite bioceramic. *J. Asian Ceram. Soc.* **2018**, *6*, 132–138. [CrossRef]

12. Tian, L.; Wang, L.; Wang, K.; Zhang, Y.; Liang, J. The Preparation and properties of porous sepiolite ceramics. *Sci. Rep.* **2019**, *9*, 2–10. [CrossRef] [PubMed]
13. Biswas, N.; Samanta, A.; Podder, S.; Ghosh, C.K.; Ghosh, J.; Das, M.; Mallik, A.K.; Mukhopadhyay, A.K. Phase pure, high hardness, biocompatible calcium silicates with excellent anti-bacterial and biofilm inhibition efficacies for endodontic and orthopaedic applications. *J. Mech. Behav. Biomed. Mater.* **2018**, *86*, 264–283. [CrossRef] [PubMed]
14. Ingole, V.H.; Sathe, B.; Ghule, A.V. Bioactive ceramic composite material stability, characterization, and bonding to bone. In *Fundamental Biomaterials: Ceramics*; Elsevier Ltd.: Amsterdam, The Netherlands, 2018; pp. 273–296.
15. Ni, S.; Chang, J. In vitro degradation, bioactivity, and cytocompatibility of calcium silicate, dimagnesium silicate, and tricalcium phosphate bioceramics. *J. Biomater. Appl.* **2009**, *24*, 139–158. [CrossRef] [PubMed]
16. De Aza, P.N.; Guitian, F.; De Aza, S. Bioactivity of wollastonite ceramics: In vitro evaluation. *Scr. Metall. Mater.* **1994**, *31*, 1001–1005. [CrossRef]
17. Azarov, G.M.; Maiorova, E.V.; Oborina, M.A.; Belyakov, A.V. Wollastonite raw materials and their applications (a review). *Glas. Ceram.* **1995**, *52*, 237–240. [CrossRef]
18. Golovanova, O.A. Experimental Modeling of Formation of the Basic Mineral Phases of Calcifications. *Russ. J. Inorg. Chem.* **2018**, *63*, 1541–1545. [CrossRef]
19. Fadeeva, T.V.; Golovanova, O.A. Physicochemical Properties of Brushite and Hydroxyapatite Prepared in the Presence of Chitin and Chitosan. *Russ. J. Inorg. Chem.* **2019**, *64*, 847–856. [CrossRef]
20. Huang, Q.W.; Wang, L.P.; Wang, J.Y. Mechanical properties of artificial materials for bone repair. *J. Shanghai Jiaotong Univ.* **2014**, *19*, 675–680. [CrossRef]
21. Juhasz, J.A.; Best, S.M. Bioactive ceramics: Processing, structures and properties. *J. Mater. Sci.* **2012**, *47*, 610–624. [CrossRef]
22. Guglielmi, M.; Kickelbick, G.; Martucci, A. *Sol-Gel Nanocomposites*; Aegerter, M., Prassas, M., Eds.; Springer: New York, NY, USA, 2014.
23. Ros-Tárraga, P.; Murciano, Á.; Mazón, P.; Gehrke, S.A.; De Aza, P.N. In vitro behaviour of sol-gel interconnected porous scaffolds of doped wollastonite. *Ceram. Int.* **2017**, *43*, 11034–11038. [CrossRef]
24. Lin, K.; Chang, J.; Chen, G.; Ruan, M.; Ning, C. A simple method to synthesize single-crystalline β-wollastonite nanowires. *J. Cryst. Growth* **2007**, *300*, 267–271. [CrossRef]
25. Singh, N.B. Hydrothermal synthesis of β-dicalcium silicate (β-Ca_2SiO_4). *Prog. Cryst. Growth Charact. Mater.* **2006**, *52*, 77–83. [CrossRef]
26. Solonenko, A.P.; Blesman, A.I.; Polonyankin, D.A. Preparation and in vitro apatite-forming ability of hydroxyapatite and β-wollastonite composite materials. *Ceram. Int.* **2018**, *44*, 17824–17834. [CrossRef]
27. Palakurthy, S.; Samudrala, R.K. In vitro bioactivity and degradation behaviour of β-wollastonite derived from natural waste. *Mater. Sci. Eng. C* **2019**, *98*, 109–117. [CrossRef] [PubMed]
28. Ismail, H.; Shamsudin, R.; Abdul Hamid, M.A. Effect of autoclaving and sintering on the formation of β-wollastonite. *Mater. Sci. Eng. C* **2016**, *58*, 1077–1081. [CrossRef]
29. Bakan, F.; Laçin, O. A novel low temperature sol—gel synthesis process for thermally stable nano crystalline hydroxyapatite. *Powder Technol.* **2013**, *233*, 295–302. [CrossRef]
30. Wang, P.; Li, C.; Gong, H.; Jiang, X.; Wang, H.; Li, K. Effects of synthesis conditions on the morphology of hydroxyapatite nanoparticles produced by wet chemical process. *Powder Technol.* **2010**, *203*, 315–321. [CrossRef]
31. Xu, J.L.; Khor, K.A.; Kumar, R. Physicochemical differences after densifying radio frequency plasma sprayed hydroxyapatite powders using spark plasma and conventional sintering techniques. *Mater. Sci. Eng. A* **2007**, *457*, 24–32. [CrossRef]
32. Ergun, C. Enhanced phase stability in hydroxylapatite/zirconia composites with hot isostatic pressing. *Ceram. Int.* **2011**, *37*, 935–942. [CrossRef]
33. Viswanathan, V.; Laha, T.; Balani, K.; Agarwal, A.; Seal, S. Challenges and advances in nanocomposite processing techniques. *Mater. Sci. Eng. R Rep.* **2006**, *54*, 121–285. [CrossRef]
34. Orru, R.; Licheri, R.; Locci, A.M.; Cincotti, A.; Cao, G. Consolidation/synthesis of materials by electric current activated/assisted sintering. *Mater. Sci. Eng. R Rep.* **2009**, *63*, 127–287. [CrossRef]

35. Tokita, M. Spark Plasma Sintering (SPS) Method, Systems, and Applications. In *Handbook of Advanced Ceramics: Materials, Applications, Processing and Properties*; Somiya, S., Ed.; Elsevier Inc.: Amsterdam, The Netherlands, 2013; pp. 1149–1178.
36. Simonenko, T.L.; Kalinina, M.V.; Simonenko, N.P.; Simonenko, E.P.; Glumov, O.V.; Mel'nikova, N.A.; Murin, I.V.; Shichalin, O.O.; Papynov, E.K.; Shilova, O.A. Spark plasma sintering of nanopowders in the CeO_2-Y_2O_3 system as a promising approach to the creation of nanocrystalline intermediate-temperature solid electrolytes. *Ceram. Int.* **2018**, *44*, 19879–19884. [CrossRef]
37. Simonenko, E.P.; Simonenko, N.P.; Simonenko, T.L.; Grishin, A.V.; Tal'skikh, K.Y.; Gridasova, E.A.; Papynov, E.K.; Shichalin, O.O.; Sevastyanov, V.G.; Kuznetsov, N.T. Sol-gel synthesis of SiC@$Y_3Al_5O_{12}$ composite nanopowder and preparation of porous SiC-ceramics derived from it. *Mater. Chem. Phys.* **2019**, *235*, 121734. [CrossRef]
38. Papynov, E.K.; Shichalin, O.O.; Mayorov, V.Y.; Modin, E.B.; Portnyagin, A.S.; Tkachenko, I.A.; Belov, A.A.; Gridasova, E.A.; Tananaev, I.G.; Avramenko, V.A. Spark Plasma Sintering as a high-tech approach in a new generation of synthesis of nanostructured functional ceramics. *Nanotechnol. Russ.* **2017**, *12*, 49–61. [CrossRef]
39. Wang, H.; He, Z.; Li, D.; Lei, R.; Chen, J.; Xu, S. Low temperature sintering and microwave dielectric properties of $CaSiO_3$–Al_2O_3 ceramics for LTCC applications. *Ceram. Int.* **2014**, *40*, 3895–3902. [CrossRef]
40. Chen, S.; Zhou, X.; Zhang, S.; Li, B.; Zhang, T. Low temperature preparation of the β-$CaSiO_3$ ceramics based on the system CaO-SiO_2-BaO-B_2O_3. *J. Alloys Compd.* **2010**, *505*, 613–618. [CrossRef]
41. Harabi, A.; Chehlatt, S. Preparation process of a highly resistant wollastonite bioceramics using local raw materials. *J. Therm. Anal. Calorim.* **2013**, *111*, 203–211. [CrossRef]
42. Papynov, E.K.; Shichalin, O.O.; Mayorov, V.Y.; Modin, E.B.; Portnyagin, A.S.; Gridasova, E.A.; Agafonova, I.G.; Zakirova, A.E.; Tananaev, I.G.; Avramenko, V.A. Sol-gel and SPS combined synthesis of highly porous wollastonite ceramic materials with immobilized Au-NPs. *Ceram. Int.* **2017**, *43*, 8509–8516. [CrossRef]
43. Papynov, E.K.; Shichalin, O.O.; Modin, E.B.; Mayorov, V.Y.; Portnyagin, A.S.; Kobylyakov, S.P.; Golub, A.V.; Medkov, M.A.; Tananaev, I.G.; Avramenko, V.A. Wollastonite ceramics with bimodal porous structures prepared by sol–gel and SPS techniques. *RSC Adv.* **2016**, *6*, 34066–34073. [CrossRef]
44. Papynov, E.K.; Mayorov, V.Y.; Portnyagin, A.S.; Shichalin, O.O.; Kobylyakovt, S.P.; Kaidalova, T.A.; Nepomnyashiy, A.V.; Sokol'nitskaya, T.A.; Zub, Y.L.; Avramenko, V.A. Application of carbonaceous template for porous structure control of ceramic composites based on synthetic wollastonite obtained via Spark Plasma Sintering. *Ceram. Int.* **2015**, *41*, 1171–1176. [CrossRef]
45. Papynov, E.K.; Shichalin, O.O.; Buravlev, I.Y.; Portnyagin, A.S. Reactive spark plasma synthesis of porous bioceramic wollastonite. *Russ. J. Inorg. Chem.* **2020**, *65*, 263–270. [CrossRef]
46. Papynov, E.K.; Shichalin, O.O.; Apanasevich, V.I.; Afonin, I.S.; Evdokimov, I.O.; Mayorov, V.Y.; Portnyagin, A.S.; Agafonova, I.G.; Skurikhina, Y.E.; Medkov, M.A. Synthetic $CaSiO_3$ sol-gel powder and SPS ceramic derivatives: "In vivo" toxicity assessment. *Prog. Nat. Sci. Mater. Int.* **2019**, *29*, 569–575. [CrossRef]
47. Dudina, D.V.; Mukherjee, A.K. Reactive spark plasma sintering: Successes and challenges of nanomaterial synthesis. *J. Nanomater.* **2013**, *2013*, 625218. [CrossRef]
48. Simonenko, E.P.; Simonenko, N.P.; Papynov, E.K.; Shichalin, O.O.; Golub, A.V.; Mayorov, V.Y.; Avramenko, V.A.; Sevastyanov, V.G.; Kuznetsov, N.T. Preparation of porous SiC-ceramics by sol–gel and spark plasma sintering. *J. Sol-Gel Sci. Technol.* **2017**, *82*, 748–759. [CrossRef]
49. Kosyanov, D.Y.; Yavetskiy, R.P.; Vorona, I.O.; Shichalin, O.O.; Papynov, E.K.; Vornovskikh, A.A.; Kuryavyi, V.G.; Vovna, V.I.; Golokhvast, K.S.; Tolmachev, A.V. Transparent 4 at% Nd^{3+}:$Y_3Al_5O_{12}$ ceramic by reactive spark plasma sintering. *AIP Conf. Proc.* **2017**, *1874*, 1–5.
50. Papynov, E.K.; Shichalin, O.O.; Skurikhina, Y.E.; Turkutyukov, V.B.; Medkov, M.A.; Grishchenko, D.N.; Portnyagin, A.S.; Merkulov, E.B.; Apanasevich, V.I.; Geltser, B.I.; et al. ZrO_2-phosphates porous ceramic obtained via SPS-RS "in situ" technique: Bacteria test assessment. *Ceram. Int.* **2019**, *45*, 13838–13846. [CrossRef]
51. Wang, H.H.; Li, X.R.; Fei, G.Q.; Mou, J. Synthesis, morphology and rheology of core-shell silicone acrylic emulsion stabilized with polymerisable surfactant. *Express Polym. Lett.* **2010**, *4*, 670–680. [CrossRef]
52. Papynov, E.K.; Mayorov, V.Y.; Palamarchuk, M.S.; Bratskaya, S.Y.; Avramenko, V.A. Sol–gel synthesis of porous inorganic materials using "core–shell" latex particles as templates. *J. Sol-Gel Sci. Technol.* **2013**, *68*, 374–386. [CrossRef]

53. Suchanek, W.; Yoshimura, M. Processing and properties of hydroxyapatite-based biomaterials for use as hard tissue replacement implants. *J. Mater. Res.* **1998**, *13*, 94–117. [CrossRef]
54. Ravaglioli, A.; Krajewski, A. *Bioceramics*, 1st ed.; Springer: Amsterdam, The Netherlands, 1992.
55. Peppas, N.A. *Handbook of Biomaterial Properties*, 1st ed.; Black, J., Hastings, G., Eds.; Springer: New York, NY, USA, 1998; Volume 65.
56. Callejas-Díaz, A.; Fernández-Pérez, C.; Ramos-Martínez, A.; Múñez-Rubio, E.; Sánchez-Romero, I.; Vargas Núñez, J.A. Impact of Pseudomonas aeruginosa bacteraemia in a tertiary hospital: Mortality and prognostic factors. *Med. Clin.* **2018**, *152*, 83–89. [CrossRef] [PubMed]

© 2020 by the authors. Licensee MDPI, Basel, Switzerland. This article is an open access article distributed under the terms and conditions of the Creative Commons Attribution (CC BY) license (http://creativecommons.org/licenses/by/4.0/).

Article

In Vitro Study on the Effect of a New Bioactive Desensitizer on Dentin Tubule Sealing and Bonding

Minh N. Luong, Laurie Huang, Daniel C. N. Chan and Alireza Sadr *

Department of Restorative Dentistry, University of Washington, Seattle, WA 98195, USA; mndluong@uw.edu (M.N.L.); laurie.whuang@gmail.com (L.H.); dcnchan@uw.edu (D.C.N.C.)
* Correspondence: arsadr@uw.edu

Received: 18 April 2020; Accepted: 25 May 2020; Published: 2 June 2020

Abstract: Bioactive mineral-based dentin desensitizers that can quickly and effectively seal dentinal tubules and promote dentin mineralization are desired. This in vitro study evaluated a novel nanohydroxyapatite-based desensitizer, Predicta (PBD, Parkell), and its effect on bond strength of dental adhesives. Human dentin discs (2-mm thick) were subjected to 0.5 M EDTA to remove the smear layer and expose tubules, treated with PBD, and processed for surface and cross-sectional SEM examination before and after immersion in simulated body fluid (SBF) for four weeks (ISO 23317-2014). The effects of two dental desensitizers on the microshear bond strength of a universal adhesive and a two-step self-etch system were compared. SEM showed coverage and penetration of nanoparticles in wide tubules on the PBD-treated dentin at the baseline. After four weeks in SBF, untreated dentin showed amorphous mineral deposits while PBD-treated dentin disclosed a highly mineralized structure integrated with dentin. Desensitizers significantly reduced microshear bond strength test (MSBS) of adhesives by 15–20% on average, depending on the bonding protocol. In conclusion, PBD demonstrated effective immediate tubules sealing capability and promoted mineral crystal growth over dentin and into the tubules during SBF-storage. For bonding to desensitizer-treated dentin, a two-step self-etching adhesive or universal bond with phosphoric acid pretreatment are recommended.

Keywords: bioactive desensitizer; hypersensitivity; SEM; nanoparticle; nanohydroxyapatite; microshear bond strength; phosphoric acid; self-etch

1. Introduction

Dentin hypersensitivity (DH) has become a more frequently encountered issue in diagnostic and therapeutic clinical situations. DH is identified by a typical short sharp pain on the exposed dentin because of the thermal, evaporative, tactile, osmotic, or chemical stimuli. The common clinical manifestations of DH include exposed dentin from abrasion, erosion or exposed root surface. Although the mechanism of pain transmission and sensitivity in dentin is still under speculation, the most accepted theory is the hydrodynamic mechanism of sensitivity. It is postulated that the sudden flow of tubular fluid within the tubules in the presence of irritating stimulus results in the activation of sensory nerve ending thereby inducing pain or sensitivity [?,?]. The density and size of patent dentinal tubules are critical factors to the severity of DH in patients.

A standard treatment for DH has not yet been established [?]. Two main methods utilized for the treatment of DH are occlusion of the dentin tubules and interference with the sensitivity of the mechanoreceptor. A plethora of products have been developed with the aim to reduce the discomfort from DH. Dentin desensitizers have been introduced for DH treatment mainly through tubular occlusion [?,?,?,?]. Calcium phosphate mineral-based desensitizers have attracted considerable interest in recent years thanks to their biocompatible properties yielding an effective dentinal tubule occlusion and reduction in dentin permeability. They also possess the possibility of mineral crystal growth in the oral environment [? ? ?]. The composition of calcium and phosphate in a material may exhibit the ability to form hydroxyapatite (HA; $Ca_{10}(PO_4)_6(OH)_2$) as a final product. HA and its variations are the main mineral component in human tooth and their biocompatibility have rendered them as clinically practical choices [?].

Dental adhesives have been utilized in the treatment of restoring teeth in combination with DH. The component in the bonding agent provides the simultaneous conditioning and priming for the underlying dentin, resulting in superior bond strength and reduced DH [? ?]. However, the sealing ability of adhesives on the dentin surface could be affected by an incomplete micro-emulsion polymerization of the hydrophilic primer and hydrophobic bonding [?]. Therefore, in order to increase the effectiveness of treatment, it would be necessary to desensitize dentin exhibiting DH prior to the placement of a restoration. It was reported that one-step self-etching adhesives and dentin desensitizers could significantly relieve DH immediately and over a month after treatment [?]. The usage of dentin desensitizer in the combination with the bonding agent can alleviate the postoperative sensitivity with the composite restoration [?]. Nevertheless, its effect on bonding performance using different bonding systems remains to be evaluated, as the desensitized treated dentin may be unfavorable for bonding [? ?]. The objective of this in vitro study was to evaluate the morphological characteristic of the new nanohydroxyapatite-based desensitizer with the proprietary composition of calcium, sodium, phosphate, and silica when applied on the tooth surface and the effect of this material on the bond strength. The null hypotheses were that the novel desensitizer could modify the morphological characteristic of the dentin of the extracted teeth and did not interfere with the bonding performance of the tested adhesives.

2. Materials and Methods

The extracted human teeth were used in this study in accordance with the guideline of the Ethics Committee of the University of Washington Human Subject Division and the Declaration of Helsinki, which allow the use of extracted deidentified human teeth. Forty molars were collected during treatment plan and patient care and stored at 4 °C in water with 0.02% thymol.

The occlusal cusps were removed by cutting perpendicularly to the long axis of the tooth 2–3 mm from the tip of the cusp to expose superficial midcoronal dentin using a low-speed water-cooled diamond saw (Isomet; Buehler, Lake Bluff, IL, USA). The roots were then sectioned at the cement-enamel junction to obtain 2-mm-thick dentin slices. The occlusal side of each disc was polished under running water by 600-grit paper (3M, St. Paul, MN, USA) for 30 s in circular motion to create a standardized smear layer. The surfaces were observed under the digital stereomicroscope (MU1000, AmScope, Irvine, CA, USA) to eliminate the cracks, residual enamel, coarseness and pollutants.

The composition and application of desensitizers and adhesives according to the manufacturer's instruction were listed in Table ??.

Table 1. Materials used in the study.

Materials	Composition	Application
Teethmate Desensitizer (TMD) (Kuraray Noritake Dental Inc, Tokyo, Japan)	Powder: tetracalcium phosphate (TTCP) and dicalcium phosphate anhydrous (DCPA) Liquid: water	Mix powder and water within 30 s, apply slurry with microapplicator 15 s, rub 60 s rinse with water spray.
Predicta Bioactive Desensitizer gel (PBD, Parkell, Edgewood, NY, USA)	Calcium, phosphate, nanohydroxyapatite	Rinse surface with warm water, dry with absorbent cotton. Paint a coat and rub gently the liquid in 10–20 s using an applicator. Wipe the product off using a cotton pledget before air blowing.

Table 1. Cont.

Materials	Composition	Application
Parkell Etching (Parkell, Edgewood, NY, USA)	32% PA Gel (H$_3$PO$_4$)	10–15 s etch then rinse.
Parkell Universal Adhesive PBOND (Parkell, Edgewood, NY, USA)	Acetone, Ethyl Alcohol, 10-MDP, 2-Hydroxylethyl methacrylate, 2-Propenoic acid	Mix well, dispense 1–3 drops. Rub the liquid using an applicator, keep moist for 20 s. Blow gently for 10 s and light cure for 10 s.
Clearfil SE 2 (Kuraray Noritake Dental, Japan)	Primer: MDP, HEMA, camphorquinone, hydrophilic dimethacrylate, N,N-diethanol p-toluidine, water. Bond: MDP, Bis-GMA, HEMA, camphorquinone, hydrophobic dimethacrylate, N,N-diethanol p-toluidine, silanated colloidal silica.	With the applicator, prime for 20 s. Apply bond and light cure for 10 s
Z100 Restorative (3M ESPE, St. Paul, MN, USA)	Bis-GMA, TEDGMA, Zirconium/Silica filler	Build the resin cylinder then light cure for 40 s.

2.1. Morphological Observation

Ten dentin discs were pretreated with 0.5 M EDTA (pH 7.4) for 2 min to completely remove the smear layer and widely open the dentinal tubules before applying the deionized water. Half of the disc surfaces were subjected to the Predicta Bioactive Desensitizer (PBD, Parkell, Edgewood, NY, USA) whereas the other half served as control (no PBD). Five discs were immediately processed for SEM examination while the other five were stored for four weeks in simulated body fluid (SBF) for bioactivity test (specified by the ISO 23317:2014) (Figure ??A).

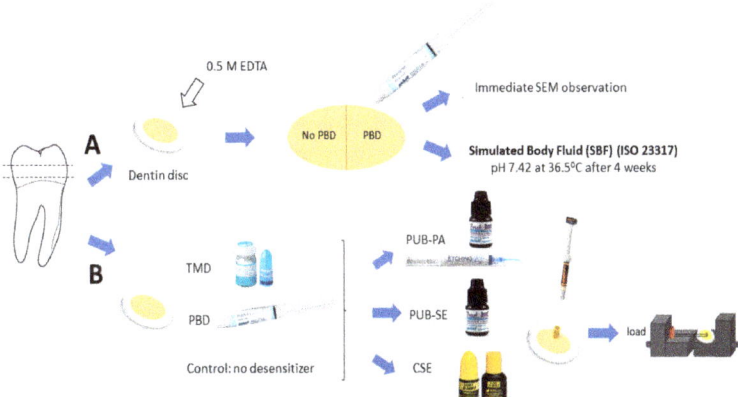

Figure 1. Methodology: Dentin discs were produced from human molar for morphological observation (**A**) and microshear bond strength test (**B**). For (**A**): Ten dentin discs were pretreated with EDTA (pH 7.4) to remove the smear layer and open dentinal tubules before rinsing. Half of the disc surfaces were subjected to Predicta Bioactive Desensitizer (PBD) and the other half served as control (no PBD). Five discs were immediately processed for SEM examination while the other five were stored for four weeks in simulated body fluid (SBF) for bioactivity test (specified by the ISO 23317:2014). For (**B**): Thirty dentin discs were allocated into three groups (n = 10): Teethmate Desensitizer (TMD) treatment, PBD treatment and no desensitizer as control. The universal adhesive (PUB) in self-etch (SE) or with phosphoric acid etching (PA)) and two-step self-etch adhesive Clearfil SE Bond 2 (CSE) were used. Small cylinders of the hybrid composite were bonded on the dentin surface and the wire loop bond test was performed at a crosshead speed of 1 mm/min.

2.2. Observation after Application of PBD

The specimens were dried in a desiccator without heating for 24 h. The discs were then superficially scored using a diamond bur and delicately fractured by finger pressure revealing the cross-section of dentinal tubules without creating an additional smear layer. The boundary between applied and non-applied dentin surface was targeted for visualization under SEM (JEOL JSM-6010LA, Tokyo, Japan). After gold-sputter coating, the micromorphology was examined at magnifications of 1000×, 4000×, 5000×, and 10,000×. Furthermore, in order to anticipate the bioactivity of PBD-treated dentin in the oral environment, the dentin discs were further placed in dust-free SBF solution incubated at 37 °C for four weeks in a transparent bottle.

2.3. Microshear Bond Strength Test

Thirty dentin discs were randomly distributed into three groups (n = 10): Teethmate Desensitizer (TMD, Kuraray Noritake Dental Inc, Tokyo, Japan) treatment, Predicta Bioactive Desensitizer (PBD) treatment and no treatment (control). The universal adhesive PBOND (PUB, Parkell, Edgewood, NY, USA) in self-etch (SE) or with phosphoric acid etching (PA, Parkell Etching Gel, Parkell, Edgewood, NY, USA) and two-step self-etch adhesive Clearfil SE Bond 2 (CSE, Kuraray Noritake Dental) were used in the experiment of microshear bond strength test (MSBS). Figure ?? represented the method of the study.

Prior to the irradiation, three pieces of Tygon tubes (Saint-Gobain Performance Plastic, Nagano, Japan) with an internal diameter of 0.8 mm and 1.0 mm in height were fixed on the dentin surface of each slice. The adhesive was light-cured for 10 s at 600 mW/cm^2 (Yoshida Light Curing Unit, Japan). The hybrid composite Z100 (A2 Shade, 3M ESPE, St. Paul, MN, USA) was carefully placed inside the tubing lumens before irradiation for 40 s. After 1 h-storage at room temperature (25 °C), the tygon tubes were removed using a sharp blade to obtain the cylinders of resin with the size of 0.8 mm in diameter and 1.0 mm in height bonded to the dentin surface. The specimens were then stored in distilled water in an incubator at 37 °C.

After storage, an MSBS test was conducted. The dentin slice was attached to the testing apparatus (Mecmesin, Compact Force Gauge, CFG+, UK) with a thin layer of cyanoacrylate glue (Model Repair II Blue, Dentsply-Sankin, Tochigi, Japan) and tested in a microshear tester (Bisco, Inc., Schaumburg, IL, USA). A 0.2 mm-diameter steel wire was looped at the bottom of the resin cylinder, in contact with the lower half-circle of the cylinder and the dentin surface. The wire loop was maintained according to the shear stress orientation at the bonding interface. The force was applied at a crosshead speed of 1 mm/min until failure then the MSBS data were recorded (Figure ??B).

2.4. Evaluation of Failure Mode

After the MSBS test, the mode of failure was determined for each specimen under the digital stereomicroscope. The fractured surfaces were classified as followed: adhesive failures between resin and dentin, mixed failures that were partially adhesive and partially cohesive, and cohesive failures that occurred entirely within the bonding resin or entirely in dentin.

2.5. Statistical Analysis

The distribution of data in each group was normal (Kolmogorov–Smirnov $p > 0.05$), parametric tests were performed. The MSBS values were statistically analyzed using two-way ANOVA with application mode and desensitizer treatment as factors. The statistical procedures were analyzed at 0.05 significance level using IBM SPSS Statistics 23 Software.

3. Results

Immediately after the application of PBD, SEM micrographs showed good coverage by PBD of the widely open dentinal tubule in EDTA-demineralized dentin as clearly observed in the comparison of the untreated and treated side (Figure ??). The cross-sectional images showed that the nanoparticles form PBD have penetrated into the dentinal tubules and further sealed the nanospaces exposed through the collagen network and lateral branches in the tubule walls of demineralized dentin. Intratubular dentin seemed to be densely penetrated on the treated side. The depth of penetration was estimated to be 5–10 μm from the dentin surface. The thickness of the formed layer as a result of the PBD application was approximately 1 μm.

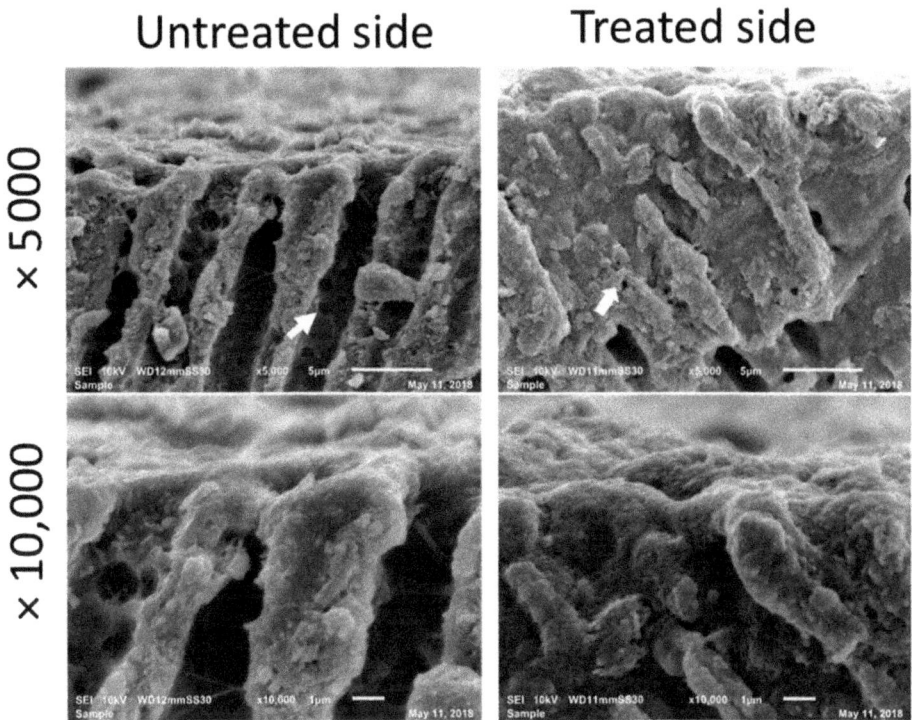

Figure 2. Cross-sectional SEM images immediately after application of PBD. The patency of dentinal tubules and open lateral branch orifice (arrow) have been well sealed on the treated side.

After four weeks of SBF immersion, SEM images of dentin treated with EDTA and PBD disclosed a highly dense and homogenous structure of dentin measuring at least 20 μm in thickness. In this structure, a surface layer was not easily distinguished from the bulk of subsurface dentin, indicating integrated mineralization of the surface zone through both mineral deposition on the surface and into the demineralized dentin, forming a highly mineralized unified structure whereas there was no such unification on untreated sides (Figure ??).

The results of MSBS of each experimental group (mean and SD) were listed in Table ?? and Figure ??. Two-way ANOVA indicated that both the use of desensitizers and application mode were significant factors in MSBS and their interaction was significant ($p < 0.05$). Generally, both PBD and TMD significantly reduced bond strength to dentin by 15–20% on average; however, the effect was adhesive dependent. For PUB, the groups using phosphoric acid etching (PUB-PA) consistently showed higher bond strength compared to the ones using self-etch (PUB). Bonding to PBD and TMD with either PUB-PA and CSE were comparable ($p > 0.05$).

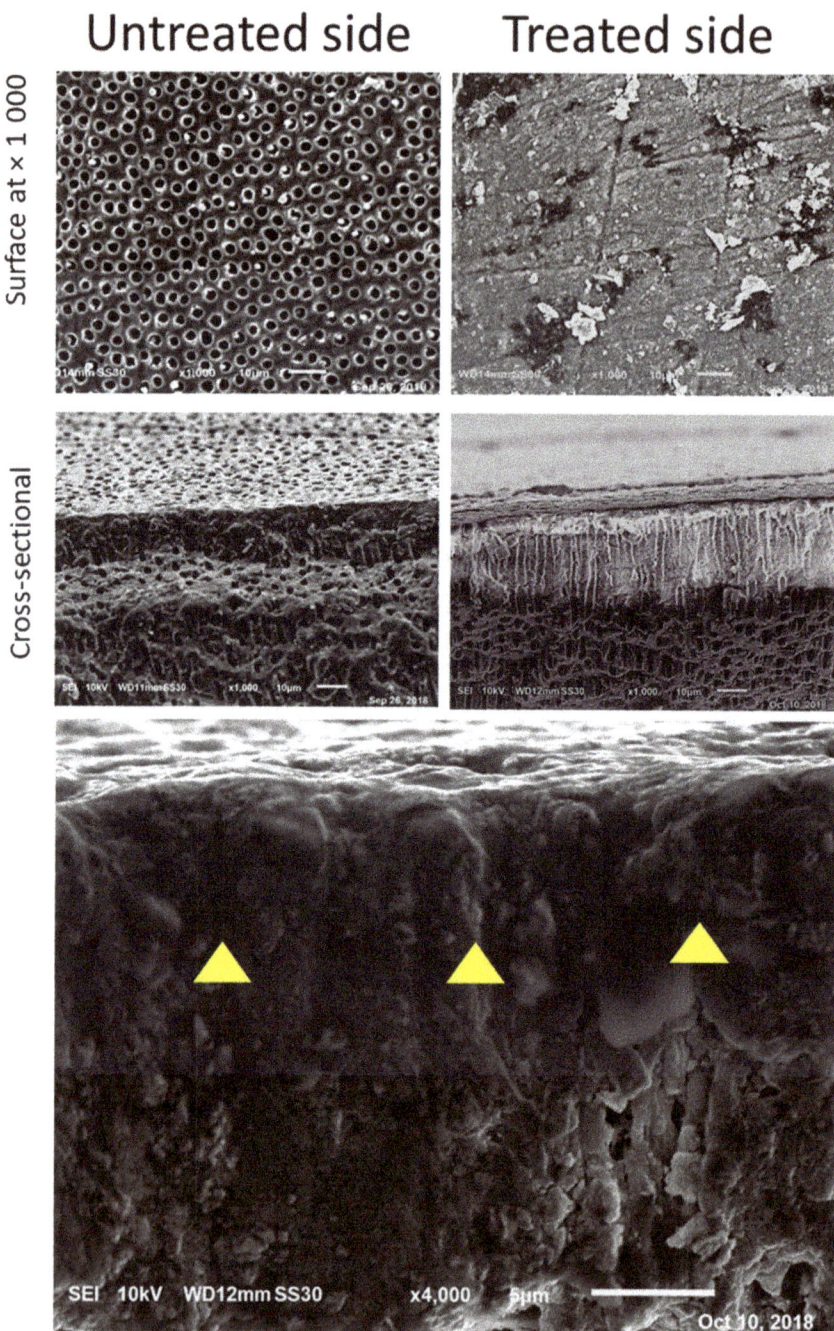

Figure 3. SEM images four weeks after application of PBD and immersion in SBF. PBD-treated dentin demonstrates a solid layer of mineral formation while no coverage is distinguished in the untreated dentin away from the PBD-treated dentin.

Table 2. Mean microshear bond strength of the experimental groups (MPa).

Mean ± SD.	TMD	PBD	Control
PUB-SE	25.5 ± 4.7	19.5 ± 3.7	24.9 ± 4.0
PUB-PA	31.8 ± 5.8	30.2 ± 5.9	37.9 ± 5.0
CSE	29.2 ± 5.7	34.2 ± 3.7	40.3 ± 4.3

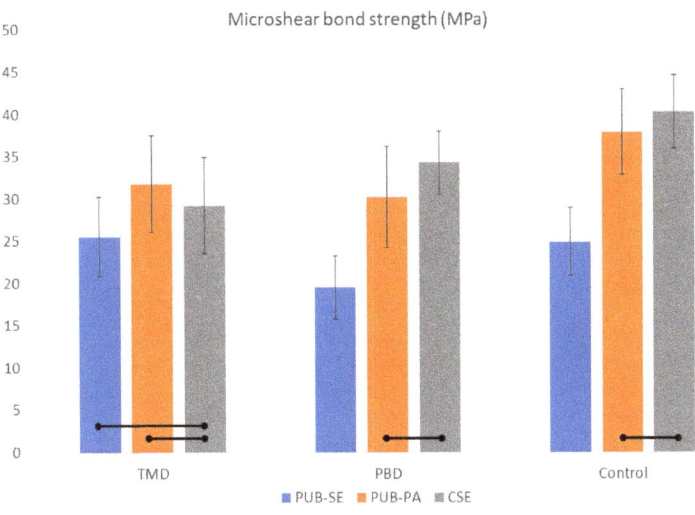

Figure 4. Bond strength of the experiment group in MPa. Bars indicate no significant difference ($p < 0.05$). Two-way ANOVA indicated that both the use of desensitizers and application mode were significant factors in MSBS and their interaction was significant ($p < 0.05$). Predicta (PBD) and Teethmate (TMD) significantly reduced bond strength to dentin by 15%–20% on average; however, the effect depended on the adhesives. For Universal Adhesive (PUB), the groups using phosphoric acid etching (PUB-PA) consistently showed higher bond strength compared to the ones using self-etch (PUB). Bonding to PBD and TMD with either PUB-PA and Clearfil SE 2 (CSE) were comparable ($p > 0.05$).

4. Discussion

Desensitizing agents occlude the dentinal tubules at the tubular orifice or within the dentinal tubules to prevent the fluid flow and therefore decrease the pain sensation by counteracting the hydrodynamic mechanism of DH. The effectiveness of various desensitizing agent to mitigate the DH has been reported in previous studies. The clinical efficacy has been reported in the application on vital abutment teeth prepared to receive full coverage or porcelain fused to metal restoration [? ?]. The pre-impression sealing of the dentin was recommended for tooth preparation on vital teeth to reduce the pressure transmitted to the pulp chamber during the crown cementation [?].

In this study, immediately after the application of PBD, the untreated dentin showed patent dentinal tubules. At the area adjacent to the PBD-treated dentin, random mineral deposits were observed on the untreated dentin. It is assumed that these deposits have formed as a result of free PBD nanoparticles adjacent to the PBD-treated dentin (Figure ??).

For evaluation of the mineral forming ability of PBD-treated dentin, SBF immersion proceeded for four weeks. The SBF was used to simulate the clinical situation where dentin interacts with body fluids, such as saliva from the oral environment and dentinal fluid and plasma from the pulpal side. The SBF is an acellular and protein-free simulated fluid with ion concentration nearly equal to those of human blood plasma. SBF was preferred over artificial saliva formulations for the bioactivity experiment in this study since the ISO test has standardized the SBF for bioactivity test and apatite formation ability, while there is not a comparable standard established for artificial saliva formulations. Due to the similarity between bone and dentin in terms of mineral composition and structure, the bioactivity test for bone implant devices would be applicable to a desensitizer material applied to dentin. The bioactive material is expected to form a layer of apatite on their surface [?]. In this study, PBD demonstrated bioactivity as defined by the ISO 23317-2014 document. Considering the composition of SBF, the forming mineral in the ISO experiment is expected to be apatite.

The SEM micrographs confirmed that the dentin regions exposed for the experiment were rich in open dentin tubules, and tubular sealing was achieved immediately by PBD. The mineral growth matured through the SBF-storage period. Fine precipitates are observed inside the tubules close to the surface whereas the group without PBD, there were patent tubules. The principal mineral content in PBD (calcium and phosphate, as disclosed by the manufacturer) could stimulate the mineralization of the dentine tubules. During immersion in SBF, different processes occur resulting in structural and chemical change at the surface of dentin: leaching, degradation, and precipitation. Calcium ions dissolve from the bioactive substance into the body fluid. The nucleation of hydroxyapatite is possible because the surrounding fluid is supersaturated with respect to hydroxyapatite due to the dissolution of the calcium ions. The process of formation and growth of the hydroxyapatite layer continues by the reactions of calcium, phosphate and hydroxide ions [?]. In the oral cavity, the supersaturated calcium phosphate salts contained in the saliva could precipitate to form a less soluble compound HA continuously [?].

The study compared the bond strength of adhesives to dentin treated with PBD and TMD. The main components of TMD are TTCP and DCPA, which are eventually converted to hydroxyapatite in an aqueous environment. This conversion is predicated upon the dissolution-precipitation reaction mechanism [?]. TTCP and DCPA dissolve and supply Ca^{2+} and PO_4^{3-}. The supersaturation of the apatite-contained solution prompted the precipitation of HA crystals. TMD has been investigated previously in vitro with a superior tubule occlusion rate compared to silver diamine fluoride and resin-containing oxalate [? ? ?]. SEM images of TMD showed all tubules occluded with crystalline precipitates, with similar precipitation on the intertubular surface [?]. The application of TMD occluded dentinal tubules and reduced dentin permeability by 92% regardless of the exposed collagen network [?]. A comparable tubule occlusion was found in the novel PBD, therefore its sealing ability for treatment DH is promising.

As far as bonding performance is concerned, dentin desensitizers significantly decreased the bond strength of bonding agents. These findings are in alignment with the results of previous work [? ? ?]. The application of desensitizers before the bonding system significantly reduced shear bond strengths to sound dentin. In the three-step bonding system, the protocol included acid etching, rinsing, and drying of dentin before applying the primer. Self-etching primers succeeded in eliminating the dentin-conditioning steps before priming. Nevertheless, the majority of self-etching primers have a milder acidity. Therefore, the lower bond strengths to sound dentin might be accounted for the desensitizer deposition blocking the dentin tubule orifices and intertubular diffusion channels. The nanoparticles layer of the desensitizer dispersed on the dentin could hinder the adhesives to interact with the demineralized substrate, resulting in reducing the bond strength [?].

The literature has reported that the dentin treated with mineral-based desensitizers did not show a reduction in the bond strength when treated with a two-step self-etching adhesive material [?]. The reason for this phenomenon was explained by the interaction created between the desensitizer, the smear layer, and dentin. This was in contrast with the result in this present study where bond strength in PUB-SE mode after application of PBD and TMD was reduced compared to the control group.

The functional monomer in self-etch adhesives could contribute to the superficial micromechanical interlocking through hybridization of dentin substrate. The exposed HA crystal that remains around collagen is expected to be beneficial, as a substrate for chemical bonding. A mineral-rich substrate allow a more substantial chemical interaction with the functional monomer and protect the collagen against hydrolysis thus prevent from premature degradation [? ?].

It was of interest to compare the bond strength characteristics of two products PBD and TMD in a combination of adhesives. The result indicated that bonding to PBD with PUB-PA and CSE is comparable to that to TMD. The bonding performance of the PBD-treated specimen was corroborated with SEM images obtained by the application of PUB on dentin. PUB with acid etching exhibited resin tag formation, which could explain higher values in groups bonded in PA mode. Nonetheless, it was reported that the sequence of application of etching and desensitizer affected the bonding performance [? ?]. When dentin was acid-etched before the application of oxalate desensitizer [?] or Novamin [?], the bonding performance was not compromised. However, in this application sequence, the formation of resin tags might be affected by tubule occlusion with calcium-containing desensitizer pastes [?]. The clinical decision of whether to etch dentin or not after desensitizer application should be made considering the potential increase in permeability of dentin following etching. The interaction of PBD and phosphoric acid etched dentin, as application of the acid prior to treatment with PBD and the universal bond, warrants further investigation.

Successful bonding to dentin may further reduce the effects of hydraulic pressure for the treatment of the post-operative DH [? ?]. The use of the universal bonding system with acid phosphoric etching or the two-step self-etching is recommended following the usage of mineral-based desensitizer to achieve a high bond strength.

The study highlights the biomimetic remineralization shown by desensitizing agent PBD. It could be of interest to compare investigate the characteristics of PBD in vivo. The present laboratory test could be to some extent a predictive tool for the clinical efficacy of desensitizing materials, while the effect of adhesive systems on the survival rate of restorations following desensitizing treatment should be further studied.

5. Conclusions

In this in vitro study, the newly developed dentin desensitizer demonstrated effective dentinal tubules sealing capability and promoted mineral crystal growth over dentin and into the tubules during simulated body fluid storage. For bonding to the desensitizer-treated dentin, dentin treatment by two-step self-etching system or universal bond with phosphoric acid is recommended.

Author Contributions: Conceptualization: A.S.; methodology: M.N.L., L.H.; image interpretation: A.S., M.N.L.; formal analysis: A.S.; resources: A.S. and D.C.N.C.; writing—original draft preparation: M.N.L.; writing review and editing: A.S. All authors have read and agreed to the published version of the manuscript.

Funding: This research and APC charges were supported by the Biomimetics Biomaterials Biophotonics Biomechanics & Technology (B4T), Department of Restorative Dentistry, University of Washington, Seattle, WA, USA.

Acknowledgments: Authors are grateful to the manufacturers of materials Parkell and Kuraray Noritake Dental for their donation of the experimental materials.

Conflicts of Interest: The authors declare no conflict of interest with regard to the authorship of this manuscript.

References

1. Brännström, M.; Åström, A. A Study on the Mechanism of Pain Elicited from the Dentin. *J. Dent. Res.* **1964**, *43*, 619–625. [CrossRef] [PubMed]
2. Orchardson, R.; Gillam, D.G. Managing dentin hypersensitivity. *J. Am. Dent. Assoc.* **2006**, *137*, 990–998. [CrossRef] [PubMed]
3. Orchardson, R.; Gangarosa, L.P.; Holland, G.R.; Pashley, D.H. Towards a standard code of practice for evaluating the effectiveness of treatments for hypersensitive dentine. *Arch. Oral Biol.* **1994**, *39*, 121S–124S. [CrossRef]
4. Miglani, S.; Aggarwal, V.; Ahuja, B. Dentin hypersensitivity: Recent trends in management. *J. Conserv. Dent.* **2010**, *13*, 218. [CrossRef] [PubMed]

5. Porto, I.C.C.M.; Andrade, A.K.M.; Montes, M.A.J.R. Diagnosis and treatment of dentinal hypersensitivity. *J. Oral Sci.* **2009**, *51*, 323–332. [CrossRef] [PubMed]
6. Han, L.; Okiji, T. Dentin tubule occluding ability of dentin desensitizers. *Am. J. Dent.* **2015**, *28*, 90–94.
7. Ishihata, H.; Kanehira, M.; Finger, W.J.; Takahashi, H.; Tomita, M.; Sasaki, K. Effect of two desensitizing agents on dentin permeability in vitro. *J. Appl. Oral Sci.* **2017**, *25*, 34–41. [CrossRef]
8. Endo, H.; Kawamoto, R.; Takahashi, F.; Takenaka, H.; Yoshida, F.; Nojiri, K.; Takamizawa, T.; Miyazaki, M. Evaluation of a calcium phosphate desensitizer using an ultrasonic device. *Dent. Mater. J.* **2013**, *32*, 456–461. [CrossRef]
9. Siso, S.H.; Dönmez, N.; Kahya, D.S.; Uslu, Y.S. The effect of calcium phosphate-containing desensitizing agent on the microtensile bond strength of multimode adhesive agent. *Niger. J. Clin. Pract.* **2017**, *20*, 964–970.
10. Thanatvarakorn, O.; Nakashima, S.; Sadr, A.; Prasansuttiporn, T.; Ikeda, M.; Tagami, J. In vitro evaluation of dentinal hydraulic conductance and tubule sealing by a novel calcium-phosphate desensitizer. *J. Biomed. Mater. Res. Part B Appl. Biomater.* **2013**, *101*, 303–309. [CrossRef]
11. Cunha-Cruz, J.; Wataha, J.C.; Zhou, L.; Manning, W.; Trantow, M.; Bettendorf, M.M.; Heaton, L.J.; Berg, J. Treating dentin hypersensitivity: Therapeutic choices made by dentists of the northwest PRECEDENT network. *J. Am. Dent. Assoc.* **2010**, *141*, 1097–1105. [CrossRef] [PubMed]
12. Fu, B.; Shen, Y.; Wang, H.; Hannig, M. Sealing ability of dentin adhesives/desensitizer. *Oper. Dent.* **2007**, *32*, 496–503. [CrossRef]
13. Grégoire, G.; Joniot, S.; Guignes, P.; Millas, A. Dentin permeability: Self-etching and one-bottle dentin bonding systems. *J. Prosthet. Dent.* **2003**, *90*, 42–49. [CrossRef]
14. Tay, F.R.; Pashley, D.H. Have dentin adhesives become too hydrophilic? *J. Can. Dent. Assoc.* **2003**, *69*, 726–732. [PubMed]
15. Yu, X.; Liang, B.; Jin, X.; Fu, B.; Hannig, M. Comparative in vivo study on the desensitizing efficacy of dentin desensitizers and one-bottle self-etching adhesives. *Oper. Dent.* **2010**, *35*, 279–286. [CrossRef]
16. Sengun, A.; Koyuturk, A.E.; Sener, Y.; Ozer, F. Effect of desensitizers on the bond strength of a self-etching adhesive system to caries-affected dentin on the gingival wall. *Oper. Dent.* **2005**, *29*, 176–181.
17. Pashley, D.H.; Carvalho, R.M. Dentine permeability and dentine adhesion. *J. Dent.* **1997**, *25*, 355–372. [CrossRef]
18. Seara, S.F.; Erthal, B.S.; Ribeiro, M.; Kroll, L.; Pereira, G.D.S. The influence of a dentin desensitizer on the microtensile bond strength of two bonding systems. *Oper. Dent.* **2002**, *27*, 154–160.
19. Plant, C.G.; Browne, R.M.; Knibbs, P.J.; Britton, A.S.; Sorahan, T. Pulpal effects of glass ionomer cements. *Int. Endod. J.* **1984**, *17*, 51–59. [CrossRef]
20. Pashley, D.H. Smear layer: Physiological considerations. *Oper. Dent. Suppl.* **1984**, *3*, 13–29.
21. Richardson, D.; Tao, L.; Pashley, D.H. Dentin permeability: Effects of crown preparation. *Int. J. Prosthodont.* **1991**, *4*, 219–225. [PubMed]
22. Siriphannon, P.; Kameshima, Y.; Yasumori, A.; Okada, K.; Hayashi, S. Formation of hydroxyapatite on $CaSiO_3$ powders in simulated body fluid. *J. Eur. Ceram. Soc.* **2002**, *22*, 511–520. [CrossRef]
23. Rahaman, M.N.; Day, D.E.; Sonny Bal, B.; Fu, Q.; Jung, S.B.; Bonewald, L.F.; Tomsia, A.P. Bioactive glass in tissue engineering. *Acta Biomater.* **2011**, *7*, 2355–2373. [CrossRef]
24. Larsen, M.J.; Pearce, E.I.F. Saturation of human saliva with respect to calcium salts. *Arch. Oral Biol.* **2003**, *48*, 317–322. [CrossRef]
25. Ishikawa, K.; Takagi, S.; Chow, L.C.; Suzuki, K. Reaction of calcium phosphate cements with different amounts of tetracalcium phosphate and dicalcium phosphate anhydrous. *J. Biomed. Mater. Res.* **1999**, *46*, 504–510. [CrossRef]
26. Thanatvarakorn, O.; Nakashima, S.; Sadr, A.; Prasansuttiporn, T.; Thitthaweerat, S.; Tagami, J. Effect of a calcium-phosphate based desensitizer on dentin surface characteristics. *Dent. Mater. J.* **2013**, *32*, 615–621. [CrossRef]
27. Garcia, R.N.; Giannini, M.; Takagaki, T.; Sato, T.; Matsui, N.; Nikaido, T.; Tagami, J. Effect of dentin desensitizers on resin cement bond strengths. *RSBO* **2016**, *12*, 14–22. [CrossRef]
28. Hashimoto, M.; Ohno, H.; Sano, H.; Tay, F.R.; Kaga, M.; Kudou, Y.; Oguchi, H.; Araki, Y.; Kubota, M. Micromorphological changes in resin-dentin bonds after 1 year of water storage. *J. Biomed. Mater. Res.* **2002**, *63*, 306–311. [CrossRef]

29. Tay, F.R.; Pashley, D.H.; Suh, B.I.; Carvalho, R.M.; Itthagarun, A. Single-step adhesives are permeable membranes. *J. Dent.* **2002**, *30*, 371–382. [CrossRef]
30. Yang, H.; Pei, D.; Chen, Z.; Lei, J.; Zhou, L.; Huang, C. Effects of the application sequence of calcium-containing desensitising pastes during etch-and-rinse adhesive restoration. *J. Dent.* **2014**, *42*, 1115–1123. [CrossRef]
31. De Munck, J.; Van Landuyt, K.; Peumans, M.; Poitevin, A.; Lambrechts, P.; Braem, M.; Van Meerbeek, B. A critical review of the durability of adhesion to tooth tissue: Methods and results. *J. Dent. Res.* **2005**, *84*, 118–132. [CrossRef] [PubMed]
32. Frankenberger, R.; Tay, F.R. Self-etch vs. etch-and-rinse adhesives: Effect of thermo-mechanical fatigue loading on marginal quality of bonded resin composite restorations. *Dent. Mater.* **2005**, *21*, 397–412. [CrossRef] [PubMed]

© 2020 by the authors. Licensee MDPI, Basel, Switzerland. This article is an open access article distributed under the terms and conditions of the Creative Commons Attribution (CC BY) license (http://creativecommons.org/licenses/by/4.0/).

Communication

Towards the Development of a Novel Ex Ovo Model of Infection to Pre-Screen Biomaterials Intended for Treating Chronic Wounds

Elena García-Gareta [1,2,]*, Justyna Binkowska [1], Nupur Kohli [1,3] and Vaibhav Sharma [1]

[1] Regenerative Biomaterials Group, The RAFT Institute & The Griffin Institute, Northwick Park & Saint Mark's Hospitals, Harrow, London HA1 3UJ, UK; j.m.binkowska@gmail.com (J.B.); n.kohli@imperial.ac.uk (N.K.); sharmav@raft.ac.uk (V.S.)
[2] Division of Biomaterials and Tissue Engineering, Eastman Dental Institute, University College London, London WC1X 8LD, UK
[3] Biomechanics Research Group, Department of Mechanical Engineering, Imperial Collage London, South Kensington Campus, London SW7 2AZ, UK
* Correspondence: garciae@raft.ac.uk

Received: 11 February 2020; Accepted: 18 May 2020; Published: 2 June 2020

Abstract: This communication reports preliminary data towards the development of a live ex vivo model of persistent infection that is based on the chick embryo chorioallantoic membrane (CAM), which can be used for pre-screening biomaterials with antimicrobial properties for their antimicrobial and angiogenic potential. Our results showed that it was possible to infect chicken embryos with *Staphylococcus aureus*, one of the main types of bacteria found in the persistent infection associated with chronic wounds, and maintain the embryos' survival for up to 48 h. Survival of the embryos varied with the dose of bacteria inoculum and with the use and time of streptomycin application after infection. In infected yet viable embryos, the blood vessels network of the CAM was maintained with minimal disruption. Microbiological tests could confirm embryo infection, but quantification was difficult. By publishing these preliminary results, we hope that not only our group but others within the scientific community further this research towards the establishment of biomimetic and reproducible ex vivo models of persistent infection.

Keywords: chronic wounds; infection; chick embryo CAM; ex ovo; biomaterials

---

1. Introduction

Chronic wounds are described as severe non-healing wounds that constitute a significant burden to healthcare systems [1]. According to a retrospective health economics study, in the U.K. the National Health System (NHS) treated 2.2 million wounds in the year 2012/2013, with an estimated cost of GBP 5 billion. In the USA, about 6.5 million patients are affected by chronic wounds and an estimated USD 25 billion is spent annually on treating these wounds [1]. Predictably, these figures will rise with an ageing population and the increasing prevalence of long-term conditions such as diabetes [1]. Chronic wounds do not progress through the overlapping healing stages of haemostasis, inflammation, proliferation, and remodelling, as these wounds are often infected and lack the angiogenic supply for the healing process to begin [1,2]. Current treatments include surgery, dressings, drugs, and devices, among others, and often a combination of them is necessary to achieve healing [1]. Therefore, there is a clinical need for new effective treatments with the ultimate goal of regeneration rather than repair. Tissue engineering is a rapidly progressing field that offers exciting potential to find new and regenerative treatments for chronic wounds.

With the advent of tissue engineering, biomedical engineers, material scientists, and cell biologists are researching novel systems for management of these difficult wounds. The research strategies for treating chronic wounds vary from stem cell-based therapies and gene therapies to tissue-engineered biomaterial scaffolds [3,4]. The rationale for using tissue-engineered scaffolds is to control the infection, restore the angiogenic response, and allow new tissue formation to take place while providing a temporary extracellular matrix (ECM). To develop biomaterials that can treat the underlying infection and promote angiogenesis, the candidate materials need to go through rigorous in vitro and in vivo testing. In the particular case of chronic wounds, currently there is not a chronic wound model in existence that can mimic the in vivo scenario for pre-screening biomaterials. Such a model should mimic the persistent infection associated with chronic wounds. Furthermore, this model should allow testing of the angiogenesis potential of the candidate biomaterials.

The extraembryonic chorioallantoic membrane (CAM) of the chick embryo is easily accessible, highly vascularised, and presents an immunodeficient environment [5]. Therefore, it has been widely used in various aspects of research such as biology, pharmacy, oncology, and tissue regeneration [5,6]. It has also been used as an ex vivo model to study pathogenesis and the invasion capacity of different bacteria such as *Klebsiella pneumoniae*, fungi such as *Cryptococcus gatii* or *Candida albicans*, and viruses such as the fowlpox-causing viruses [7–10]. In these studies, the CAM was inoculated with the pathogen of interest, and after subsequent infection, the researchers assessed the survival and morphological features of the CAM and embryo. Thus far, only *in ovo* systems have been used. However, *ex ovo* systems may offer advantages over *in ovo* systems. Although *in ovo* CAM assays are very popular, a significant variation exists in in ovo techniques as a result of a lack of standardisation. Furthermore, in ovo systems often develop contamination from the eggshell dust, and therefore are inefficient in maintaining sterility of the embryo [6]. Recently, research into ex ovo techniques has seen the development of efficient, reproducible, and cost-effective CAM assays that are becoming increasingly popular in the fields of biomaterials and tissue engineering [6].

We are currently developing a novel ex vivo model of persistent infection using ex ovo chick embryo CAM for testing the antimicrobial potential of tissue-engineered materials followed by their subsequent angiogenesis potential. In the present preliminary study, we try to answer the initial question of whether it is possible to infect chick embryos/CAM with bacteria and maintain their survival ex ovo for up to 48 h. For this purpose, we used *Staphylococcus aureus*, one of the main types of bacteria found in the persistent infection associated with chronic wounds [11], and antibiotics (i.e., streptomycin) to maintain and/or enhance embryo survival.

2. Materials and Methods

2.1. S. aureus Strains

A clinical strain and a control antibiotic-susceptible strain (ATCC 29213) of *S. aureus* were kindly donated by The Hillingdon Hospitals NHS Foundation Trust (Uxbridge, UK). Donated *S. aureus* samples were in mannitol salt agar (MSA) plates.

2.2. Luria-Bertani (LB) Agar Plates

LB agar powder (17.5 g, 22700025, Thermo Fisher Scientific, Loughborough, UK) was dissolved in 500 mL of deionised water, autoclaved, and cooled down to approximately 50 °C. Working in a biosafety cabinet, the agar was poured into Petri dishes (around 25 mL of agar per dish). Plates were partially covered with the lids and allowed to cool down and dry. Finally, the lids were closed, and the plates stored with the agar side up at 4 °C until used (for a maximum of 4 weeks).

2.3. S. aureus Culture/Colony Isolation

The agar plate(s) were taken out of the fridge to warm up to room temperature. Using a sterile swab, a single colony of *S. aureus* from the sample MSA plate (clinical or control strain) was picked and streaked onto a fresh LB agar plate, which was incubated for up to 24 h at 37 °C. The same procedure was carried out to maintain culture of *S. aureus* in LB agar plates but transferring individual colonies from LB agar plates to fresh ones.

2.4. Streptomycin Sensitivity Assay

Filter paper (Whatman, WHA1001150, Sigma-Aldrich, Dorset, UK) was cut into 6 mm discs and autoclaved prior to use. Streptomycin solutions (streptomycin sulfate salt, S6501, Merck, Feltham, UK) ranging from 100 mg/mL to 5 µg/mL were prepared and stored at 4 °C under sterile conditions.

A standard EUCAST (European Committee on Antimicrobial Susceptibility Testing) disc diffusion method [12] was used as follows. Agar plates were taken out of the fridge several minutes before the procedure. Isolated colonies of *S. aureus* were picked from an LB plate (see Section 2.3) using a sterile swab, transferred into a sterile tube containing 1 mL of sterile water, and vortexed to suspend the bacteria in the water. The Optical Density (OD) at 600 nm of the prepared bacterial suspension was measured. If required, the suspension was diluted to obtain 4×10^8 cfu/mL. A sterile swab was dipped in the final suspension, and excess fluid was removed and swabbed onto a fresh LB agar plate to obtain confluent bacterial growth. Bacteria were let to set for a few minutes before applying the streptomycin-soaked discs onto the plate; using sterile tweezers, filter paper discs were dipped in the required streptomycin solution, excess fluid was removed, and then the soaked discs were placed onto the plates, which were incubated for 24 h at 37 °C. Finally, the diameters of the zones of inhibition were measured using a ruler.

2.5. Ex Ovo Culture System

The use of chicken embryos in our study did not require ethical approval as per the guidelines of The Institutional Animal Care and Use Committee (IACUC) and the National Institutes of Health, USA, and the Animals Scientific Procedures Act (ASPA), U.K., which state that a chicken embryo that has not reached the 14th day of its gestation period would not experience pain and can therefore be used for experimentation without any ethical restrictions or prior protocol approval [6].

Fertilized chicken eggs obtained from a local producer in Middlesex (Quedgeley, UK) were kept in a specialized incubator at 38 °C and ≈45% relative humidity for 3 days, when the embryos were transferred to a shell-less culture system consisting of a glass–cling film set-up routinely used in our laboratory [6]. Briefly, Pyrex glasses of 8 cm diameter were autoclaved and filled up to three-quarters with sterile water, and then a clean cling film layer (pre-sterilised with 70% Industrial Methylated Spirit (IMS) and dried) was placed inside the glasses with the bottom of the cling film touching the water. Rubber bands secured the cling film on the glasses. After 3 days in the specialized incubator, eggs were wiped with cytosol wipes, cracked open against the sharp edge of a triangular block, and the contents were immediately transferred to the glass-cling film set up. The yolk sac and the embryo were identified and assessed for viability by looking for a beating heart. The glasses were then covered with a Petri dish, transferred to the incubator, and grown for 6 days at 38 °C and 80%–90% humidity.

2.6. Ex Ovo CAM Infection with S. aureus

At day 9, the CAMs were infected using 2 doses of 50 µL of *S. aureus* bacterial inoculum, applied topically avoiding dispersing the inoculum very close to the embryo (Figure 1A). The inoculation spots were marked on the cover Petri dish lid (Figure 1). Control CAMs were inoculated with two doses of 50 µL of deionised water. The ex ovo cultures were covered with the corresponding Petri dish lid and incubated at 38 °C and 80%–90% humidity. Viability of the embryos was monitored daily. Macroscopic and microscopic images (using a GT vision stereo microscope, GXM-XTL3T101, GT Vision Ltd, Stansfield, UK) of the embryos were taken regularly. Allantoic fluid samples were taken either from embryos that died during the experiment or that were sacrificed if still viable. The allantoic fluid was collected (≈1 mL) at relevant time points, diluted 5–6-fold, plated, and incubated at 37 °C for 24 h to examine the presence of infection.

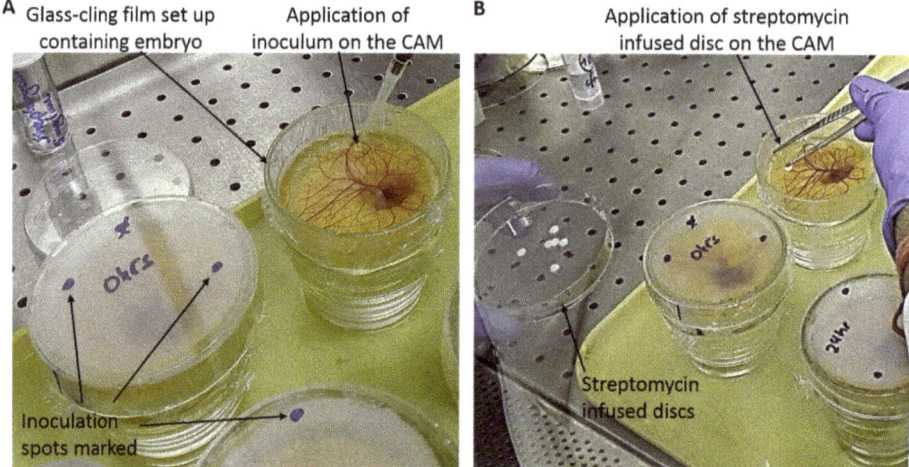

Figure 1. Photos taken during the experimental procedure of this study showing (**A**) application of *Staphylococcus aureus* bacterial inoculum on the chorioallantoic membrane (CAM) of a viable embryo contained in a glass–cling film set-up and inoculation spots marked on the cover Petri dish lids, and (**B**) application of streptomycin-infused 6 mm diameter filter paper discs on the CAM using sterile tweezers.

2.7. Preparation and Application of Streptomycin-Infused Discs on CAM

A solution of 5 mg/mL of streptomycin in sterile water was prepared, and 6 mm diameter filter paper discs were dipped into it. Streptomycin-infused discs (×4) were gently applied onto the CAM using sterile tweezers (Figure 1B). The ex ovo cultures were immediately covered with the corresponding Petri dish lid and incubated at 38 °C and 80%–90% humidity.

A control for streptomycin activity was run alongside the ex ovo cultures by using the standard EUCAST method described before. Four streptomycin-infused discs were applied onto the LB agar plate, which was incubated at 37 °C for up to 24 h.

3. Results and Discussion

3.1. Streptomycin Sensitivity Assay

An assay to determine the streptomycin concentration that yielded the desired level of bactericidal effect was carried out, where *S. aureus* was tested against several concentrations of streptomycin using the standard EUCAST disc diffusion method [12]. *S. aureus* from both a clinical strain and a control antibiotic susceptible strain were used. Results showed that the minimum streptomycin concentration necessary to yield a noticeable bactericidal effect on both strains tested was 5 mg/mL. Below this concentration, the bactericidal effect was faint, whereas above it the effect was hardly enhanced (Figure 2). Therefore, 5 mg/mL was chosen to carry out our study.

3.2. Embryo Survival after Infection of CAM in the Absence of Streptomycin

According to the literature, an inoculum of 10^8 cfu/mL or higher would result in embryo death [7,8]. A quick test confirmed this previous finding, where our embryos died within 24 h following CAM inoculation with 10^8–10^9 cfu/mL inoculum (results not shown). A study by Gow et al. using an inoculum of 10^5 cfu/mL showed embryo death within 24 h. However, in the cited study, the infecting pathogen was the fungus *Candida albicans* [10]. In our study, we lowered the bacterial load in the inoculum to 10^5 cfu/mL for investigating embryo survival following CAM infection.

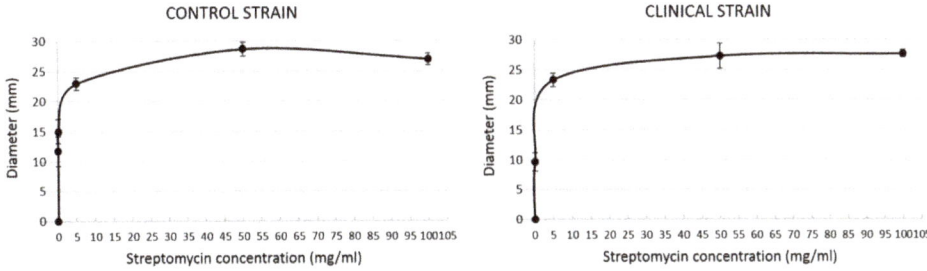

Figure 2. Results from the streptomycin sensitivity assay. Graphs show average ± standard deviation ($n = 3$ per streptomycin concentration).

Results showed that embryo survival decreased to 50% 24 h after inoculation and continued to decrease following another 24 h of incubation (Figure 3A). By day 3 (72 h), infected embryos were dead. Our control embryos (non-inoculated CAM) survived over the 72 h experiment (Figure 3A). On examining the allantoic fluid, the presence of bacterial colonies was observed in infected embryos, suggesting the presence of infection within the CAM as well (Figure 3B,C).

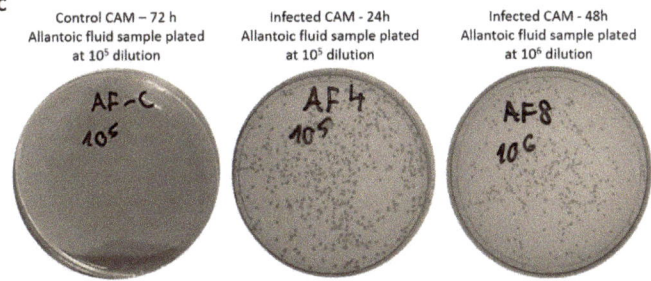

Figure 3. (**A**) Percentage of embryo survival after CAM infection (infected CAM) with 10^5 cfu/mL inoculum compared to non-infected CAMs (control CAM). Results show average ($n = 3$ for control and $n = 6$ for infected CAM at the beginning of the experiment). (**B**) *S. aureus* load in allantoic fluid samples of infected CAMs. X means that there were too many colonies and a count could not be performed. (**C**) Representative images of Luria-Bertani (LB) plates showing *S. aureus* colony formation after plating samples of the allantoic fluid from control and infected CAMs.

These results suggest that it is possible to infect chick embryos/CAM with *S. aureus* (10^5 cfu/mL inoculum) and maintain their survival for up to 48 h. However, the percentage of embryo survival under these conditions is approximately 30%. Therefore, the next question to answer is whether survival rate could be increased with the use of antibiotics, the aminoglycoside streptomycin in our study.

3.3. Embryo Survival after Infection of CAM in the Presence of Streptomycin

An increase in embryo survival was observed in the presence of the antibiotic streptomycin when applied at the time of inoculation (0 h) (Figure 4A). On injecting the CAMs with the antibiotic 24 h after inoculation, survival increased after 24 h incubation; however, after further 24 h of incubation, the embryos died. Examination of the allantoic fluid showed clear presence of infection in the embryos/CAM when the antibiotic was applied 24 h after inoculation, whereas no infection could be seen in the control group or when streptomycin was applied at the time of inoculation (Figure 4B).

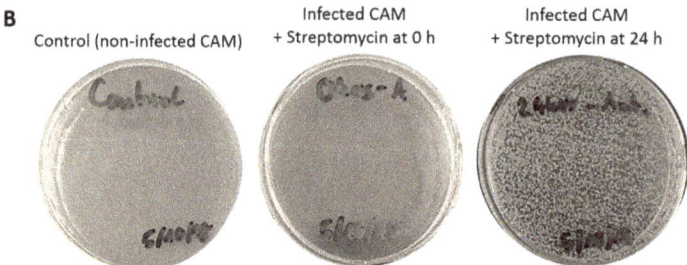

Figure 4. (**A**) Percentage of embryo survival after CAM infection (infected CAM) with 10^5 cfu/mL inoculum compared to non-infected CAMs (control CAM). Streptomycin was applied either at the time of inoculation (0 h) or 24 h after inoculation (24 h). Results show average (n = 2 per group). (**B**) Representative images of LB plates showing *S. aureus* colony formation after plating allantoic fluid samples (10^5 dilution). No colonies were observed in the control group plates. As can be observed, there were too many colonies to perform a count and subsequent quantification, which could only be done using a higher dilution.

These results suggest that embryo survival depends on the inoculum dose, the use of antibiotics, and the time of their application on the infected CAM. Further work on this project intends to look into a higher range for these variables to find optimum conditions. Nevertheless, the answer to the initial question of whether it is possible to infect chick embryos/CAM with bacteria and maintain their survival for up to 48 h was affirmative, which sets the ground for developing an ex vivo model of persistent infection.

3.4. CAM Morphology of Infected Embryos

Another aim of this study was to investigate CAM morphology of infected embryos to understand how infection with *S. aureus* affects the CAM blood vessel network, which could have an effect when studying the angiogenic potential of candidate materials. The CAM is highly vascularised with both mature vessels and capillaries and is easily accessible for orthotopic implantation of biomaterials without initiating an embryo immune reaction [6]. If this network of vessels and capillaries was badly disrupted due to infection, testing of biomaterials on it would not be possible.

Non-infected CAMs showed a well-defined network of blood vessels with no obvious pathological features (Figure 5A,G). On the other hand, CAMs with signs of infection showed various degrees of blood vessel damage, which was mainly observed on the periphery of the CAM, as well as throughout the network (Figure 5B,E). In some instances, the blood vessel network appeared quite disrupted, making it not suitable for further biomaterial testing (Figure 5E). However, some infected embryos showed minimal blood vessel network disruption (Figure 5B). Microscopic analysis showed a cloudy background for infected CAMs (Figure 5H) compared to non-infected CAMs (Figure 5G). Application of streptomycin at the time of inoculation seemed to clear infection because no evident signs of it could be observed for the majority of these CAMs (Figure 5C,D) compared to applying the antibiotic 24 h after inoculation (Figure 5E). When embryo death occurred, the CAM blood vessel network appeared completely destroyed (Figure 5F).

These results suggest that under certain conditions it is possible to infect CAMs with *S. aureus* and keep the embryo viable while maintaining the network of CAM blood vessels with minimal disruption. In the future, it would be useful to investigate the way in which to quantify blood vessel damage to establish a threshold of viability for biomaterial testing.

We acknowledge that the system presented here is, at the moment, fragile, where we only managed to keep a low percentage of infected embryos alive for up to 48 h. We also acknowledge variation of outcomes and reproducibility. The CAM model in itself is a highly biological variable, even in non-infected eggs [6]. Therefore, the differences observed in our study may stem from this intrinsic feature of the CAM model, rather than inappropriate methodology. Further experiments with increased biological replicates may improve reproducibility.

It is worth mentioning that extraction of allantoic fluid was technically difficult and invasive to the embryo. Although information on the dynamics of infection spread would be useful, manipulation of the embryo during the experiment could introduce additional contamination and stress, increasing the chances of death due to the experimental procedure versus the infection itself. However, this can be resolved by extracting the allantoic fluid after sacrificing the embryo, as we did in our study, at given time intervals. The analysis of allantoic fluid in this preliminary study was largely exploratory and we could not anticipate the final bacterial concentration in the embryo because no reports of similar experiments exist. In future work, we do intend to take into account this study's observations when quantifying bacterial burden. Notwithstanding, the results presented still allow comparison of bacterial burden between CAMs with different survival outcomes.

In addition, more research and optimisation are needed to test a wider range of conditions, such as the ones mentioned earlier (i.e., inoculum dose, the use of antibiotics, and the time of their application on the infected CAM), in order to find the optimum conditions to achieve consistent and reproducible viability of infected embryos. Nevertheless, this system is more representative of the biological situation in chronic wounds where living tissues are ridden with bacterial infection. Moreover, it is very cost-effective.

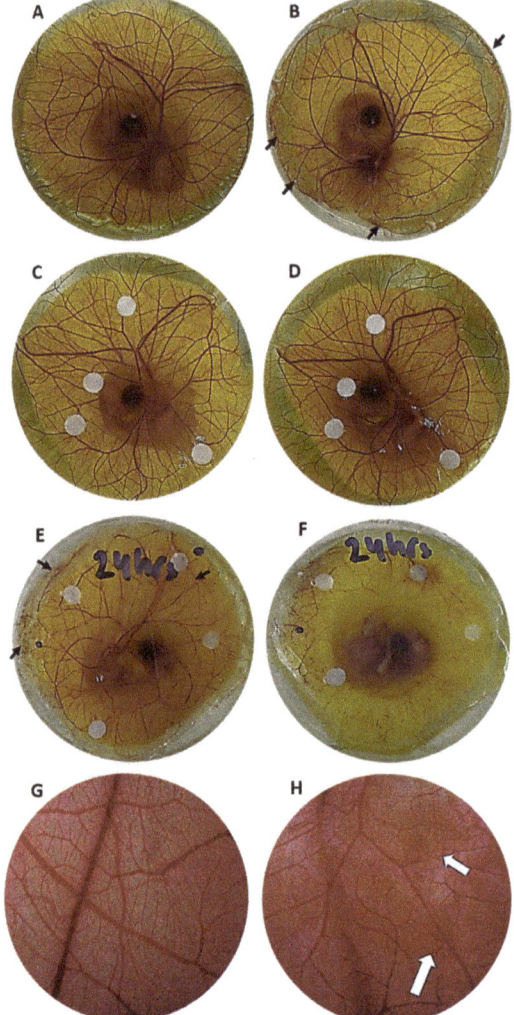

Figure 5. Macroscopic images of the CAM with and without bacterial infection, where black arrows point at vessel damage due to the bacterial invasion. (**A**) Non-infected CAM incubated for 48 h. (**B**) CAM infected with 10^5 bacterial inoculum and incubated for 48 h. (**C**) CAM infected with 10^5 bacterial inoculum with antibiotic applied at the time of infection (0 h) and incubated for 24 h. (**D**) Same embryo as C after 48 h incubation. (**E**) CAM infected with 10^5 bacterial inoculum with antibiotic applied 24 h post-infection and incubated for 24 h. (**F**) Same embryo as (**E**) after 48 h incubation (embryo is dead and network of blood vessels is completely destroyed). (**G**) Stereomicroscopy image of non-infected CAM at 24 h showing a mostly clear background and well-defined blood vessels. (**H**) Stereomicroscopy image of infected CAM at 24 h post-infection showing a cloudy background (white arrows point at yellowish cloudy spots) and in some areas blood vessels appear less well-defined and fainter than in non-infected CAM. Please note that with the purpose of protecting the live ex vivo cultures, photos were taken with the Petri dish lid on, but the lid might have been taken off if the embryos were already dead or sacrificed. Additionally, the lid might have been moved (without opening) to capture the morphological changes in the CAM/embryo.

4. Conclusions

The main conclusions from this study are (1) it was possible to infect chicken embryos with *S. aureus*, one of the main types of bacteria found in the persistent infection associated with chronic wounds, and maintain the embryos' survival for up to 48 h; (2) survival of the embryos varied with the dose of bacteria inoculum and with the use and time of application after infection of an antibiotic (streptomycin in this study); (3) in infected yet viable embryos, the blood vessels network of the CAM was maintained with minimal disruption; and (4) microbiological tests can confirm infection of the embryo, but quantification is difficult.

In summary, we report here preliminary data towards the development of a live ex vivo model of persistent infection that can be used for pre-screening biomaterials intended for treating chronic wounds for their antimicrobial and angiogenic potential. This model is relatively simple, quick, and low-cost, and mimics the in vivo situation more closely than traditionally used antimicrobial tests using agar plates and dilution assays. In addition, keeping in accordance with the principles of the National Centre for the Replacement Refinement and Reduction of Animals in Research (NC3R's), this model does not require administrative procedures for obtaining ethics committee approval for animal experimentation. Nevertheless, we do acknowledge that further work is still needed towards further optimisation to study the interaction of biomaterials with two different strains of bacteria commonly found in chronic wounds, in order to assess their antimicrobial and angiogenic properties. By publishing these preliminary results, we hope that not only our group but others within the scientific community further this research towards the establishment of biomimetic and reproducible ex vivo models of persistent infection. Such models would not only be useful in the research of treatments for chronic wounds, but also in those applications where infection is an important factor to take into account when investigating new therapies.

Author Contributions: Conceptualization, V.S. and N.K.; methodology, V.S., N.K., and J.B.; investigation and data acquisition, J.B., V.S., and N.K.; data analysis E.G.-G.; writing—original draft preparation, E.G.-G.; writing—review and editing, E.G.-G., J.B., N.K., and V.S.; supervision, V.S. and E.G.-G.; funding acquisition, E.G.-G. All authors have read and agreed to the published version of the manuscript.

Funding: This work was supported by the Restoration of Appearance and Function Trust (RAFT, U.K., registered charity number 299811).

Conflicts of Interest: The authors declare no conflict of interest.

References

1. Price, A.; Naik, G.; Harding, K. "Skin repair technology". In *Biomaterials for Skin Repair and Regeneration*; Woodhead Publishing: Cambridge, UK, 2019; pp. 27–57.
2. García-Gareta, E. Introduction to biomaterials for skin repair and regeneration. In *Biomaterials for Skin Repair and Regeneration*; Woodhead Publishing: Cambridge, UK, 2019; pp. xiii–xxvii.
3. Davison-Kotler, E.; Sharma, V.; Kang, N.V.; García-Gareta, E. A Universal Classification System of Skin Substitutes Inspired by Factorial Design. *Tissue Eng. Part B Rev.* **2018**, *24*, 279–288. [CrossRef]
4. Shevchenko, R.V.; James, S.L.; James, S.E. A review of tissue-engineered skin bioconstructs available for skin reconstruction. *J. R. Soc. Interface* **2010**, *7*, 229–258. [CrossRef]
5. Yuan, Y.J.; Xu, K.; Wu, W.; Luo, Q.; Yu, J.L. Application of the chick embryo chorioallantoic membrane in neurosurgery disease. *Int. J. Med. Sci.* **2014**, *11*, 1275–1281. [CrossRef]
6. Kohli, N.; Sawadkar, P.; Ho, S.; Sharma, V.; Snow, M.; Powell, S.; Woodruff, M.A.; Hook, L.; García-Gareta, E. Pre-screening the intrinsic angiogenic capacity of biomaterials in an optimised *ex ovo* chorioallantoic membrane model. *J. Tissue Eng.* **2020**, *11*. [CrossRef]
7. Adam, R.; Mussa, S.; Lindemann, D.; Oelschlaeger, T.A.; Deadman, M.; Ferguson, D.J.; Moxon, R.; Schroten, H. The avian chorioallantoic membrane in ovo—A useful model for bacterial invasion assays. *Int. J. Med. Microbiol.* **2002**, *292*, 267–275. [CrossRef]
8. Nnadi, E.N.; Enweani, I.B.; Ayanbimpe, G.M. Infection of Chick Chorioallantoic Membrane (CAM) as a Model for the Pathogenesis of Cryptococcus gattii. *Med. Mycol. J.* **2018**, *59*, E25–E30. [CrossRef]

9. Woodruff, A.M.; Goodpasture, E.W. The susceptibiliry of the chorio-allantoic membrane of chick embryos to infection with the fowl-pox virus. *Am. J. Pathol.* **1931**, *7*, 209–222.
10. Gow, N.A.R.; Knox, Y.; Munro, C.A.; Thompson, W.D. Infection of chick chorioallantoic membrane (CAM) as a model for invasive hyphal growth and pathogenesis of Candida albicans. *Med. Mycol.* **2003**, *41*, 331–338. [CrossRef] [PubMed]
11. Serra, R.; Grande, R.; Butrico, L.; Rossi, A.; Settimio, U.F.; Caroleo, B.; Amato, B.; Gallelli, L.; de Franciscis, S. Chronic wound infections: the role of Pseudomonas aeruginosa and *Staphylococcus aureus*. *Expert Rev. Anti. Infect. Ther.* **2015**, *13*, 605–613. [CrossRef] [PubMed]
12. European Committee on Antimicrobial Susceptibility Testing. *European Committee on Antimicrobial Susceptibility Testing Breakpoint Tables for Interpretation of MICs and Zone Diameters*; European Committee on Antimicrobial Susceptibility Testing: Växjö, Sweden, 2018.

© 2020 by the authors. Licensee MDPI, Basel, Switzerland. This article is an open access article distributed under the terms and conditions of the Creative Commons Attribution (CC BY) license (http://creativecommons.org/licenses/by/4.0/).

Article

Substituted Nano-Hydroxyapatite Toothpastes Reduce Biofilm Formation on Enamel and Resin-Based Composite Surfaces

Andrei C. Ionescu [1,*], Gloria Cazzaniga [1], Marco Ottobelli [1], Franklin Garcia-Godoy [2] and Eugenio Brambilla [1]

1. Oral Microbiology and Biomaterials Laboratory, Department of Biomedical, Surgical and Dental Sciences, University of Milan, via Pascal 36, 20133 Milan, Italy; gloria.cazzaniga@yahoo.it (G.C.); marco.ottobelli@unimi.it (M.O.); eugenio.brambilla@unimi.it (E.B.)
2. Bioscience Research Center and Clinical Research, College of Dentistry, University of Tennessee Health Science Center, 875 Union Avenue, Memphis, TN 38163, USA; fgarciagodoy@gmail.com
* Correspondence: andrei.ionescu@unimi.it; Tel.: +39-0250319007

Received: 3 April 2020; Accepted: 21 May 2020; Published: 1 June 2020

Abstract: *Background*: Toothpastes containing nano-hydroxyapatite (n-HAp) substituted with metal ions provide calcium and phosphate ions to dental hard tissues, reducing demineralization, and promoting remineralization. Few data are available about the effect of these bioactive compounds on oral microbiota. *Methods*: This in vitro study evaluated the influence of two commercially-available substituted n-HAp-based toothpastes (α: Zn-carbonate substituted n-HAp; β: F, Mg, Sr-carbonate substituted n-HAp) on early colonization (EC, 12 h) and biofilm formation (BF, 24 h) by oral microbiota. Controls were brushed with distilled water. Artificial oral microcosm and *Streptococcus mutans* biofilms were developed using human enamel and a resin-based composite (RBC) as adherence surfaces. Two test setups, a shaking multiwell plate and a modified drip-flow reactor (MDFR), were used to simulate clinical conditions during the night (low salivary flow and clearance) and daytime, respectively. Energy-dispersive X-ray spectrometry (EDS) was used to evaluate specimens' surfaces after toothpaste treatment. Fluoride release from β toothpaste was evaluated. Viable adherent biomass was quantified by MTT assay, and biofilms' morphology was highlighted using confocal microscopy. *Results*: EDS showed the presence of remnants from the tested toothpastes on both adherence surfaces. β toothpaste showed significantly lower EC and BF compared to control using the artificial oral microcosm model, while α toothpaste showed lower EC and BF compared to control, but higher EC and BF compared to β toothpaste. The effect shown by β toothpaste was, to a minimal extent, due to fluoride release. Interestingly, this result was seen on both adherence surfaces, meaning that the tested toothpastes significantly influenced EC and BF even on RBC surfaces. Furthermore, the effect of toothpaste treatments was higher after 12 h than 24 h, suggesting that toothbrushing twice a day is more effective than brushing once. *Conclusions*: The efficacy of these treatments in reducing microbial colonization of RBC surfaces may represent a promising possibility in the prevention of secondary caries.

Keywords: fluoride(s); biofilm(s); *Streptococcus mutans*; bioreactor(s); enamel; composite materials; biomaterials; hydroxyapatite; nanostructured materials; dental

---

1. Introduction

Biofilms colonizing dental surfaces represent the habitat of a wide range of microorganisms. In healthy conditions, this microbial community provides fundamental benefits to host surfaces, including homeostasis of critical environmental parameters such as pH, as well as protection against colonization by exogenous species. However, loss of this equilibrium can result in dysbiosis, leading to the onset of oral diseases, including dental caries [1,2].

Mechanical biofilm control can be beneficially associated with the use of active compounds to reduce biofilm formation, positively affecting the biological equilibrium of the ecosystem [3]. The search for new bioactive compounds to prevent dental caries development and progression has led researchers to focus their attention on the use of nanotechnologies, especially hydroxyapatite (HAp), metals, and metal oxide nanoparticles [4–6]. Nanotechnology applications offer the opportunity to modulate the formation of dental biofilms using nanoparticles with bioactive effects. Indeed, their possible use in biofilm control is related to some peculiar characteristics that materials acquire when nano-sized: shape, surface-to-volume ratio, chemical reactivity, and ability to interact with the bacterial cell wall directly [6–8]. Considerable efforts have been made to clarify the transfer mechanisms of such nanoparticles through biofilms, indicating that nanoparticles can diffuse through water channels inside biofilms, interacting both with the microbial cells and with the hard tissue surface.

Toothpastes containing nano-hydroxyapatite (n-HAp) and its precursors can provide calcium and phosphate ions to dental hard tissues, reducing demineralization, and promoting remineralization. N-HAp may also form a protective layer on dental hard tissues with chemical characteristics very close to natural tissues, showing biomimetic properties [9,10]. Biomimetic n-HAp or other calcium phosphate-based nanosystems have been shown to actively shift the ionic balance toward remineralization [11,12]. Recently, n-HAp-based toothpastes containing metal ions such as Zn, F, Mg, and Sr have been introduced to improve the remineralization processes. The small quantity of metal ions with which nHAp is doped modulates the properties of the compound already from its deposition, influencing crystal growth and length, and influencing its solubility and strength. They are firmly embedded inside the lattice and are not supposed to be released. Most studies, therefore, provide data about the remineralization potential of these compounds on surface lesions, with few reporting their influence on early microbial colonization (EC) and biofilm formation (BF) [13–16].

Resin-based composites (RBCs) provide an example of artificial surfaces intensely colonized by cariogenic biofilms due to their lack of buffering effect [17]. This characteristic has been put into relation with the high occurrence of secondary caries in the sound dental tissues surrounding an RBC restoration [17]. The possibility of influencing microbial colonization of RBCs, therefore, can be seen as a means to prevent the occurrence of such lesions. For this reason, RBCs are a potential target of preventive measures exploiting the potential of nanotechnologies in order to reduce secondary caries occurrence.

The aim of this study was to evaluate the in vitro effect of two n-Hap-based toothpastes on early colonization (EC) (12 h) and biofilm formation (BF) (24 h) of human enamel and RBC surfaces. The study was performed by aerobically culturing the monospecies *Streptococcus mutans* and an artificial oral microcosm using both a static and a continuous flow setup. The aim of the two models was to assess the effect of the tested toothpastes under environments simulating clinical conditions during the night (low salivary flow and clearance) and day (high salivary flow and clearance). The null hypothesis was that toothpaste treatment would not influence EC and BF on sound enamel surfaces in the tested models. An additional null hypothesis was that toothpaste treatment would not influence EC and BF on RBC surfaces in the tested models.

2. Materials and Methods

2.1. Preparation of Specimens

Reagents, culture media, and disposables used in this study were obtained from E. Merck AG (Darmstadt, Germany). A total of 110 human teeth extracted for clinical reasons were obtained and stored at −20 °C until use. The Institutional Review Board of the University of Milan approved the protocol (codename: SALTiBO-2017). Enamel disks 6.0 mm in diameter and 1.5 mm thick were cut from the teeth surfaces by means of custom-made water-cooled diamond trephine burs (Indiam, Carrara,

Italy). A total of 213 enamel disks were prepared for the study, using this procedure to obtain more than one disk from each tooth.

A total of 213 disks (6.0 mm diameter and 1.5 mm thickness) of a nanohybrid resin-based composite (RBC, Clearfil Majesty ES-2, shade A2, Kuraray Europe GmbH, Hattersheim am Main, Germany) were obtained by packing an excess of uncured material into a custom-made PTFE mold. The top and bottom surfaces of the RBC were covered with a cellulose acetate strip (Mylar) and condensed against a glass plate by applying a load of 1 kg for 20 s. The specimens were then irradiated for 40 s by placing the tip of a light-curing unit (MiniLED, Satelec, Acteon Group, Merignac, France, 420–480 nm emission, 1250 mW/cm^2 irradiance) into direct contact with the acetate strip. To remove most of leachates, RBC specimens were then stored under light-proof conditions in phosphate-buffered saline (PBS) for 6 days at 37 °C, with the buffer replaced twice a day.

Both enamel and RBC disks were polished using grinding paper (1000, 2000, and 4000 grit) to obtain a surface roughness with Ra < 0.2 µm. They were then randomly divided into three groups (α, β, and control, n = 71/group, Figure 1).

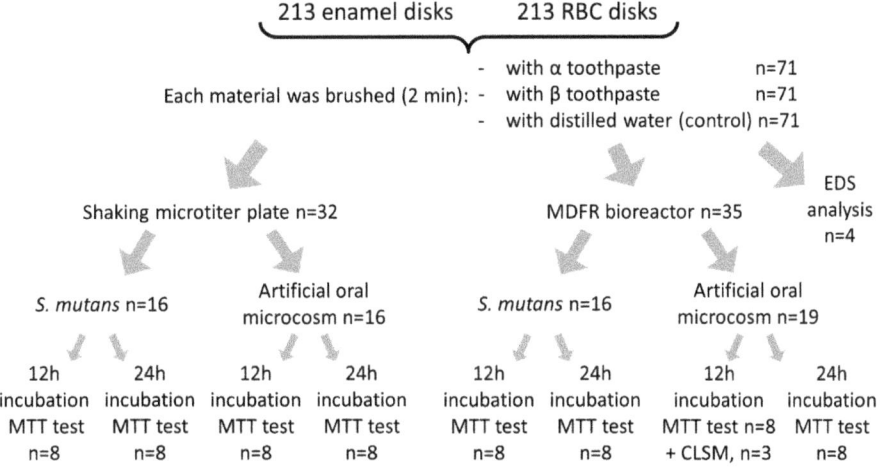

Figure 1. Diagram of specimens processing for the present study. After brushing with the tested toothpastes or with distilled water, enamel and RBC specimens were rinsed, sterilized, then subjected to microbiological analysis and surface imaging (SEM, CLSM) and analysis (EDS). The microbiological analysis included two test setups, two microbiological models, and two incubation times.

2.2. Toothpastes

Two toothpastes containing n-HAp were tested (Table 1). The toothpastes were inserted into test tubes coded by the Greek letter corresponding to the first two groups (α, β). In this way, the experimenters performing microbiological procedures were blinded regarding their composition. Each toothpaste was mixed to a slurry using one part of toothpaste and two parts of distilled water. Then, enamel and RBC disks of groups α and β were manually brushed by a single operator with the corresponding toothpaste for 2 min and rinsed with distilled water for 1 min. Disks belonging to the control group were brushed with distilled water for 2 min, then rinsed for 1 min with distilled water. All disks were then sterilized using a chemical peroxide-ion plasma sterilizer (STERRAD, ASP, Irvine, CA, USA).

Table 1. Label, manufacturer and composition of the tested substituted n-Hap-containing toothpastes.

Label	Name	Company	Composition
Toothpaste α	Biorepair Total Protection Plus	Coswell Funo (BO), Italy	Purified water, zinc carbonate hydroxyapatite, glycerin, sorbitol, hydrated silica, silica, aroma, cellulose gum, tetrapotassium pyrophosphate, sodium myristoyl sarcosinate, sodium methyl cocoyl taurate, sodium saccharin, citric acid, phenoxyethanol, benzyl alcohol, sodium benzoate.
Toothpaste β	Biosmalto Caries, Abrasion and Erosion	Curasept Healthcare S.p.A. Saronno (VA), Italy	Purified water, glycerin, hydrated silica, fluoride-hydroxyapatite, Mg-Sr-carbonate hydroxyapatite conjugated with chitosan, cellulose gum, xylitol, cocamidopropyl betaine, xantham gum, aroma, acesulfame K, ethylhexylglycerin, phenoxyethanol, sodium benzoate, citric acid.

2.3. Surface Imaging and Analysis

Scanning electron microscopy (SEM) and energy-dispersive X-ray spectroscopy (EDS) analysis were performed (n = 4/group, Figure 1) using a TM4000Plus Tabletop scanning electron microscope (Hitachi, Schaumburg, IL, USA) equipped with an EDS probe (Q75, Bruker, Berlin, Germany) to investigate the presence of toothpaste residues on the specimen surface. Dry specimens were observed in surface-charge reduction mode without sputter-coating, using an accelerating voltage of 15 KV. Three randomly selected fields were acquired for each specimen at 500× magnification and were analyzed using the EDS probe in full-frame mode using an acquisition time of 150 s. SEM micrographs (backscattered electrons mode) and EDS elemental maps of the surfaces at 5000× magnification were also obtained. The acquired data represent the elemental composition of the ≈1 μm superficial layer.

2.4. Saliva Collection

Whole saliva was collected from five healthy volunteers, according to a previously described protocol [18]. They refrained from oral hygiene for 24 h, did not have any active dental disease, and did not have antibiotic therapy for at least three months prior to the experiment. Chilled test tubes were used for saliva collection. Saliva was then pooled, heated to 60 °C for 30 min and centrifuged (27,000× g, 4 °C, 30 min). The sterile supernatant was collected into sterile tubes and stored at −20 °C. Saliva was thawed at 37 °C for 1 h before use. The Institutional Review Board of the University of Milan approved the use of saliva (codename: SALTiBO-2017), and written, informed consent was obtained from all volunteers.

2.5. Microbiological Procedures

Two microbiological models were used in this study: a monospecific *S. mutans* culture and an artificial oral microcosm based on mixed oral flora (Figure 1).

A pure suspension of *S. mutans* strain ATCC 35668 was obtained as described elsewhere [19]. Briefly, Mitis Salivarius Bacitracin agar (MSB agar) plates were inoculated with the *S. mutans* strain and incubated aerobically in a 5% CO_2-supplemented atmosphere at 37 °C for 48 h. After transferring an inoculum in Brain Heart Infusion (BHI) broth and further incubating in a 5% CO_2-supplemented atmosphere at 37 °C for 12 h, a pure culture of the microorganism was obtained. Cells were harvested by centrifugation (2200× g, 19 °C, 5 min), washed twice with sterile phosphate-buffered saline (PBS), and resuspended in the same buffer., The suspension was sonicated using low-energy output (7W for 30 s, B-150 Sonifier, Branson, Danbury, CT, USA) to disperse bacterial chains. Finally, the suspension was adjusted to an optical density of 1.0 on the McFarland scale, corresponding to a concentration of approximately 6.0×10^8 cells/mL.

A mixed oral flora inoculum was obtained from fresh, pooled saliva expectorated from the same five donors previously described, filtered through sterile glass wool, vigorously stirred for 2 min, and immediately used. Two test setups were used in this study: a shaking multiwell plate and a modified drip-flow reactor (MDFR).

2.5.1. Shaking Multiwell Plate

Disks were placed in 48-well sterile plates and incubated at 37 °C for 24 h in sterile saliva; then, the saliva was removed by gentle aspiration. Each well was inoculated with 100 µl of either *S. mutans* suspension or mixed oral flora, and 900 µl of sterile modified artificial saliva medium. The medium composition was the following: 10.0 g/L sucrose, 2.5 g/L mucin (type II, porcine gastric), 2.0 g/L bacteriological peptone, 2.0 g/L tryptone, 1.0 g/L yeast extract, 0.35 g/L NaCl, 0.2 g/L KCl, 0.2 g/L $CaCl_2$, 0.1 g/L cysteine hydrochloride, 0.001 g/L hemin, and 0.0002 g/L vitamin K1 [20]. Plates were then incubated aerobically in an orbital shaker at 37 °C and 100 rpm. After either 12 h (colonization) or 24 h (biofilm formation), adherent viable biomass assessment was performed.

2.5.2. MDFR

The bioreactor used in the present study was a modification of a commercially available drip-flow reactor (DFR 110, BioSurface Technologies; Bozeman, MT, USA) according to previously described [21]. The modified design (MDFR) allowed the placement of customized PTFE trays on the bottom of the flow cells to submerge the specimen surfaces in the flowing medium. Specimens from each group were randomly divided into the eight flow cells of two identical MDFRs. After sterilization (STERRAD), the MDFR was assembled inside a sterile hood, and a salivary pellicle was obtained on the specimen surface after incubation with sterile saliva at 37 °C for 24 h. Saliva was then removed, each flow-cell was inoculated with 10 mL of either *S. mutans* suspension or mixed oral flora suspension to allow bacterial adherence. After 4 h, a peristaltic pump (RP-1k; Rainin, Emeryville, CA, USA) provided the constant flow of sterile modified artificial saliva medium through the flow cells (9.6 mL/h). For each microbiological inoculum, the amount of viable adherent biomass was evaluated after either 12 h or 24 h of further incubation in aerobic conditions (Figure 1).

2.6. Viable Biomass Assessment

Viable and metabolically active biomass adherent to the specimen surface was assessed using a tetrazolium-based assay as described previously [21]. In brief, a tetrazolium salt stock solution was prepared by dissolving 5 mg/mL 3-(4,5)-dimethylthiazol-2-yl-2,5-diphenyltetrazolium bromide (MTT) in sterile PBS; a phenazinium salt stock solution was prepared by dissolving 0.3 mg/mL of N-methylphenazinium methyl sulphate (PMS) in sterile PBS. The solutions were stored at 2 °C in light-proof vials until the day of the experiment when a fresh measurement solution (FMS) was prepared by diluting 1:10 v/v of MTT stock solution and 1:10 v/v of PMS stock solution in sterile PBS. A lysing solution (LS) was prepared by dissolving 10% v/v of sodium dodecyl sulphate (SDS) and 50% v/v dimethylformamide in distilled water and stored at 2 °C until the day of the experiment when it was warmed at 37 °C for 2 h before use. After the specified incubation times, the orbital shaker and the MDFR medium flow was halted, flow cells and plates were opened, the specimens were carefully removed and immediately placed into Petri plates containing sterile PBS at 37 °C. They were gently washed three times with sterile PBS to remove non-adherent cells and finally placed inside the wells of 48-well plates containing 300 µL of FMS each. The plates were incubated at 37 °C under light-proof conditions for 3 h. During incubation, electron transport across the microbial plasma membrane and, to a lesser extent, microbial redox systems, converted the yellow salt to insoluble purple formazan crystals. The conversion at the cell membrane level was facilitated by the intermediate electron acceptor (PMS). The unreacted FMS was gently removed by aspiration, and the formazan crystals were dissolved by adding 300 µL of LS to each well. The plates were stored for an additional 1 h under light-proof conditions at room temperature; 100 µL of the solution was then transferred into

96-well plates. The absorbance of the solution was measured using a spectrophotometer (Genesys 10-S, Thermo Spectronic, Rochester, NY, USA) at a wavelength of 550 nm; results were expressed as relative absorbance in optical density (OD) units corresponding to the amount of adherent, viable and metabolically active biomass.

2.7. Confocal Laser-Scanning Microscopy (CLSM)

A total of three specimens for each material and treatment group (Figure 1) were prepared for CLSM analysis. They were incubated using the MDFR test setup with the artificial oral microcosm model for 12 h. Then, they were gently removed from the flow cells, rinsed twice with sterile PBS, stained using the FilmTracer™ LIVE/DEAD® Biofilm Viability Kit (Invitrogen Ltd., Paisley, UK), and analyzed using confocal laser-scanning microscopy (CLSM; Eclipse Ti2 inverted CLSM, Nikon, Tokyo, Japan). Three randomly selected image stack sections were recorded for each specimen. Confocal images were obtained using a dry objective (20×; NA = 0.5) at a resolution of 2048 × 2048 pixels, with a zoom factor of 1.0 and a scan speed of 400 Hz, and digitalized using the Nikon Imaging Software (NIS)—Elements Viewer, v. 4.50. Three channels were acquired in parallel scanning; the first used an excitation at 405 nm and emission at 420–470 nm in order to subtract potential autofluorescence digitally. The other two channels had an excitation wavelength of 488 nm, and emission was acquired at 500–570 nm (green channel, live bacteria) and 610–760 nm (red channel, dead bacteria). For each image stack section, 3D-rendering reconstructions were obtained using Drishti (Ajay Limaye, Australian National University, CAN, AUS, http://sf.anu.edu.au/Vizlab/drishti/).

2.8. Fluoride Release Measurements

Measurements of fluoride release from the fluoride-hydroxyapatite-containing toothpaste (β) were performed to correlate it with the hypothesized antimicrobial activity of the toothpaste. A total of 48 enamel disks and 48 RBC disks were additionally prepared and brushed as previously specified (cf. Sections 2.1 and 2.2): one half (n = 24) was brushed with slurry from toothpaste β, and the other half was brushed with distilled water. Immediate fluoride release (t = 0) was evaluated from eight disks for each toothpaste. These disks were placed in 48-well plates containing 300 µL of sterile saliva in each well for 24 h. Disks were then discarded, 1:50 dilution of total ionic strength solution adjustment buffer, TISAB III (Orion Research Inc, Boston, MA, USA) was added, and the solution adjusted to pH = 5.5. The fluoride content of the solution was determined using the ion-selective electrode micro-method. The remaining disks were weighted with an analytical balance, then sterilized, and a monospecies *S. mutans* biofilm was allowed to develop on their surfaces after salivary pellicle formation using the MDFR as described in Section 2.5.2 for the same incubation times (12 h and 24 h). A total of eight disks for each toothpaste and time group were weighed again, and the weight of the adherent wet biomass was calculated. After that, each disk was inserted into one well of 48-well plates containing 300 µL of 10 vol% SDS in distilled water and 1:50 TISAB III (pH = 5.5), and the plates were sonicated for 5 min to allow dispersion of the biomass. Then, disks were discarded, and the fluoride content of the solution was determined.

The amount of released fluoride was calculated and displayed in parts per billion (ppb) after normalizing the fluoride readings by the disks' surfaces. The amount of fluoride incorporated in the biofilm structures was calculated and displayed in parts per million (ppm) after normalizing the fluoride readings by the amount of adherent biomass. All analyses were performed using the ion-selective electrode micro-method, as described previously [22]. In brief, a stock solution with a fluoride concentration of 1000 ppm was appropriately diluted with 10 vol% SDS in distilled water at pH = 5.5 to obtain fluoride standards with fluoride concentrations ranging from 0.0019 to 64 ppm. A calibration curve was obtained using a digital pH/mV meter (SA-720, Orion Research Inc, Boston, MA, USA). A 1:50 final dilution of TISAB III as an ionic strength adjustor was added to each standard before the analyses. A negative reference standard (0 ppm fluoride) was prepared by adding 1:50 TISAB III to the solution containing 10 vol% SDS in distilled water; this solution was also used to rinse the electrodes between measurements.

2.9. Statistical Analysis

All statistical analyses were performed using statistical software (JMP 12.0, SAS Institute, Inc., Cary, NC, USA). The OD and F data were reported as means and standard errors calculated from the natural values. The normality of distributions was preliminarily checked using Shapiro-Wilk test, and homogeneity of variances was preliminarily checked using Bartlett's test. A multi-way ANOVA model was used on OD dataset considering the factors: toothpaste (α, β, control), adherence surface (enamel, RBC), test setup (shaking multiwell plate, MDFR), microbiological model (*S. mutans*, oral microcosm), and incubation time (12 h, 24 h). A multi-way ANOVA model was used on fluoride dataset considering the factors: toothpaste (β, control), adherence surface (enamel, RBC), and incubation time (12 h, 24 h). Student-Newman-Keuls post-hoc test was used to highlight significant differences ($p < 0.05$).

3. Results

3.1. Surface Imaging and Analysis

To investigate the presence of toothpaste residues on enamel and RBC surfaces as well as the composition of the tested toothpastes, specimens were observed using SEM-EDS in surface-charge reduction mode without sputter-coating, using an accelerating voltage of 15 KV (Figure 2). EDS results showed that the elemental composition of both toothpastes was very similar, except that α toothpaste contained zinc, while β toothpaste contained strontium, magnesium, and fluoride (Figure 3). Enamel specimens showed a significant increase in surface zinc content after treatment with α toothpaste, and a significant increase in strontium and magnesium content after treatment with β toothpaste (Table 2). Very interestingly, EDS detected the presence of calcium and phosphate on the surfaces of RBC specimens treated with both toothpastes, as well as traces of zinc on specimens treated with α toothpaste and magnesium on specimens treated with β toothpaste (Figure 2, Table 2).

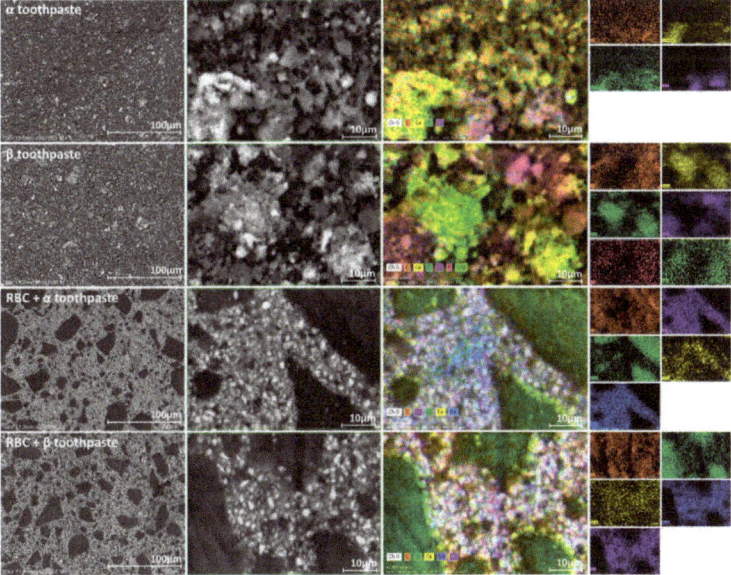

Figure 2. The panel, to be read horizontally, depicts the SEM backscattered electrons micrographs (500×

and 5000×, respectively) and EDS elemental maps (5000×) of the specimen surfaces. The EDS maps are additionally presented as single-channel maps to better identify the topographical presence of each element. The first two rows of the panel display the surfaces of the vacuum-dried toothpastes tested in this study: α toothpaste (containing Zn-carbonate substituted n-HAp) and β toothpaste (containing F, Mg, Sr-carbonate substituted n-HAp). Their aspect is very similar, showing silica microparticles (identified by Si signal) and clusters of n-HAp (identified by Ca and P signals). Zn and Sr signals were below the detection limits in mapping mode and were not displayed. The last two rows of the panel represent the surfaces of the tested restorative material after treatment with the toothpastes: RBC + α toothpaste and RBC + β toothpaste. The RBC composition included silica particles (identified by Si signal), and alumina and barium glass micro and nanoparticles (identified by Al and Ba signals). Ca signal was displayed as a marker of toothpaste remnants. It is noteworthy that toothpaste remnants could be associated with alumina and barium glass fillers rather than silica particles.

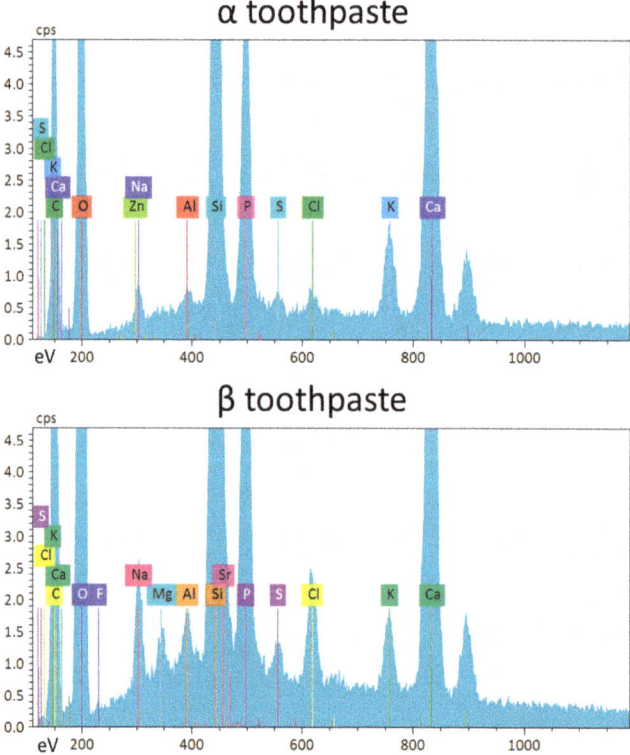

Figure 3. EDS spectra acquired from vacuum-dried tested toothpastes surfaces (α toothpaste containing Zn-carbonate substituted n-HAp, and β toothpaste containing F, Mg, Sr-carbonate substituted n-HAp). Strong Ca and P signals are identified belonging to the n-HAp, as well as the signals corresponding to the corresponding doping elements (Zn in α toothpaste and F, Mg, Sr, in β toothpaste). High counts of Si were also detected in both toothpastes, together with relatively low counts of Al and other elements. The relative amounts of n-HAp doping elements and other elements such as Al and S are below the conventionally considered detection limit of EDS (about 1 wt%). However, the presence of such elements is shown by peaks that were clearly identifiable on all acquired spectra. In this sense, the acquisition of several spectra over the surfaces of many specimens, and the use of statistical analysis on acquired data demonstrating low variability in signals among the different acquisitions (cf. Table 2) allows overcoming, to some extent, this detection limit, improving the performances of EDS analysis.

Table 2. Surface elemental composition of the experimental groups and vacuum-dried toothpastes (α toothpaste containing Zn-carbonate substituted n-HAp and β toothpaste containing F, Mg, Sr-carbonate substituted n-HAp)), as assessed by EDS. Data are presented as mean (±1 SD); different letters indicate, for each element, significant differences between groups (Tukey test, $p < 0.05$).

Element	Control Enamel	Control RBC	α Toothpaste	β Toothpaste	Enamel+ α Toothpaste	Enamel+ β Toothpaste	RBC+ α Toothpaste	RBC+ β Toothpaste
C	6.19(0.32) c	33.36(0.49) a	26.08(2.01) b	25.53(1.10) b	5.49(0.75) c	6.12(0.20) c	32.77(1.86) a	33.96(0.34) a
O	38.21(1.38) a,b	33.04(0.33) c,d	43.75(3.23) a	42.69(0.38) a	40.12(2.93) a,b	36.06(1.78) b,c,d	32.89(2.21) d	32.63(0.60) c,d
Ca	38.71(0.90) a	0.00(0.00) c	12.12(2.22) b	9.80(0.40) b	36.97(2.30) a	39.13(0.74) a	1.26(0.12) c	0.30(0.13) c
P	16.31(0.45) a	0.00(0.00) c	4.95(0.95) b	4.93(0.42) b	15.78(1.39) a	16.59(1.20) a	0.76(0.17) c	0.12(0.06) c
Al	0.03(0.04) c	1.46(0.04) a,b	0.29(0.05) b,c	0.32(0.14) b,c	0.09(0.04) c	0.07(0.05) c	1.31(0.78) a,b	1.87(0.03) a
Si	0.08(0.02) d	20.50(0.32) a	9.36(1.81) c	12.12(0.59) b	0.49(0.15) d	0.39(0.04) d	20.23(0.69) a	20.24(0.64) a
Ba	0.00(0.00) c	11.37(0.56) a	0.00(0.00) c	0.00(0.00) c	0.00(0.00) c	0.00(0.00) c	10.52(0.39) b	10.74(0.09) a,b
F	0.00(0.00) b	0.00(0.00) b	0.00(0.00) b	0.18(0.08) a	0.00(0.00) b	0.04(0.02) b	0.00(0.00) b	0.00(0.00) b
Mg	0.18(0.01) b,c	0.00(0.00) d,e	0.00(0.00) d,e	0.35(0.01) a	0.13(0.03) c,d	0.28(0.11) a,b	0.00(0.00) e	0.03(0.01) c,d,e
Zn	0.05(0.06) c	0.00(0.00) c	0.27(0.05) a	0.00(0.00) c	0.17(0.04) a,b	0.05(0.06) c	0.09(0.03) b,c	0.00(0.00) c
Sr	0.00(0.00) c	0.00(0.00) c	0.00(0.00) c	0.93(0.09) a	0.00(0.00) c	0.38(0.10) b	0.00(0.00) c	0.00(0.00) c
Na	0.02(0.03) d	0.00(0.00) d	1.09(0.01) a	1.38(0.01) a	0.52(0.06) b,c	0.71(0.23) b	0.09(0.10) d	0.21(0.04) c,d
S	0.00(0.00) b	0.00(0.00) b	0.25(0.06) a	0.30(0.00) a	0.00(0.00) b	0.00(0.00) b	0.00(0.00) b	0.00(0.00) b
Cl	0.21(0.05) b,c	0.03(0.02) d	0.30(0.07) b	1.08(0.13) a	0.24(0.03) b	0.22(0.01) b,c	0.07(0.01) d	0.09(0.02) c,d
K	0.00(0.00) c	0.00(0.00) c	1.56(0.02) a	0.92(0.06) b	0.00(0.00) c	0.00(0.00) c	0.07(00.09) c	0.02(0.00) c

3.2. Microbiological Results

The effect of the tested toothpastes on EC and BF, expressed as mean OD values ± 1 standard error, according to test setup are shown in Figures 4 and 5. ANOVA results showed an overall significant influence of the tested toothpastes on both EC and BF, with a significant interaction between the factors, toothpaste, and microbiological model ($p = 0.017$). No significant interactions were found between toothpaste and incubation time, adherence surface, or test setup, meaning that the toothpastes had the same effect on microbial viability both in EC and BF, independently from the type of bioreactor used. Most importantly, the tested toothpastes had the same effect on microbial viability, both on enamel and RBC surfaces. On the contrary, ANOVA highlighted a significant interaction ($p < 0.0001$) among adherence surface, test setup, microbiological model, and incubation time.

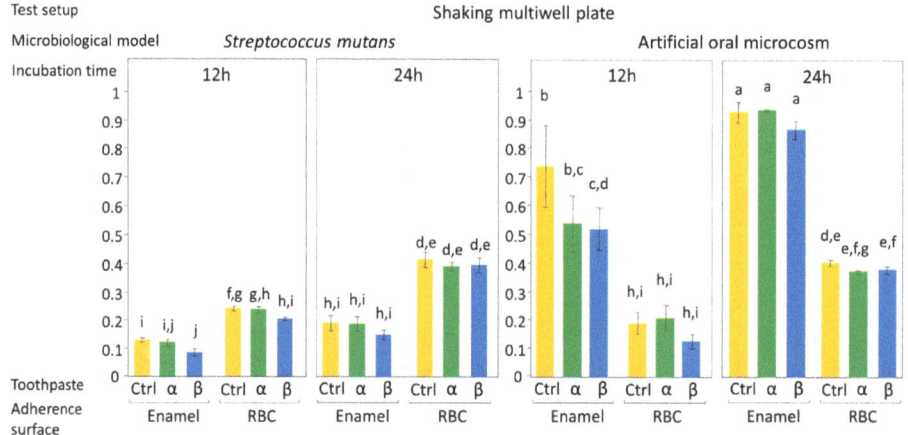

Figure 4. Biofilm formation on the surface of the tested specimens using the shaking multiwell plate test setup, according to the microbiological model (*S. mutans* monospecific biofilm or artificial oral microcosm, aerobically grown) and to the incubation time (12 h or 24 h). Sucrose-enriched sterile modified artificial saliva medium was used in all experiments. Low shear stress on specimens' surfaces was obtained by an orbital incubator, to simulate oral conditions during the night. Moreover, the closed system setup allows a progressive increase in microorganism catabolites and antimicrobial agents released from the surfaces. Results of viable biomass assay are expressed as mean OD ± SE. Different superscript letters indicate significant differences between groups (student's test, $p < 0.05$). α toothpaste contains Zn-carbonate substituted n-HAp; β toothpaste contains F, Mg, Sr-carbonate substituted n-HAp while Ctrl group was brushed with distilled water. β toothpaste significantly reduced the early colonization of the artificial oral microcosm on enamel surfaces when compared to the control.

Considering the results of the post-hoc test on the toothpaste factor, the artificial oral microcosm model showed significantly lower EC and BF on group β disks compared to the control, with group α disks showing lower EC and BF compared to control, but higher EC and BF compared to β toothpaste. Surprisingly, this pattern was found on both tested adherence surfaces, demonstrating an influence of toothpastes on EC and BF even over RBC surfaces. In the monospecific *S. mutans* model, the toothpastes did not significantly influence EC and BF.

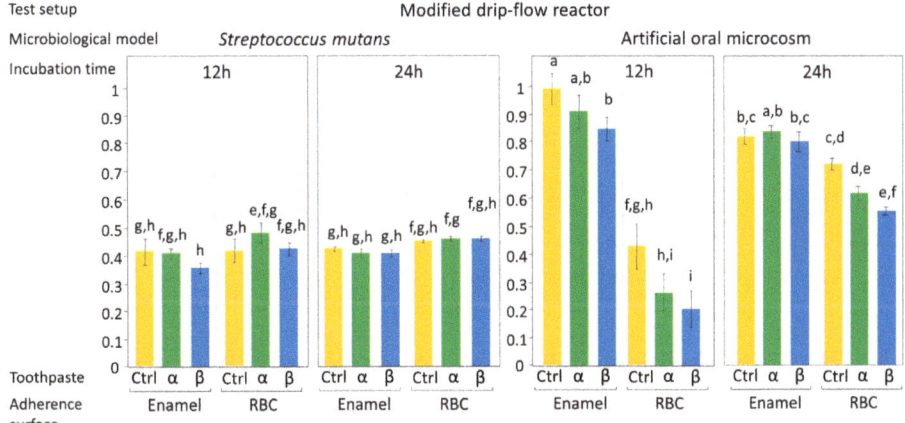

Figure 5. Biofilm formation on the surface of the tested specimens using the MDFR bioreactor test setup, according to the microbiological model (*S. mutans* monospecific biofilm or artificial oral microcosm, aerobically grown) and to the incubation time (12 h or 24 h). Sucrose-enriched sterile modified artificial saliva medium was used in all experiments, being pumped through the flow-cells of the bioreactor. High hydrodynamic stress conditions that occur during the daytime can thus be simulated. An elution of microorganism catabolites and antimicrobial agents released from the surfaces can also occur. Results of viable biomass assay are expressed as mean OD ± SE. Different superscript letters indicate significant differences between groups (student's test, $p < 0.05$). α toothpaste contains Zn-carbonate substituted n-HAp; β toothpaste contains F, Mg, Sr-carbonate substituted n-HAp while Ctrl group was brushed with distilled water. β toothpaste significantly reduced early colonization (enamel and RBC) and biofilm formation (RBC) of the artificial oral microcosm when compared to the control. Interestingly, the adherence surface showed a more considerable influence than toothpaste treatment on early colonization and biofilm formation, independent of the microbiological model or test setup applied. The effect of the toothpaste on RBC surfaces was not expected and opens the possibility to control microbial colonization on RBCs, and, ultimately, secondary caries prevention, by such treatments.

Considering the role of the adherence surface, the oral microcosm model showed significantly higher EC and BF on enamel than on the tested RBC in both test setups (shaking multiwell plate and MDFR). *S. mutans* showed significantly higher EC and BF on the tested RBC than on enamel in the shaking multiwell plate test setup and significantly higher BF on the tested RBC than on enamel in the MDFR.

Regarding incubation time, as expected, there was an overall significant increase in biomass over time. The artificial oral microcosm grown in the MDFR on enamel surfaces showed no significant difference in viable adherent biomass between 12 and 24 h incubation ($p = 0.0711$). Likewise, *S. mutans* grown in the MDFR on both adherence surfaces showed no significant difference in viable adherent biomass between 12 and 24 h incubation ($p = 0.5665$ and $p = 0.7691$ for enamel and RBC, respectively).

From the test setup point of view, MDFR showed an overall significantly higher EC and BF than the shaking multiwell plate, except for BF of artificial oral microcosm on enamel surfaces.

Considering the microbiological model, the artificial oral microcosm showed higher EC and BF than *S. mutans* on enamel surfaces in both test setups, whereas *S. mutans* showed higher EC and BF than the artificial oral microcosm in the shaking multiwell plate on RBC surfaces.

A set of additional enamel and RBC specimens were made and treated with the toothpastes or with the control, as described in the Methods section, to provide a morphological view of the microbial colonization. The most biologically meaningful result was found to be the one obtained using the MDFR test setup and the artificial oral microcosm model for 12 h (EC, Figure 5). Therefore, this particular situation was furthherly investigated using confocal laser-scanning microscopy and live/dead staining. The results are illustrated in Figure 6. Enamel specimens showed a uniform microbial layer colonizing all surfaces, with multilayered structures starting to develop. The aspect is in keeping with the viable biomass results, showing early colonization of β toothpaste < α toothpaste = control. On enamel specimens treated with β toothpaste, a slightly higher amount of dead cells and microcolonies can be seen. RBC specimens showed much lower colonization than enamel, showing early colonization of β toothpaste < α toothpaste < control. The surfaces of the specimens provided an intense red fluorescence that was not wholly subtracted by parallel scanning in the near-UV 405 nm channel. This artifact was not due to dead cells, but rather to autofluorescence coming from the RBC, and also to an affinity of the propidium iodide dye with RBC surfaces, as previously shown [22].

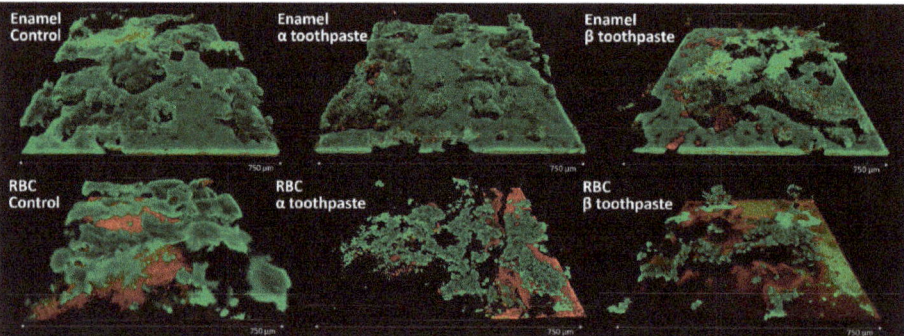

Figure 6. CLSM results of the adherence surfaces treated with the control and the tested toothpastes using the MDFR test setup and the artificial oral microcosm model for 12 h. Scans were analyzed 3D reconstructions obtained using Drishti software. Enamel surfaces provided a much higher early colonization than RBC surfaces. In the background of the enamel control specimen, a central microcolony shows a long tail detached from the surface and oriented horizontally downstream (to the right). This feature is typical of biofilms that develop on surfaces in the presence of relatively high shear stress and is also a means to colonize downstream surfaces rapidly. This feature demonstrates the good morphological resemblance of bioreactor-grown biofilms with in vivo ones. Enamel specimens treated with β toothpaste showed a higher amount of dead cells compared to the other groups. RBC specimens treated with β toothpaste showed the lowest early colonization overall, consistently with viable biomass results.

3.3. Fluoride Release Results

The fluoride release from the disks treated with the fluoride-hydroxyapatite containing toothpaste, and the fluoride content of the biofilms after 12 and 24 h are displayed in Figure 7. A significant albeit shallow release of fluoride was seen from enamel specimens treated with β toothpaste. Contrary to RBC specimens, enamel specimens showed biofilms containing significant amounts of fluoride at both incubation times. After 12 h, treatment with β toothpaste significantly decreased fluoride presence in the overlying biofilm compared to the control ($p < 0.001$).

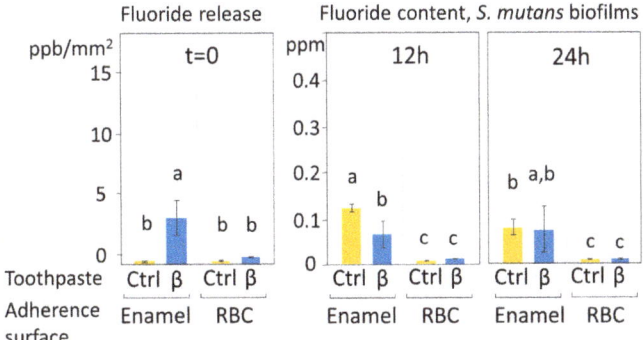

Figure 7. Immediate fluoride release (ppb/mm^2 ± SE) after disks' brushing and fluoride presence (ppm ± SE) in the biofilms grown over the disks' surfaces after 12 and 24 h. Different superscript letters indicate significant differences between groups (Student's test, $p < 0.05$). β toothpaste contains F, Mg, Sr-carbonate substituted n-HAp while Ctrl group was brushed with distilled water. β toothpaste did not increase the baseline fluoride presence in biofilms grown over enamel surfaces despite showing a low but significant immediate fluoride release. The reduction in fluoride presence seen in enamel specimens after 12 h of incubation when compared to the control may be due to an uptake of fluoride by the tested toothpaste.

4. Discussion

Nano-hydroxyapatite is considered one of the most promising bioactive materials with a broad spectrum of applications in preventive and restorative dentistry. N-HAp particles have been demonstrated to have a similar structure to natural enamel apatite crystals and can remineralize dental hard tissues [15,23–25]. Non-substituted n-HAp is also used for desensitizing and anti-erosion purposes [26,27]. The synthesis of nano-scaled zinc carbonate or magnesium-strontium carbonate hydroxyapatite ($ZnCO_3$/n-HAp, Mg-$SrCO_3$/n-HAp) represented significant progress in this field, allowing the introduction of n-HAp-based toothpastes with remineralization and antibacterial potential [28]. Different substituted n-Hap, coupled to other compounds with antibacterial activity, such as Zn PCA, chitosan, and fluoride, have been developed to achieve the goal of the remineralization of dental hard tissues [4,29–32]. Nevertheless, very few data are available in the literature about the antibacterial activity of these compounds, which is important in caries prevention because of the impact of oral biofilms in the dynamics of carious lesion formation.

Biofilms produce organic acids that cause the demineralization of the hard tissues. Moreover, biofilms can act as a reservoir of calcium and phosphate for remineralization, but also as a diffusion barrier for other compounds, such as fluoride, thus reducing the efficacy of some caries-preventive agents [16]. From a microbiological point of view, n-HAp-containing products might allow adsorbed nanoparticles to interact with bacterial cells, reducing adherence to hard tissue surfaces, thereby reducing biofilm formation.

The current in vitro study compared the effect of two substituted n-HAp-based toothpastes on early colonization and biofilm formation. Teeth specimens treated with the α toothpaste showed no significant differences in EC and BF compared to the control and β group, whereas specimens treated with β toothpaste showed a significantly lower EC and BF compared to the control (Figures 4 and 5). Both toothpaste formulations contained doping agents having antibacterial activity, namely zinc in α toothpaste and fluoride and strontium in β toothpaste, suggesting that n-HAp doping agents, rather than n-HAp itself, play a crucial role in determining the antibacterial properties of these compounds. Furthermore, fluoride and strontium have been shown to exhibit synergistic antibacterial performance [33,34]. EDS data (Table 2) showed that the zinc and fluoride concentrations were in the same range, while strontium in the β toothpaste was four times higher. A possible explanation of these results could be that the association of fluoride and strontium was more effective in reducing EC and BF than zinc alone at the tested concentrations. Nevertheless, fluoride release (Figure 7), which was only found on brushed enamel surfaces, did not help in explaining the reduced EC and BF by β toothpaste on RBC surfaces. Furthermore, despite a significant initial release of fluoride, the amount of this ion was not increased in biofilms grown over enamel and RBC surfaces treated with the β toothpaste, confirming the fact that biofilms do not tend to accumulate fluoride [16].

EDS data also showed the presence of calcium and phosphate on the surfaces of RBC specimens treated with both toothpastes, as well as traces of zinc on specimens treated with α toothpaste and magnesium on specimens treated with β toothpaste (Figure 2 and Table 1). It is noteworthy that toothpaste remnants could be associated with alumina and barium glass fillers rather than silica particles. This observation, needing further confirmation, opens the possibility for such compounds to actively interact with RBC surfaces with the effect of reducing or preventing secondary caries occurrence.

Regarding the influence of the adherence surface on the antibacterial activity of substituted n-HAp, as the surface features can deeply influence biological behavior, the roughness of all specimens was reduced below 0.2 µm Ra [35] to exclude the possible influence of this parameter on EC and BF. The two tested surfaces were selected considering their buffering capacity: a buffering surface (enamel) and a non-buffering one (RBC). The non-buffering surface was selected because it has been demonstrated to promote the growth of biofilms with a high concentration of cariogenic bacteria [17]. Our data showed that the adherence surfaces did not influence the effect of the tested toothpastes on EC and BF. This result was observed for both *S. mutans* and the oral microcosm, indicating that, unexpectedly, both substituted n-HAps had the same effect on buffering and non-buffering surfaces independent of their chemical composition, suggesting a similar interaction of the active components with the specimen surface. EDS results confirm the presence of active compounds from both toothpastes on both adherence surfaces. This presence was seen to a larger extent on enamel compared to RBC. However, the presence of such active compounds on the tested RBC surface, as seen in the present study, is very interesting from the perspective of controlling EC and BF on these surfaces, with the final aim of reducing secondary caries occurrence. Indeed, small but accumulative effects can prevent dysbiotic changes in dental biofilms and help maintain a beneficial oral microbiome [36]. In this sense, the use of this kind of active principles might be an efficient means of reducing secondary caries occurrence by actively promoting our natural microbiota and reducing the impact of the drivers of dysbiosis, such as, for instance, the presence of a non-buffering surface.

Two test setups were used, shaking multiwell plates and a drip-flow bioreactor to simulate clinical situations as close as possible, avoiding the variabilities that characterize in vivo studies. The first test setup was designed to evaluate EC and BF under low hydrodynamic stress conditions, for instance, during the night. Moreover, the first setup favors a progressive increase in microorganism catabolites and antimicrobial agents released from the surfaces. The second one simulated relatively high hydrodynamic stress conditions that occur during the daytime. *S. mutans* showed lower EC and BF in the first test setup compared to the second one, probably due to the limited amount of nutrients available and the accumulation of metabolic acids in the culture medium.

From the microbiological model point of view, the approach simulated as much as possible the clinical situation by testing an oral microcosm in addition to a monospecific *S. mutans* biofilm. The oral microcosm was used to evaluate the mixed oral flora developing on hard tissues, while the monospecific *S. mutans* model was used to assess the behavior of the cariogenic part of flora when treated with the tested compounds. In our study, *S. mutans* showed a higher affinity for RBC surfaces than for enamel, which is probably related to the buffering capacity of the tested surfaces being a crucial factor for both EC and BF.

Still, the microbiological models and incubation conditions may not be an accurate replica of all the complex interactions taking place in the oral environment. However, in vitro simulations are useful since they can push microenvironmental conditions to levels that might be difficult to control, or even unethical if replicated in vivo. The culture medium was made to have a high content in sucrose, therefore promoting the growth of fermentative Streptococcus as well as the acidogenic species in the mixed inoculum. The aim was to provide a shift in microbial composition towards a pathogenic biofilm, to test the activity of the toothpastes in a harsh environment. An active principle such as substituted n-HAp is especially useful if it can provide a controlled release of ions over time and, especially, in response to defined environmental conditions. A fermentative biofilm can reach pH values lower than 4.0, promoting the demineralization of enamel (critical pH ≈ 5.5). n-HAp is known to be very stable at neutral pH. Therefore, an acidogenic microbial challenge as the one produced in the present setups helped in understanding the effect of the toothpastes when challenged by pathogenic biofilms.

Considering incubation time, 12 h was selected as toothbrushing twice a day is the most frequently advised routine, while a 24 h incubation can provide a sufficient amount of time to evaluate BF. The results showed that the most relevant differences between the tested toothpaste groups were found on *S. mutans* EC, which means after 12 h of incubation. In a translational sense, this may suggest that the tested toothpastes provided the best effect when used twice a day, rather than a single time.

5. Conclusions

A toothpaste treatment based on substituted n-HAp showed an in vitro antibacterial effect both on natural and RBC surfaces, which was more pronounced after 12 h than after 24 h. On RBC surfaces, toothpaste remnants seemed to show selective affinity with some particles, such as alumina and barium glass fillers, rather than silica particles. This observation is worthy of future investigations aimed at clarifying the interactions between substituted n-HAp and RBC fillers from the perspective of secondary caries prevention. This phenomenon provides a basis to design clinical studies aimed to confirm in vivo the preventive efficacy of these active principles.

Author Contributions: Conceptualization, A.C.I. and E.B.; Data curation, A.C.I., G.C. and M.O.; Formal analysis, M.O., F.G.-G., and E.B.; Funding acquisition, E.B.; Investigation, A.C.I. and G.C.; Methodology, A.C.I., G.C. and M.O.; Project administration, E.B.; Resources, E.B.; Software, G.C. and M.O.; Supervision, A.C.I. and E.B.; Validation, A.C.I.; Visualization, E.B., F.G.-G.; Writing—original draft, A.C.I., G.C. and M.O.; Writing—review & editing, F.G.-G., and E.B. All authors have read and agreed to the published version of the manuscript.

Funding: This research received no external funding.

Acknowledgments: The authors are grateful to Massimo Tagliaferro and to Nanovision S.R.L. for providing the SEM-EDS equipment.

Conflicts of Interest: The authors declare no conflict of interest.

References

1. Flemming, T.F.; Beikler, T. Control of oral biofilms. *Periodontology 2000* **2011**, *55*, 9–15. [CrossRef]
2. Rosier, B.T.; Marsh, P.D.; Mira, A. Resilience of the oral microbiota in health: Mechanisms that prevent dysbiosis. *J. Dent. Res.* **2018**, *97*, 371–380. [CrossRef] [PubMed]
3. Marsh, P.D. Plaque as a biofilm: Pharmacological principles of drug delivery and action in the sub-and supragingival environment. *Oral Dis.* **2003**, *9*, 16–22. [CrossRef] [PubMed]
4. Allaker, R.P. The use of nanoparticles to control oral biofilm formation. *J. Dent. Res.* **2010**, *89*, 1175–1186. [CrossRef] [PubMed]
5. Besinis, A.; De Peralta, T.; Handy, R.D. Inhibition of biofilm formation and antibacterial properties of a silver nano-coating on human dentine. *Nanotoxicology* **2014**, *8*, 745–754. [CrossRef]
6. Hannig, M.; Hannig, C. Nanomaterials in preventive dentistry. *Nat. Nanotechnol.* **2010**, *5*, 565–569. [CrossRef]
7. Allaker, R.P.; Memarzadeh, K. Nanoparticles and the control of oral infections. *Int. J. Antimicrob. Agents* **2014**, *43*, 95–104. [CrossRef]
8. Beyth, N.; Houri-Haddad, Y.; Domb, A.; Khan, W.; Hazan, R. Alternative antimicrobial approach: Nano-antimicrobial materials. *Evid.-Based Complement. Altern. Med.* **2015**, *2015*, 246012. [CrossRef]
9. Cochrane, N.J.; Cai, F.; Huq, N.L.; Burrow, M.F.; Reynolds, E.C. New approaches to enhanced remineralization of tooth enamel. *J. Dent. Res.* **2010**, *89*, 1187–1197. [CrossRef]
10. Nongonierma, A.B.; Fitzgerald, R.J. Biofunctional properties of caseinphosphopeptides in the oral cavity. *Caries Res.* **2012**, *46*, 234–267. [CrossRef]

11. Grychtol, S.; Basche, S.; Hannig, M.; Hannig, C. Effect of cpp/acp on initial bioadhesion to enamel and dentin in situ. *Sci. World J.* **2014**, *2014*, 512682. [CrossRef]
12. Najibfard, K.; Ramalingam, K.; Chedjieu, I.; Amaechi, B.T. Remineralization of early caries by a nano-hydroxyapatite dentifrice. *J. Clin. Dent.* **2011**, *22*, 139–143. [PubMed]
13. Lelli, M.; Putignano, A.; Marchetti, M.; Foltran, I.; Mangani, F.; Procaccini, M.; Roveri, N.; Orsini, G. Remineralization and repair of enamel surface by biomimetic Zn-carbonate hydroxyapatite containing toothpaste: A comparative in vivo study. *Front. Physiol.* **2014**, *5*, 333–340. [CrossRef] [PubMed]
14. Lu, K.L.; Meng, X.C.; Zhang, J.X.; Li, X.Y.; Zhou, M.L. Inhibitory effect of synthetic nano-hydroxyapatite on dental caries. *Key Eng. Mater.* **2007**, *336*, 1538–1541. [CrossRef]
15. Tschoppe, P.; Zandim, D.L.; Martus, P.; Kielbassa, A.M. Enamel and dentine remineralization by nano-hydroxyapatite toothpastes. *J. Dent.* **2011**, *39*, 430–437. [CrossRef]
16. Zhang, M.; He, L.B.; Exterkate, R.A.M.; Cheng, L.; Li, J.Y.; ten Cate, J.M.; Crielaard, W.; Deng, D.M. Biofilm layers affect the treatment outcomes of NaF and nano-hydroxyapatite. *J. Dent. Res.* **2015**, *94*, 602–607. [CrossRef]
17. Nedeljkovic, I.; De Munck, J.; Slomka, V.; Van Meerbeek, B.; Teughels, W.; Van Landuyt, K.L. Lack of buffering by composites promotes shift to more cariogenic bacteria. *J. Dent. Res.* **2016**, *95*, 875–881. [CrossRef]
18. Guggenheim, B.; Giertsen, E.; Schupbach, P.; Shapiro, S. Validation of an in vitro biofilm model of supragingival plaque. *J. Dent. Res.* **2001**, *80*, 363–370. [CrossRef]
19. Brambilla, E.; Ionescu, A.C.; Cazzaniga, G.; Ottobelli, M.; Samaranayake, L.P. Levorotatory carbohydrates and xylitol subdue *Streptococcus mutans* and *Candida albicans* adhesion and biofilm formation. *J. Basic Microbiol.* **2016**, *56*, 480–492. [CrossRef]
20. Ionescu, A.; Brambilla, E.; Wastl, D.S.; Giessibl, F.J.; Cazzaniga, G.; Schneider-Feyrer, S.; Hahnel, S. Influence of matrix and filler fraction on biofilm formation on the surface of experimental resin-based composites. *J. Mater. Sci. Mater. Med.* **2015**, *26*, 58. [CrossRef]
21. Cazzaniga, G.; Ottobelli, M.; Ionescu, A.C.; Paolone, G.; Gherlone, E.; Ferracane, J.L.; Brambilla, E. In vitro biofilm formation on resin-based composites after different finishing and polishing procedures. *J. Dent.* **2017**, *67*, 43–52. [CrossRef] [PubMed]
22. Ionescu, A.C.; Brambilla, E.; Hahnel, S. Does recharging dental restorative materials with fluoride influence biofilm formation? *Dent. Mater.* **2019**, *35*, 1450–1463. [CrossRef] [PubMed]
23. Huang, S.B.; Gao, S.S.; Yu, H.Y. Effect of nano-hydroxyapatite concentration on remineralization of initial enamel lesion in vitro. *Biomed. Mater.* **2009**, *4*, 034104. [CrossRef] [PubMed]
24. Li, L.; Pan, H.; Tao, J.; Xu, X.; Mao, C.; Gu, X.; Tang, R. Repair of enamel by using hydroxyapatite nanoparticles as the building blocks. *J. Mater. Chem.* **2008**, *18*, 4079–4084. [CrossRef]
25. Roveri, N.; Battistella, E.; Bianchi, C.L.; Foltran, I.; Foresti, E.; Iafisco, M.; Lelli, M.; Naldoni, A.; Palazzo, B.; Rimondini, L. Surface enamel remineralization: Biomimetic apatite nanocrystals and fluoride ions different effects. *J. Nanomater.* **2009**, *2009*, 746383. [CrossRef]
26. Aykut-Yetkiner, A.; Attin, T.; Wiegand, A. Prevention of dentine erosion by brushing with anti-erosive toothpastes. *J. Dent.* **2014**, *42*, 856–861. [CrossRef]
27. De Melo Alencar, C.; de Paula, B.L.F.; Ortiz, M.I.G.; Magno, M.B.; Silva, C.M.; Maia, L.C. Clinical efficacy of nano-hydroxyapatite in dentin hypersensitivity: A systematic review and meta-analysis. *J. Dent.* **2019**, *82*, 11–21. [CrossRef]
28. Orsini, G.; Procaccini, M.; Manzoli, L.; Giuliodori, F.; Lorenzini, A.; Putignano, A. A double-blind randomized-controlled trial comparing the desensitizing efficacy of a new dentifrice containing carbonate/hydroxyapatite nanocrystals and a sodium fluoride/potassium nitrate dentifrice. *J. Clin. Periodontol.* **2010**, *37*, 510–517. [CrossRef]
29. Brambilla, E.; Ionescu, A.; Cazzaniga, G.; Edefonti, V.; Gagliani, M. The influence of antibacterial toothpastes on in vitro *Streptococcus mutans* biofilm formation: A continuous culture study. *Am. J. Dent.* **2014**, *27*, 160–166.
30. Hannig, C.; Basche, S.; Burghardt, T.; Al-Ahmad, A.; Hannig, M. Influence of a mouthwash containing hydroxyapatite microclusters on bacterial adherence in situ. *Clin. Oral Investig.* **2013**, *17*, 805–814. [CrossRef]
31. Kong, M.; Chen, X.G.; Xing, K.; Park, H.J. Antimicrobial properties of chitosan and mode of action: A state of the art review. *Int. J. Food Microbiol.* **2010**, *144*, 51–63. [CrossRef] [PubMed]
32. Li, X.; Wang, J.; Joiner, A.; Chang, J. The remineralisation of enamel: A review of the literature. *J. Dent.* **2014**, *42*, S12–S20. [CrossRef]

33. Dabsie, F.; Gregoire, G.; Sixou, M.; Sharrock, P. Does strontium play a role in the cariostatic activity of glass ionomer?: Strontium diffusion and antibacterial activity. *J. Dent.* **2009**, *37*, 554–559. [CrossRef] [PubMed]
34. Guida, A.; Towler, M.R.; Wall, J.G.; Hill, R.G.; Eramo, S. Preliminary work on the antibacterial effect of strontium in glass ionomer cements. *J. Mater. Sci. Lett.* **2003**, *22*, 1401–1403. [CrossRef]
35. Bollen, C.M.; Lambrechts, P.; Quirynen, M. Comparison of surface roughness of oral hard materials to the threshold surface roughness for bacterial plaque retention: A review of the literature. *Dent. Mater.* **1997**, *13*, 258–269. [CrossRef]
36. Marsh, P.D. In sickness and in health—What does the oral microbiome mean to us? An ecological perspective. *Adv. Dent. Res.* **2018**, *29*, 60–65. [CrossRef]

© 2020 by the authors. Licensee MDPI, Basel, Switzerland. This article is an open access article distributed under the terms and conditions of the Creative Commons Attribution (CC BY) license (http://creativecommons.org/licenses/by/4.0/).

Article

Anti-Bacterial Properties and Biocompatibility of Novel SiC Coating for Dental Ceramic

Samira Esteves Afonso Camargo [1], Azeem S. Mohiuddeen [1], Chaker Fares [2], Jessica L. Partain [2], Patrick H. Carey IV [2], Fan Ren [2], Shu-Min Hsu [1], Arthur E. Clark [1] and Josephine F. Esquivel-Upshaw [1,*]

1. Department of Restorative Dental Sciences, Division of Prosthodontics, University of Florida College of Dentistry, Gainesville, FL 32610, USA; scamargo@dental.ufl.edu (S.E.A.C.); AMohiuddeen@dental.ufl.edu (A.S.M.); shuminhsu@ufl.edu (S.-M.H.); BCLARK@dental.ufl.edu (A.E.C.)
2. Department of Chemical Engineering, University of Florida Herbert Wertheim College of Engineering, Gainesville, FL 32611, USA; c.fares@ufl.edu (C.F.); jpartain3@ufl.edu (J.L.P.); careyph@ufl.edu (P.H.C.IV); fren@che.ufl.edu (F.R.)
* Correspondence: JESQUIVEL@dental.ufl.edu

Received: 14 April 2020; Accepted: 9 May 2020; Published: 20 May 2020

Abstract: A 200 nm plasma-enhanced chemical vapor-deposited SiC was used as a coating on dental ceramics to improve anti-bacterial properties for the applications of dental prosthesis. A thin SiO_2 (20 nm) in the same system was deposited first, prior to SiC deposition, to improve the adhesion between SiC to dental ceramic. Silane and methane were the precursors for SiC deposition, and the SiO_2 deposition employed silane and nitrous oxide as the precursors. SiC antimicrobial activity was evaluated on the proliferation of biofilm, *Streptococcus sanguinis*, and *Streptococcus mutans* on SiC-coated and uncoated dental ceramics for 24 h. The ceramic coating with SiC exhibited a biofilm coverage of 16.9%, whereas uncoated samples demonstrated a significantly higher biofilm coverage of 91.8%, measured with fluorescence and scanning electron microscopic images. The cytotoxicity of the SiC coating was evaluated using human periodontal ligament fibroblasts (HPdLF) by CellTiter-BlueCell viability assay. After 24 h of HPdLF cultivation, no obvious cytotoxicity was observed on the SiC coating and control group; both sets of samples exhibited similar cell adhesion and proliferation. SiC coating on a ceramic demonstrated antimicrobial activity without inducing cytotoxic effects.

Keywords: antimicrobial; ceramic; coating; silicon carbide; cytotoxicity

1. Introduction

Improving the dental restoration's lifetime through careful selection of materials has been the focus of many previous studies, where traditionally different alloys and metal systems have been evaluated for biocompatibility and antibacterial properties [1–10]. Although the majority of these approaches have used methods involving the bulk system, use of thin films can offer an easier avenue to maintaining the integrity of the restoration while grafting unique properties to selective areas where the film is applied. One of the novel thin films of interest is silicon carbide due to the ease of deposition, lack of reactivity to the oral environment, and high strength [11–15] of this material. Silicon carbide has found significant success in two major areas: semiconductors and ceramics. Silicon carbide can easily be deposited by plasma deposition systems to form conformal films with tight control over thickness from nanometers to micrometers [16–20]. Silicon carbide may present a biosafe path to protect restorative surfaces from bacterial adhesion and degradation without compromising the bulk properties of traditional dental material technology.

In recent reports, the development and optimization of SiC-based coatings to minimize surface corrosion and wear on glass ceramic veneers has been demonstrated [14,21]. Hsu et al. [21] reported that

SiC-coated dental ceramics demonstrated a significant reduction in corrosion when placed in caustic environments that represent the extreme conditions of the oral cavity. Chen et al. [22] demonstrated that the thickness of the SiC coating can also be optimized to match any tooth shade required within the dental field.

As new dental materials such as SiC are developed, analyzing their anti-bacterial and biocompatibility properties is imperative. The biofilms that adhere to dental materials can be described as a micro-ecosystem formed by various species of microorganisms, surrounded by a protein extracellular matrix and polysaccharides generated by them [23]. These biofilms are responsible for plaque formation, leading to dental caries, periodontal disease, peri-implantitis, and enamel demineralization. They can adhere to both biotic and abiotic surfaces, such as prostheses, implants, or host tissues [24]. Bacterial adhesion to a substrate is a multifactorial process that involves surface properties inherent to both the bacteria and the biomaterial [25,26]. Bacteria present in the oral cavity naturally tend to adhere to ceramic materials and also to the interface between tooth and restoration [27,28]. They adhere on the cervical third of the proximal surface, and along the gingival margin where they are protected from mechanical action, creating a well-structured biofilm [29].

Understanding how these biofilms adhere to surfaces is crucial in qualifying new dental materials. On solid surfaces such as enamel, the ability to aggregate, the order of appearance of the microorganisms [30,31], and the environment [23] are essential factors in oral biofilm formation. The composition of the material, as well as the surface structure, can influence the initial bacterial adhesion and compromise dental health. Poor oral hygiene or a compromised immune system can also be contributing factors that aid colonization [32]. SiC coating provides a smoother surface after corrosion, which could minimize plaque accumulation, secondary caries, and periodontal inflammation from occurring [21].

Although SiC has been studied and applied as a viable biomaterial in the literature [15–18], no work to date has studied the feasibility of SiC as a dental material. One of the precluding tasks to determine if SiC can be utilized as a dental material is to quantify how the coating affects monomicrobial and polymicrobial biofilm adhesion, as well as the cytotoxicity for human cells.

The objective of this study was to evaluate the effect of SiC coating on monomicrobial and polymicrobial biofilm adhesion, as well as determine the viability of human cells in contact with the coating surface. Understanding the biological and mechanical behavior of dental materials in the oral environment is crucial to developing predictable restorations, and thus we proposed the hypotheses that SiC coatings on ceramic will (1) decrease monomicrobial and polymicrobial biofilm adhesion on the surface of the ceramic, and (2) exhibit biocompatibility without signs of cytotoxicity.

2. Material and Methods

2.1. Ceramic Sample Preparation

Fluorapatite glass-ceramic disks (Ivoclar Vivadent AG, Schaan, Liechtenstein, 12.6 × 2 ± 0.2 mm) were polished using silicon carbide abrasive paper (Carbimet, Buehler, Lake Bluff, IL, USA) and subsequently cleaned using the following procedure: (1) acetone cleaning in an ultrasonic bath, (2) isopropyl alcohol rinse followed by a compressed nitrogen drying step, and (3) ozone treatment to remove surface carbon contamination.

2.2. SiC Coating

After the ceramic disks were polished and cleaned, half of the disks (40/80) were coated with silicon dioxide (SiO_2) and silicon carbide (SiC) on both sides. These depositions were done within a plasma-enhanced chemical vapor deposition system (PECVD, PlasmaTherm 790, Saint Petersburg, FL, USA). The corresponding thicknesses of the deposited SiO_2 and SiC films were 20 nm and 200 nm, respectively. The SiO_2 film was utilized as an adhesion layer between the ceramic and SiC, and was optimized in a previous report [21]. The precursors for the SiO_2 film were silane (SiH_4) and nitrous oxide (N_2O). Following the SiO_2 deposition, methane (CH_4) and silane (SiH_4) were the precursors used for silicon carbide deposition. The deposition temperature for both films was 300 °C, and the deposition rates for SiO$_2$ was 330 Å/min and for SiC was 170 Å/min [14].

2.3. Bacteria Growth

To study the antimicrobial properties of the coated and uncoated disks, monomicrobial and polymicrobial strains (ATCC—American Type Culture Collection) of *Streptococcus mutans* (ATCC 35688) and *Streptococcus sanguinis* (ATCC BAA-1455) were utilized. *S. mutans* and *S. sanguinis* were used for this study because these bacteria are the early colonizers of initial supragingival biofilm in the first 8 h and are present in greater quantities in the oral biofilm. The strains were grown onto agar plates with brain heart infusion broth (BHI, Himedia) for 24 h at 37 °C. After the growth, each microbial suspension was centrifuged at 4700 rpm for 10 min (Centrifuge model MPW-350, Beckman Coulter, Indianapolis, IN, USA) to separate the supernatant and microbial suspension. The centrifuge process was performed twice to minimize the quantity of debris. After separation, the microbial suspension adjusted to 10^7 colony-forming units (CFU)/mL and was ready to be placed onto the coated and uncoated glass-ceramic disks.

2.4. Experimental Design

The two groups in this study were (i) non-coated fluorapatite glass-ceramic disks as the reference group, and (ii) SiO$_2$/SiC-coated fluorapatite glass-ceramic disks (SiC-disks). Twenty-four ceramic specimens of each group were sterilized in an autoclave (121 °C, 60 min) and then placed into individual sterile plates. A total of 12 samples from each group were used to study the effects of monomicrobial suspensions, and the remaining 12 samples were used to study polymicrobial suspensions (*S. sanguinis* and *S. mutans*). For all samples, 100 uL of mono- or polymicrobial suspension was added to each plate containing a coated or uncoated disk, along with 1 mL of BHI broth to adequately cover the full disk.

To study the biocompatibility and cytotoxicity of the SiC coating, human periodontal ligament fibroblasts (HPdLF, Lonza, Basel, Switzerland) were cultured and subsequently placed onto coated and uncoated samples. The HPdLF were cultured at 37 °C in a growth media consisting of Dulbecco's modified Eagle's medium (DMEM) with 10% fetal bovine serum and 1% penicillin/streptomycin. After the cells were cultured, they were seeded onto 16 sterilized ceramic disks from each group, each within its own sterilized well plate. The approximate concentration of cells after seeding was 20,000 cells per well.

The quantitative data were presented as the means ± standard deviations. The statistical differences were compared using one-way ANOVA and Tukey's test (Graph Prism 6.0, GraphPad Software Inc., San Diego, CA, USA). For all analyses, statistical significance was pre-set at $\alpha = 0.05$.

2.5. Characterization Techniques

2.5.1. Scanning Electron Microscopy

Non-coated, SiC-coated, and glazed disks were examined under scanning electron microscopy using a MAICE system (JEOL JSM-6400 Scanning Electron Microscope, JEOL LTD, Tokyo, Japan) to characterize surface roughness. After the coated and non-coated disks were incubated for 24 h, the culture medium was removed, and the polymicrobial biofilm adhered to the samples was fixed in a solution of 3% glutaraldehyde, 0.1 mol/L sodium cacodylate, and 0.1 mol/L sucrose for 45 min. Samples were soaked for 10 min in a buffer solution of 0.1 mol/L sodium cacodylate and 0.1 mol/L sucrose. Sample surfaces and cells were processed in serial ethanol dehydrations for 10 min each and dehydrated in hexamethyldisilazane (HDMS) before being stored in a desiccator until SEM imaging. The samples were then sputter-coated with a palladium–gold alloy (Polaron SC 7620 Sputter Coater, Quorum Technologies, Laughton, East Sussex, United Kingdom) with a thickness of 10 nm to reduce charging effects during SEM analysis (10–15 mA, under a vacuum of 130 mTorr). After this, the SEM was operated at 5 kV, spot 3 to 6 (FEI NOVA 430, Hillsboro, OR, USA).

2.5.2. Atomic Force Microscopy (AFM)

Topographies of non-coated and coated with SiC samples were evaluated using an atomic force microscopy system (Bruker/Veeco/Digital Instruments NanoScope V, Billerica, MA, USA). The AFM was operated in tapping mode, using a silicon AFM probe (RTESP-300, Bruker, Billerica, MA, USA), with radius less than 10 nm, and resonance frequency between 200 and 400 kHz.

2.5.3. Water Contact Angle Measurements

The contact angle is described as the angle between the liquid–solid interface. The water contact angle (WCA) measurements were performed on the neutralized non-coated and coated SiC surfaces (neutralized up to pH 6.0). Static contact angles were assessed by the sessile drop method with a contact angle goniometer (Krüss DSA 10, Matthews, NC, USA), equipped with video capture. The automatic dosing feature of the DSA 10 dispensed a water drop on the non-coated and coated SiC surfaces, and the needle was manually withdrawn. Images were captured after contact of a droplet with the surface by a camera leveled with the surface. Contact angle measurements were analyzed by the circle fitting profile available with the DSA 10 imaging software. Three separate measurements were made on each sample at different locations.

2.5.4. Cytotoxicity Test

Cell viability were determined by a CellTiter-BlueCell Viability Assay (Promega G808A), which was used according to the manufacturer's instructions. After 24 h, 50 µL of CellTiter-Blue dye was added to ceramic specimens for every 500 µL of culture media, and samples were incubated for 6 h at 37 °C and 5% CO_2. Sample fluorescence was analyzed using a spectrophotometer (SmartSpec Plus, Bio-Rad, Hercules, CA, USA) at a wavelength of 600 nm, which generated density optic values.

2.5.5. Colony-Forming Units

After 24 h of incubation, the ceramic samples were removed and placed inside an Eppendorf tube with Ringer's solution (Sigma-Aldrich). The monomicrobial biofilm of *S. sanguinis* and *S. mutans* was disaggregated by an ultrasonic homogenizer (Sonopuls HD 2200–Bandelin Electronic, Berlin, Germany) for 30 s. The generated microbial suspension was serially diluted (1:10), and 100 µL was seeded on BHI agar for each respective sample. After 48 h of incubation, the concentration of colony-forming units (CFU) per milliliter was determined by visual counting.

2.5.6. Fluorescence Assay

After the incubation period, bacteria adhered to the samples were stained with SYTO 9 dye (Live/Dead BacLight Bacterial Viability Kit, ThermoFisher Scientific, Waltham, MA, USA) according to the manufacturer's instructions. Fluorescence images of the live bacteria were recorded in a fluorescence microscope (Zeiss Imager-A2, Jena, Germany) and analyzed by ImageJ software. Bacteria coverage percentages were averaged over five random areas on each filter specimen ($n = 8$).

The quantitative data are presented as the means ± standard deviations. The statistical differences were compared using one-way ANOVA and Tukey's test (Graph Prism 6.0, GraphPad Software Inc., San Diego, CA, USA). For all analyses, statistical significance was pre-set at $\alpha = 0.05$.

3. Results

3.1. Atomic Force Microscopy (AFM) and Scanning Electron Microscopy (SEM)

According to Figure 1, the AFM analysis showed an irregular surface topography in non-coated samples with roughness Rq 9.58. In contrast, SiC-coated samples were slightly smoother with Rq 7.98 ($p > 0.05$). Non-coated, SiC-coated, and glazed disks demonstrated significant qualitative differences in surface roughness (Figure 2A–F).

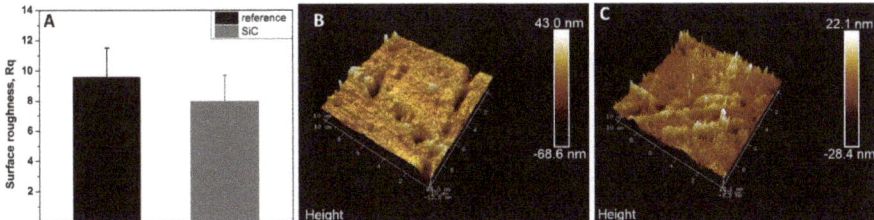

Figure 1. Roughness means values and standard deviation in non-coated (reference) and coated (ref SiC) samples (**A**). Atomic force microscopy (AFM) images of surface topography of non-coated (**B**) and coated SiC (**C**), with amplifications of 10 µm.

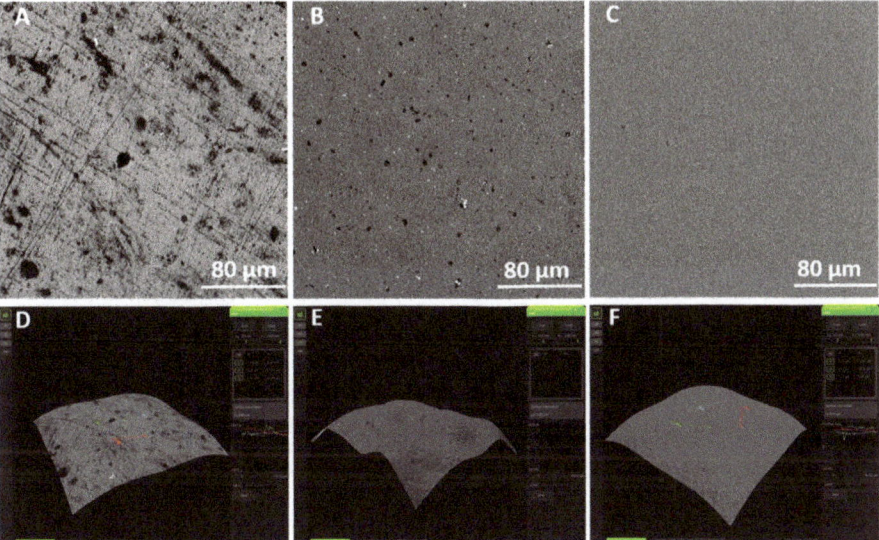

Figure 2. SEM images of roughness of non-coated and coated samples. (**A,D**) Non-coated fluorapatite disk, (**B,E**) glazed fluorapatite disk, and (**C,F**) SiC-coated fluorapatite disk.

3.2. Water Contact Angle Measurements (WCAs)

The outcomes indicate different WCAs between the non-coated and coated surfaces (Table 1). The coated SiC surface presented higher contact angles than the non-coated surface ($p > 0.05$).

Table 1. Mean of contact angle measurements on non-coated and coated SiC surfaces.

Ceramic Samples	Contact Angle (±SD)
Non-coated	46° ± 2°
Coated SiC	60° ± 1°

3.3. Bacterial Growth

After 24 h, the amount of polymicrobial biofilm (*S. mutans* and *S. sanguinis*) was less on the SiC-coated disks. The biofilm coverage on these disks was 16.9%, whereas uncoated reference samples showed a significantly higher biofilm coverage of 91.8% ($p < 0.0001$) (Figure 3).

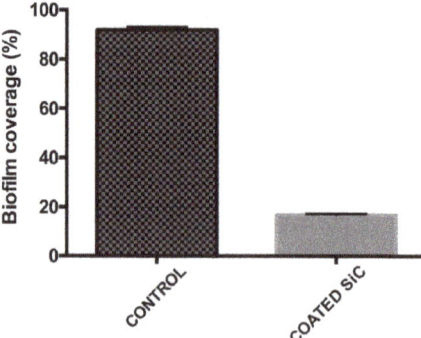

Figure 3. Live coverage of *Streptococcus mutans* and *Streptococcus sanguinis* after 24 h of culture on the non-coated (control) and coated surfaces.

The fluorescence images in Figure 4 demonstrate higher polymicrobial biofilm (S. mutans and S. sanguinis) formation on the control group than on the coated group. The SEM images confirmed the results from the live assays, showing a reduction in the number of adherent biofilms on the coated group for S. mutans and S. sanguinis after 24 h of culture (Figure 5).

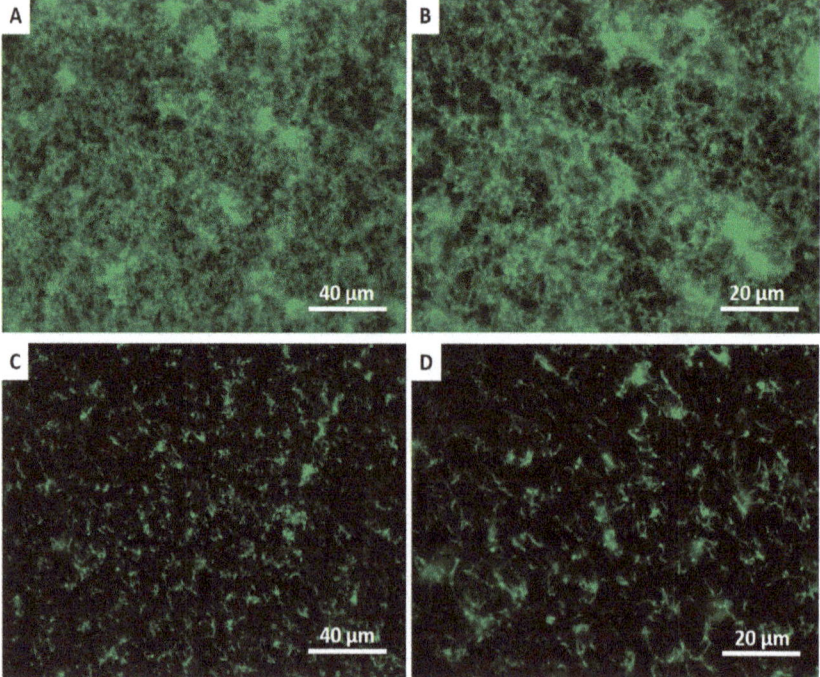

Figure 4. Live fluorescence images of *S. mutans* and *S. sanguinis* cultured for 24 h on the non-coated (**A,B**) and coated (**C,D**) surfaces. The cultures were stained with SYTO 9 to dye the living bacteria green.

There was a significant reduction in the number of CFU/mL after 24 h contact with ceramic SiC coating compared to the control group (Figure 6), showing $p = 0.0003$ for *S. sanguinis* and $p \leq 0.0001$ for *S. mutans*. The reduction of the monomicrobial biofilm of *S. sanguinis* and *S. mutans* were also similar ($p = 0.2528$).

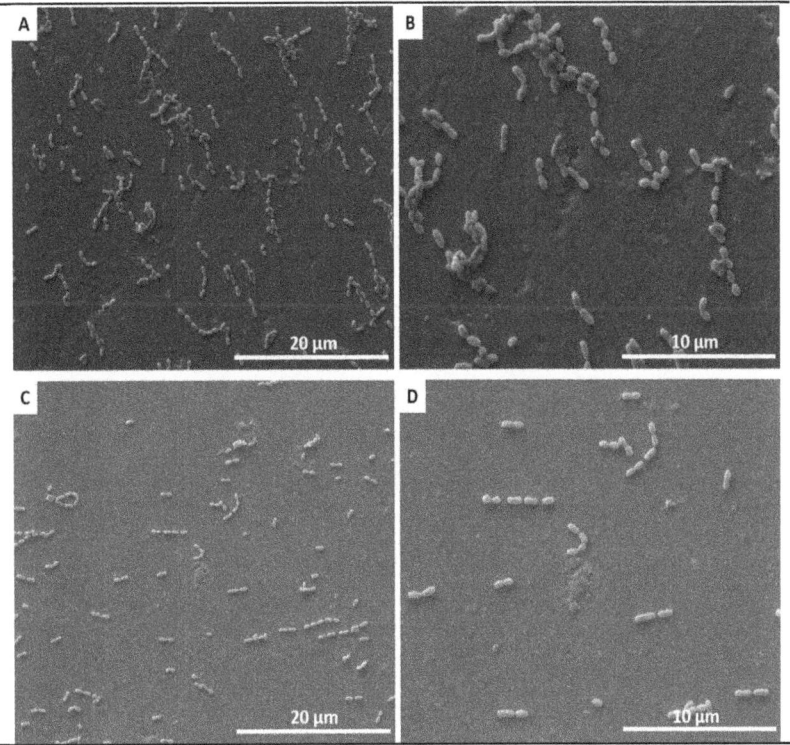

Figure 5. SEM adhesion of polymicrobial biofilm of *S. mutans* and *S. sanguinis* after 24 h of cultivation on non-coated (**A,B**) and coated (**C,D**) ceramic disks.

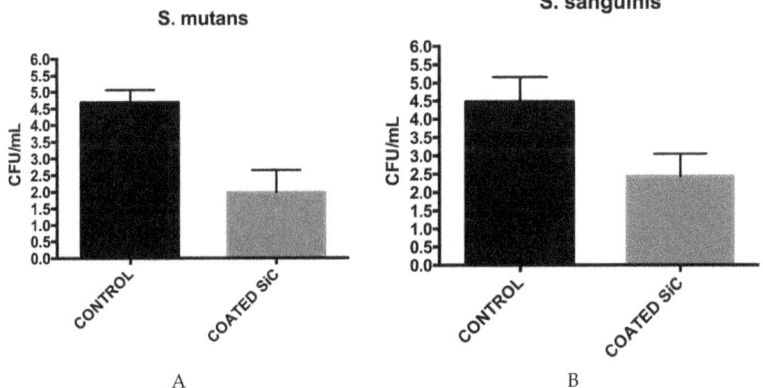

Figure 6. Mean (±SD) of colony-forming units (CFU)/mL of monomicrobial biofilms of *S. mutans* (**A**) and *S. sanguinis* (**B**) presented in the non-coated (control) and coated groups.

3.4. Biocompatibility Testing

In order to determine whether the SiC coating presented biocompatibility, the cytotoxicity was evaluated by the CellTiter-Blue assay. After 24 h of HPdLF cultivation on the samples, no obvious cytotoxicity was observed, as evidenced by the absorbance of the cells cultured on SiC coating being comparable to the control group ($p = 0.3904$) (Figure 7).

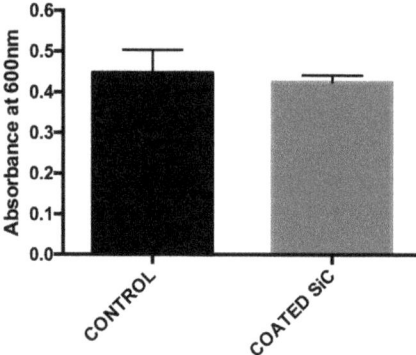

Figure 7. Cytotoxicity of the non-coated (control) and coated groups after 24 h of human periodontal ligament fibroblasts (HPdLF) culture assessed by CellTiter-Blue absorbance.

The SEM images showed the interaction of cell extensions with the SiC-coated and non-coated surfaces. SEM results demonstrated that cells adhered and covered the full surface of the disk after 24 h in culture. Additionally, the cell morphology was similar for cells cultured on SiC-coated and non-coated samples, where HPdLFs appeared oval-shaped and flattened on the surface (Figure 8).

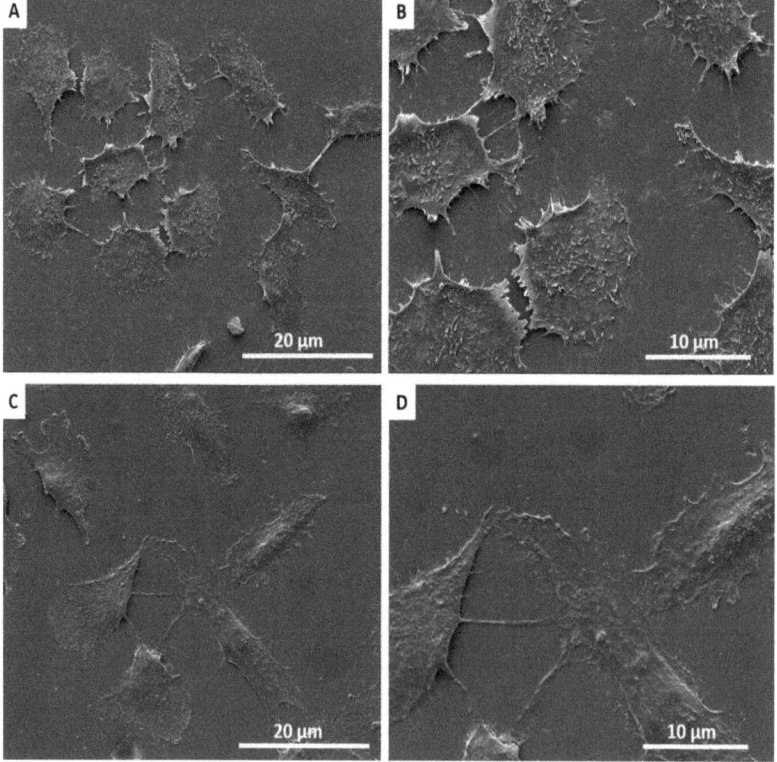

Figure 8. SEM images showing the adhesion of HPdLF on non-coated (**A**,**B**) and coated (**C**,**D**) surfaces after 24 h of incubation.

4. Discussion

The adhesion of streptococci to solid surfaces has been described as a two-step process, where the first step is influenced predominantly by macroscopic substratum properties such as surface roughness and surface free energy [33]. High substratum surface roughness has been shown to induce increased plaque formation in vivo [34,35], as the initial adhesion of microorganisms preferably starts at locations that provide shelter against shear forces. The surface roughness of differently prepared lithia disilicate ceramic restorations (pressed (Press), milled (CAD), fluorapatite layered (ZirPress/Ceram), and glazed (Ceram Glaze)) is closely related to the adherence of *S. mutans* [27]. The application of a SiC coating on the surfaces of ceramic significantly minimized biofilm adhesion on the surface. A possible explanation is that SiC altered the surface roughness of the ceramic. SEM 3D analysis of ceramic surfaces before and after coating with SiC (Figure 2A–F) revealed a distinct difference in surface roughness between the two surfaces. The SiC coating achieved planarization of the ceramic surface, essentially minimizing surface roughness that bacteria can adhere to. Even when compared to a glazed ceramic surface, the SiC coating produced a smoother surface (Figure 2B,C). This change in surface roughness was further confirmed by a difference in surface topography of coated and non-coated surfaces, as observed by AFM (Figure 1).

Numerous studies have also demonstrated that surfaces with high surface free energy foster microbial adherence in vitro and in vivo [35–38]. In addition, bacteria with high cell surface free energy appear to adhere preferentially to surfaces with high surface free energy.

Bacterial adhesion to a substrate and the initial biofilm composition is also related to surface hydrophobicity, and communication between existing microorganisms [28,37,38]. Although the literature shows that bacteria adhere better to hydrophobic surfaces, biofilm formation depends not only on the surface topography but also on the bacteria species involved. Evidence suggests that the presence of polysaccharides on the cell surface of Gram-positive bacteria, such as *S. mutans* and *S. sanguinis*, tends to make the bacterial cell more hydrophilic [35,38]. Additionally, the cell shape and size, including surface characteristics such as extracellular substances such as flagella and fimbriae, as well as the production of proteins, are believed to have a significant influence on the process of bacterial adhesion [33,34].

On the basis of contact angle measurements conducted in this study, we found that coated SiC surfaces were more hydrophobic than non-coated surfaces; however, we found lower colonized bacteria. This might suggest that SiC-coated samples can cause the detachment of bacteria from the biofilm after 24 h.

In vitro studies have been performed on bacterial adhesion to various ceramic materials [28,39–43]. In the present study, we verified the adhesion of streptococcal species and the accumulation of complex biofilms, which are of critical importance to predict the effect of biofilms on the surface ceramics. *Streptococcus mutans* and *Streptococcus sanguinis* were used for this study because these bacteria are predominant members in dental plaque, which is often described as a dual-species biofilm model [36,44]. The association of these microorganisms can interfere in the development of each other and promote different results between polymicrobial and monomicrobial biofilms.

Lower adhesion of polymicrobial biofilm of *S. mutans* and *S. sanguinis* on the surface of fluorapatite glass-ceramic coated with SiC was demonstrated. Additionally, we found a significant reduction in their monomicrobial biofilm. Two studies found significant differences in adhesion of different streptococcal species among different ceramic materials. These differences were observed in the bacterial surface coverage, as well as in the thickness of the biofilm between the tested ceramic materials. The lowest surface coverage (19.0%) and biofilm thickness (1.9 mm) were determined on the HIP Y-TZP ceramic; the highest mean values were identified with the lithium disilicate glass-ceramic (46.8%, 12.6 mm) [41,42]. On the other hand, Meier et al. [40] did not observe any significance. Our study showed that the percentage of polymicrobial biofilm coverage was five times less for the ceramic coated with SiC (16.9%) compared with the uncoated group (91.8%). Ceramic coating also presented significant reductions of CFU/mL in monomicrobial biofilms of *S. mutans* and *S. sanguinis*.

This may be related to the alteration of the surface energy and slightly smoother surface of coated SiC surfaces, which can decrease the bacteria adhesion.

Because significant differences in bacteria adhesion were identified among the groups, the hypothesis that SiC coatings on ceramic will decrease monomicrobial and polymicrobial biofilm adhesion on the surface of the ceramic has to be accepted.

Furthermore, a study by Auschill et al. [45] found that biofilms form a very thin layer on ceramic, but demonstrated the highest viability compared with biofilms on amalgam, gold, and composite resin. Dal Piva et al. [28] found a lower value of bacterial adhesion for zirconia reinforced lithium silicate (ZLS) ceramic compared to zirconia partially stabilized by yttrium (YZHT) ceramic. Lithium disilicate glass-ceramics (LDS) have lower *S. mutans* adhesion than fully stabilized zirconia (FSZ) and partially stabilized zirconia (PSZ), but the difference was not reflected in early biofilm formation [46]. Polished feldspathic ceramics showed less *Candida albicans* adherence, whereas the adherence of streptococci was greater than *C. albicans* in all conditions [39].

Ceramic materials may exhibit positive or negative results in terms of cell viability, according to the ceramic composition [37,38]. Studies have demonstrated a significant decrease in cell viability for computer-assisted design (CAD) block feldspathic ceramics [47]. Feldspathic ceramic glass and glass-ceramics based on the $3CaO \cdot P_2O_5-SiO_2-MgO$ system were considered not cytotoxic by MTT ((3-(4,5-dimethylthiazol-2-yl)-2,5-diphenyltetrazolium bromide) [39] and neutral red analysis [48], respectively. Additionally, lithia–silica glass-ceramic was biocompatible, promoting cell adhesion and proliferation after 21 days [38]. However, zirconia-reinforced lithium silicate (ZLS) ceramic showed severe initial cytotoxicity when in contact with fibroblasts for 24 h [28]. This study demonstrated that fluorapatite glass-ceramic with or without SiC coating presented no cytotoxicity on human periodontal ligament cells by CellTiter-Blue. SiC-coated surfaces seeded with HPdLF cells may foster cell survival similar to surfaces that were not coated with SiC. Additionally, cells appeared elongated and intimately attached to both ceramic surfaces by SEM images. Some cell-to-cell interaction on the surface of ceramics was observed (Figure 8). Because the biocompatibility of the materials can be evaluated by the direct interaction between the ceramic surface and the cells, we can affirm that SiC coating is biocompatible.

This study confirmed that SiC coating can decrease bacterial adhesion on the surface of ceramic veneers while maintaining biocompatibility with other cells. SiC recently demonstrated optimized color properties and corrosion resistance [21,22]. This coating can be applied to dental materials to improve their clinical performance.

5. Conclusions

A SiC coating on ceramic demonstrated enhanced bactericidal ability without inducing cytotoxic effects. The SiC coating presented bactericidal activity against *S. mutans* and *S. sanguinis* after 24 h of culture. In addition, the coating did not induce cytotoxic effects on periodontal ligament fibroblasts. These results indicate that a SiC coating can be used as a tool to prevent bacterial adhesion and improve the clinical performance of ceramic restorations. Further studies will include in vivo validation of coating performance.

Author Contributions: Conceptualization, F.R., J.F.E.-U., and A.E.C.; methodology, S.E.A.C., F.R., and J.F.E.-U.; investigation: S.E.A.C., A.S.M., C.F., and J.L.P.; data curation: S.E.A.C., S.-M.H., F.R., A.E.C., and J.F.E.-U.; writing—original draft preparation, S.E.A.C.; writing—review and editing, S.E.A.C., F.R., P.H.C.IV, and J.F.E.-U. All authors have read and agreed to the published version of the manuscript.

Funding: This work was supported by the National Institutes of Health-NIDCR (grant number R01 DE025001).

Acknowledgments: Ceramic materials were supplied by Ivoclar Vivadent. Dektak was performed at the Nanoparticle Research Facility at the University of Florida.

Conflicts of Interest: The authors report no conflicts of interest related to this study.

Ethical Statements: The human periodontal ligament fibroblasts (HPdLF) cell line was obtained from Lonza Bioscience (catalog #: CC-7049, Basel, Switzerland).

References

1. Liu, R.; Tang, Y.; Zeng, L.; Zhao, Y.; Ma, Z.; Sun, Z.; Ren, L.; Yang, K. In vitro and in vivo studies of anti-bacterial copper-bearing titanium alloy for dental application. *Dent. Mat.* **2018**, *34*, 1112–1126. [CrossRef] [PubMed]
2. Liu, R.; Memarzadeh, K.; Chang, B.; Zhang, Y.; Ma, Z.; Allaker, R.P.; Ren, L.; Yang, K. Antibacterial effect of copper-bearing titanium alloy (Ti-Cu) against Streptococcus mutans and Porphyromonas gingivalis. *Sci. Rep.* **2016**, *6*, 29985. [CrossRef] [PubMed]
3. Van Hove, R.P.; Sierevelt, I.N.; van Royen, B.J.; Nolte, P.A. Titanium-Nitride Coating of Orthopaedic Implants: A Review of the Literature. *BioMed. Res. Int.* **2015**, *9*, 485975. [CrossRef] [PubMed]
4. Vilarrasa, J.; Delgado, L.M.; Galofré, M.; Àlvarez, G.; Violant, D.; Manero, J.M.; Blanc, V.; Gil, F.J.; Nart, J. In vitro evaluation of a multispecies oral biofilm over antibacterial coated titanium surfaces. *J. Mater. Sci. Mater. Med.* **2018**, *29*, 164.
5. Pokrowiecki, R.; Zareba, T.; Szaraniec, B.; Palka, K.; Mielczarek, A.; Menaszek, E.; Tyski, S. In vitro studies of nanosilver-doped titanium implants for oral and maxillofacial surgery. *Int. J. Nanomed.* **2017**, *12*, 4285–4297. [CrossRef]
6. Mombelli, A.; Oosten, M.A.C.; Schürch, E.; Lang, N.P. The microbiota associated with successful or failing osseointegrated titanium implants. *Oral Microbiol. Immunol.* **1987**, *2*, 145–151. [CrossRef]
7. Mei, S.; Wang, H.; Wang, W.; Tong, L.; Pan, H.; Ruan, C.; Ma, Q.; Liu, M.; Yang, H.; Zhang, L.; et al. Antibacterial effects and biocompatibility of titanium surfaces with graded silver incorporation in titania nanotubes. *Biomaterials* **2014**, *35*, 4255–4265. [CrossRef]
8. Li, G.; Zhao, Q.-M.; Yang, H.-l.; Cheng, L. Antibacterial and Microstructure Properties of Titanium Surfaces Modified with Ag-Incorporated Nanotube Arrays. *Mater. Res.* **2016**, *19*, 735–740. [CrossRef]
9. Berry, C.W.; Moore, T.J.; Safar, J.A.; Henry, C.A.; Wagner, M.J. Antibacterial activity of dental implant metals. *Implant Dent.* **1992**, *1*, 59–65. [CrossRef]
10. Carey, P.H.; Ren, F.; Jia, Z.; Batich, C.D.; Camargo, S.E.A.; Clark, A.E.; Esquivel-Upshaw, J.F. Antibacterial Properties of Charged TiN Surfaces for Dental Implant Application. *Chem. Select.* **2019**, *4*, 9185–9189. [CrossRef]
11. Anggraini, L.; Isonishi, K.; Ameyama, K. Toughening and Strengthening of Ceramics Composite through Microstructural Refinement. 2016. Available online: https://aip.scitation.org/doi/abs/10.1063/1.4945458 (accessed on 19 January 2020).
12. Negita, K. Effective Sintering Aids for Silicon Carbide Ceramics: Reactivities of Silicon Carbide with Various Additives. *J. Am. Ceram. Soc.* **1986**, *69*, C-308–C-310. [CrossRef]
13. Ohji, T.; Jeong, Y.-K.; Choa, Y.-H.; Niihara, K. Strengthening and Toughening Mechanisms of Ceramic Nanocomposites. *J. Am. Ceram. Soc.* **2005**, *81*, 1453–1460. [CrossRef]
14. Rudneva, V.V.; Galevsky, G.V.; Kozyrev, N.A. Silicon nano-carbide in strengthening and ceramic technologies. *IOP Conf. Series Mater. Sci. Eng.* **2015**, *91*. [CrossRef]
15. Zhan, G.-D.; Mitomo, M.; Kim, Y.-W. Microstructural Control for Strengthening of Silicon Carbide Ceramics. *J. Am. Ceram. Soc.* **2004**, *82*, 2924–2926. [CrossRef]
16. Filatova, E.A.; Hausmann, D.; Elliott, S.D. Understanding the Mechanism of SiC Plasma-Enhanced Chemical Vapor Deposition (PECVD) and Developing Routes toward SiC Atomic Layer Deposition (ALD) with Density Functional Theory. *ACS Appl. Mater. Interfaces* **2018**, *10*, 15216–15225. [CrossRef]
17. Flannery, A.F.; Mourlas, N.J.; Storment, C.W.; Tsai, S.; Tan, S.H.; Heck, J.; Monk, D.; Kim, T.; Gogoi, B.; Kovacsa, G.T.A. PECVD silicon carbide as a chemically resistant material for micromachined transducers. *Sens. Act. A Phys.* **1998**, *70*, 48–55. [CrossRef]
18. Flannery, A.F.; Mourlas, N.J.; Storment, C.W.; Tsai, S.; Tan, S.H.; Kovacs, G.T.A. PECVD silicon carbide for micromachined transducers. In Proceedings of the International Solid State Sensors and Actuators Conference (Transducers '97), Chicago, IL, USA, 16–19 June 1997.
19. Huran, J.; Hotovy, I.; Pezoldt, J.; Balalykin, N.I.; Kobzev, A.P. RF plasma deposition of thin amorphous silicon carbide films using a combination of silan and methane. In Proceedings of the 2006 International Conference on Advanced Semiconductor Devices and Microsystems, Smolenice Castle, Slovakia, 16–18 October 2006.

20. Iliescu, C.; Poemar, D.P. PECVD Amorphous Silicon Carbide (α-SiC) Layers for MEMS Applications. In *Physics and Technology of Silicon Carbide Devices*; Hijikata, Y., Ed.; IntechOpen, 2012. Available online: https://www.intechopen.com/books/physics-and-technology-of-silicon-carbide-devices/pecvd-amorphous-silicon-carbide-sic-layers-for-mems-applications (accessed on 1 December 2019).
21. Chen, Z.; Fares, C.; Elhassani, R.; Ren, F.; Kim, M.; Hsu, S.-M.; Esquivel-Upshaw, J.F. Demonstration of SiO2/SiC-based protective coating for dental ceramic prostheses. *J. Am. Ceram. Soc.* **2019**, *102*, 6591–6599. [CrossRef]
22. Hsu, S.-M.; Ren, F.; Chen, Z.; Kim, M.; Fares, C.; Clark, A.E.; Neal, D.; Esquivel-Upshaw, J.F. Novel Coating to Minimize Corrosion of Glass-Ceramics for Dental Applications. *Materials* **2020**, *13*, 1215. [CrossRef]
23. Kolenbrander, P.E.; Palmer, R.J.Jr.; Periasamy, S.; Jakubovics, N.S. Oral multispecies biofilm development and the key role of cell-cell distance. *Nat. Rev. Microbiol.* **2010**, *8*, 471–480. [CrossRef]
24. Ammons, M.C.; Tripet, B.P.; Carlson, R.P.; Kirker, K.R.; Gross, M.A.; Stanisich, J.J.; Copié, V. Quantitative NMR metabolite profiling of methicillin-resistant and methicillin-susceptible Staphylococcus aureus discriminates between biofilm and planktonic phenotypes. *J. Proteome Res.* **2014**, *13*, 2973–2985. [CrossRef]
25. An, Y.H.; Friedman, R.J. Concise review of mechanisms of bacterial adhesion to biomaterial surfaces. *J. Biomed. Mater. Res.* **1998**, *43*, 338–348. [CrossRef]
26. Katsikogianni, M.; Missirlis, Y.F. Concise review of mechanisms of bacterial adhesion to biomaterials and of techniques used in estimating bacteria-material interactions. *Eur. Cells Mater.* **2004**, *8*, 37–57. [CrossRef] [PubMed]
27. Vo, D.T.; Arola, D.; Romberg, E.; Driscoll, C.F.; Jabra-Rizk, M.A.; Masri, R. Adherence of Streptococcus mutans on lithium disilicate porcelain specimens. *J. Prosthet. Dent.* **2015**, *114*, 696–701. [CrossRef]
28. Dal Piva, A.; Contreras, L.; Ribeiro, F.C.; Anami, L.C.; Camargo, S.; Jorge ABottino, M.A. Monolithic Ceramics: Effect of Finishing Techniques on Surface Properties, Bacterial Adhesion and Cell Viability. *Oper. Dent.* **2018**, *43*, 315–325. [CrossRef]
29. Kidd, E.A.; Fejerskov, O. What constitutes dental caries? Histopathology of carious enamel and dentin related to the action of cariogenic biofilms. *J. Dent. Res.* **2004**, *83*, C35–C38. [CrossRef]
30. Hahnel, S.; Rosentritt, M.; Handel, G.; Bürgers, R. Influence of saliva substitute films on initial Streptococcus mutans adhesion to enamel and dental substrata. *J. Dent.* **2008**, *36*, 977–983. [CrossRef]
31. Bürgers, R.; Gerlach, T.; Hahnel, S.; Schwarz, F.; Handel, G.; Gosau, M. In vivo and in vitro biofilm formation on two different titanium implant surfaces. *Clin. Oral Implants Res.* **2010**, *21*, 156–164. [CrossRef]
32. Dağistan, S.; Aktas, A.E.; Caglayan, F.; Ayyildiz, A.; Bilge, M. Differential diagnosis of denture-induced stomatitis, Candida, and their variations in patients using complete denture: A clinical and mycological study. *Mycoses* **2009**, *52*, 266–271. [CrossRef]
33. Rosentritt, M.; Behr, M.; Bürgers, R.; Feilzer, A.J.; Hahnel, S. In vitro adherence of oral streptococci to zirconia core and veneering glass-ceramics. *J. Biomed. Mater. Res. B Appl. Biomater.* **2009**, *91*, 257–263. [CrossRef]
34. Quirynen, M.; Marechal, M.; Busscher, H.J.; Weerkamp, A.H.; Darius, P.L.; Van Steenberghe, D. The influence of surface free energy and surface roughness on early plaque formation. An in vivo study in man. *J. Clin. Periodontol.* **1990**, *17*, 138–144. [CrossRef]
35. Quirynen, M.; Van der Mei, H.C.; Bollen, C.M.; Schotte, A.; Marechal, M.; Doornbusch, G.I.; Naert, I.; Busscher, H.J.; Van Steenberghe, D. An in vivo study of the influence of the surface roughness of implants on the microbiology of supra- and sub-gingival plaque. *J. Dent. Res.* **1993**, *72*, 1304–1309. [CrossRef]
36. Teughels, W.; Van Asche, N.; Sliepen, I.; Quirynen, M. Effect of material characteristics and/or surface topography on biofilm development. *Clin. Oral Impl. Res.* **2006**, *17*, 68–81. [CrossRef] [PubMed]
37. Sun, T.; Guo, M.; Cheng, Y.; Ou, L.; He, P.; Liu, X. Graded Nano Glass-Zirconia Material for Dental 37. Applications-Part II Biocompatibility Evaluation. *J. Biomed. Nanotechnol.* **2017**, *13*, 1682–1693. [CrossRef] [PubMed]
38. Daguano, J.K.M.B.; Milesi, M.T.B.; Rodas, A.C.D.; Weber, A.F.; Sarkis, J.E.S.; Hortellani, M.A.; Zanotto, E.D. In vitro biocompatibility of new bioactive lithia-silica glass-ceramics. *Mater. Sci. Eng. C Mater. Biol. Appl.* **2019**, *94*, 117–125. [CrossRef] [PubMed]
39. Contreras, L.; Dal Piva, A.; Ribeiro, F.C.; Anami, L.C.; Camargo, S.; Jorge, A.; Bottino, M.A. Effects of Manufacturing and Finishing Techniques of Feldspathic Ceramics on Surface Topography, Biofilm Formation, and Cell Viability for Human Gingival Fibroblasts. *Oper. Dent.* **2018**, *43*, 593–601. [CrossRef] [PubMed]

40. Meier, R.; Hauser-Gerspach, I.; Lüthy, H.; Meyer, J. Adhesion of oral streptococci to all-ceramics dental restorative materials in vitro. *J. Mater. Sci. Mater. Med.* **2008**, *19*, 3249–3253. [CrossRef]
41. Hahnel, S.; Leyer, A.; Rosentritt, M.; Handel, G.; Bürgers, R. Surface properties and in vitro Streptococcus mutans adhesion to self-etching adhesives. *J. Adhes. Den.* **2009**, *11*, 263–269.
42. Bremer, F.; Grade, S.; Kohorst, P.; Stiesch, M. In vivo biofilm formation on different dental ceramics. *Quintessence Int.* **2011**, *42*, 565–574.
43. Hahnel, S.; Mühlbauer, G.; Hoffmann, J.; Ionescu, A.; Bürgers, R.; Rosentritt, M.; Handel, G.; Häberlein, I. Streptococcus mutans and Streptococcus sobrinus biofilm formation and metabolic activity on dental materials. *Acta Odontol. Scan.* **2012**, *70*, 114–121. [CrossRef]
44. Zhu, B.; Macleod, L.C.; Kitten, T.; Xu, P. Streptococcus sanguinis biofilm formation & interaction with oral pathogens. *Future Microbiol.* **2018**, *13*, 915–932.
45. Auschill, T.M.; Arweiler, N.B.; Brecx, M.; Reich, E.; Sculean, A.; Netuschil, L. The effect of dental restorative materials on dental biofilm. *Eur. J. Oral Sci.* **2002**, *110*, 48–53. [CrossRef] [PubMed]
46. Viitaniemi, L.; Abdulmajeed, A.; Sulaiman, T.; Söderling, E.; Närhi, T. Adhesion and Early Colonization of S. Mutans on Lithium Disilicate Reinforced Glass-Ceramics, Monolithic Zirconia and Dual Cure Resin Cement. *Eur. J. Prosthodont. Restor. Dent.* **2017**, *25*, 228–234. [PubMed]
47. Pabst, A.M.; Walter, C.; Grassmann, L. Influence of CAD/CAM all-ceramic materials on cell viability, migration ability and adenylate kinase release of human gingival fibroblasts and oral keratinocytes. *Clin. Oral Investig.* **2014**, *18*, 1111–1118. [CrossRef]
48. Daguano, J.K.; Rogero, S.O.; Crovace, M.C.; Peitl, O.; Strecker, K.; Dos Santos, C. Bioactivity and cytotoxicity of glass and glass-ceramics based on the 3CaO·P$_2$O$_5$-SiO$_2$-MgO system. *J. Mater. Sci. Mater. Med.* **2013**, *24*, 2171–2180. [CrossRef] [PubMed]

© 2020 by the authors. Licensee MDPI, Basel, Switzerland. This article is an open access article distributed under the terms and conditions of the Creative Commons Attribution (CC BY) license (http://creativecommons.org/licenses/by/4.0/).

Article

Cerium Dioxide Particles to Tune Radiopacity of Dental Adhesives: Microstructural and Physico-Chemical Evaluation

Isadora Martini Garcia [1], Vicente Castelo Branco Leitune [1], Antonio Shigueaki Takimi [2], Carlos Pérez Bergmann [3], Susana Maria Werner Samuel [1], Mary Anne Melo [4,5,*] and Fabrício Mezzomo Collares [1,*]

1. Dental Materials Laboratory, School of Dentistry, Federal University of Rio Grande do Sul, Rua Ramiro Barcelos 2492, Rio Branco, Porto Alegre, RS 90035-003, Brazil; isadora.garcia@ufrgs.br (I.M.G.); vicente.leitune@ufrgs.br (V.C.B.L.); susana.samuel@ufrgs.br (S.M.W.S.)
2. Laboratory for Electrochemical Processes and Corrosion, Engineering School, Federal University of Rio Grande do Sul, Bento Gonçalves, 9500, Prédio 43427, Sala 216, Porto Alegre, RS 91501-970, Brazil; antonio.takimi@gmail.com
3. Laboratory of Ceramic Materials, Federal University of Rio Grande do Sul, Avenida Osvaldo Aranha 99, Porto Alegre, RS 90035-003, Brazil; bergmann@ufrgs.br
4. Division of Operative Dentistry, Department of General Dentistry, University of Maryland School of Dentistry, Baltimore, MD 21201, USA
5. Ph.D. Program in Biomedical Sciences, University of Maryland School of Dentistry, Baltimore, MD 21201, USA
* Correspondence: mmelo@umaryland.edu (M.A.M.); fabricio.collares@ufrgs.br (F.M.C.); Tel.: +1-410-706-8705 (M.A.M.); +55-51-3308-6000 (F.M.C)

Received: 16 January 2020; Accepted: 6 February 2020; Published: 11 February 2020

Abstract: The insufficient radiopacity of dental adhesives applied under composite restorations makes the radiographic diagnosis of recurrent caries challenging. Consequently, the misdiagnosis may lead to unnecessary replacement of restorations. The aims of this study were to formulate experimental dental adhesives containing cerium dioxide (CeO_2) and investigate the effects of different loadings of CeO_2 on their radiopacity and degree of conversion for the first time. CeO_2 was characterized by X-ray diffraction analysis, Fourier transforms infrared spectroscopy, and laser diffraction for particle size analysis. Experimental dental adhesives were formulated with CeO_2 as the inorganic filler with loadings ranging from 0.36 to 5.76 vol.%. The unfilled adhesive was used as a control. The studied adhesives were evaluated for dispersion of CeO_2 in the polymerized samples, degree of conversion, and radiopacity. CeO_2 presented a monoclinic crystalline phase, peaks related to Ce-O bonding, and an average particle size of around 16 µm. CeO_2 was dispersed in the adhesive, and the addition of these particles increased the adhesives' radiopacity ($p < 0.05$). There was a significant decrease in the degree of conversion with CeO_2 loadings higher than 1.44 vol.%. However, all materials showed a similar degree of conversion in comparison to commercially available adhesives. CeO_2 particles were investigated for the first time as a promising compound to improve the radiopacity of the dental adhesives.

Keywords: dental materials; dentistry; adhesives; light-curing of dental adhesives; composite resins; methylmethacrylate; oxides; cerium; polymers; dentine bonding agents

1. Introduction

Currently, resin composites and dental adhesive systems are used in restorative dentistry as primary direct restorative materials [1]. The failure of resin composites is mainly due to recurrent caries and fractures [2]. The diagnosis of recurrent caries underneath resin composite is a challenge for dentists [3]. Recurrent carious lesions may not always be seen during a clinical examination at the interface between the resin composite, and it requires radiographic evaluation for diagnosis [4,5].

Further, the application of adhesive systems can be seen radiographically as a radiolucent area, which mimics the radiographic appearance of carious lesions [6]. The radiolucency of dental adhesives can contribute to the misinterpretation of radiographic images [4,7,8]. Based on it, dentists may intervene surgically in existing composite restorations, replacing the resin [9].

A previous report has assessed and demonstrated the lack of radiopacity on many restorative dental materials [10]. In their results, all assessed materials were radiolucent and required alterations to their composition to facilitate their detection using radiographic images. The radiopacity of dental adhesives depends on their filler content, and it can be enhanced by incorporating elements with a high atomic weight as inorganic fillers [11].

Cerium dioxide (CeO_2), a rare-earth oxide found in the lanthanide series of the periodic table, has been increasingly used as a nanotherapeutic material [12]. The numerous commercial applications for CeO_2 also called ceria, include glass and glass polishing, phosphors, ceramics, catalysts, and metallurgy [13]. Cerium is found in various minerals, and its primary deposits are located in the United States (Florida and Idaho) and Brazil [14]. Interesting biological properties have been observed for both nanometric and micrometric CeO_2 [15]. Thus, as further advancement in CeO_2's applications, this compound has gained substantial interest in several innovative applications, mainly due to its redox property and catalytic activity [15,16]. In the dental field, CeO_2 was primarily used for dental ceramics since this compound stimulates the natural fluorescence found in human dental enamel [11]. The high atomic number 58 of cerium suggests that it can promote considerable attenuation of a dental X-ray beam [17,18]. The incorporation of CeO_2 in adhesives can be a valuable strategy to promote radiopacity and improve the detection of dental adhesives underneath resin composites. Therefore, the aims of this study were to formulate experimental dental adhesives containing CeO_2 and investigate the effects of different loadings of CeO_2 on their radiopacity and degree of conversion for the first time.

2. Results

Figure 1 shows the results of CeO_2 particles' characterization. In Figure 1A, the X-ray diffraction analysis of the powder shows the crystallinity pattern of a monoclinic phase of CeO_2 (powder diffraction file, ICDD-PDF 37-1468). In Figure 1B, the chemical composition analyzed by Fourier Transform Infrared Spectroscopy (FTIR) displays the peaks related to Ce–O bonds at 400 cm^{-1}. Figure 1C,D presents the results of size distribution analysis. The histogram shows a non-normal distribution of size (Figure 1C). The 10th percentile showed a particle diameter of 1.15 µm. The median (50th percentile) was 14.98 µm, the 90th percentile was 31.32 µm, and the mean particle size was around 16 µm (Figure 1D).

Figure 2 presents the qualitative assessment via micro-Raman of the materials' surfaces containing 0.36 (A) and 0.76 (B) vol.% of CeO_2. The image consists of a 2-D array of measured spectra, which means that the distribution of the chemical composition can be investigated. After mapping the polymer's surfaces, the integration of the corresponding CeO_2 peak (464 cm^{-1}) was used for analysis and graphs generation. In the graphs, from blue to yellow, more CeO_2 is identified. It was observed that the higher the load of CeO_2 in the resin, the larger the area under the curve (peak) in micro-Raman spectra for this oxide, generating more yellow regions in the graph. In these analyses, more areas in yellow were presented in the adhesive with 0.76 vol.% of CeO_2 in comparison to the group with 0.36 vol.%.

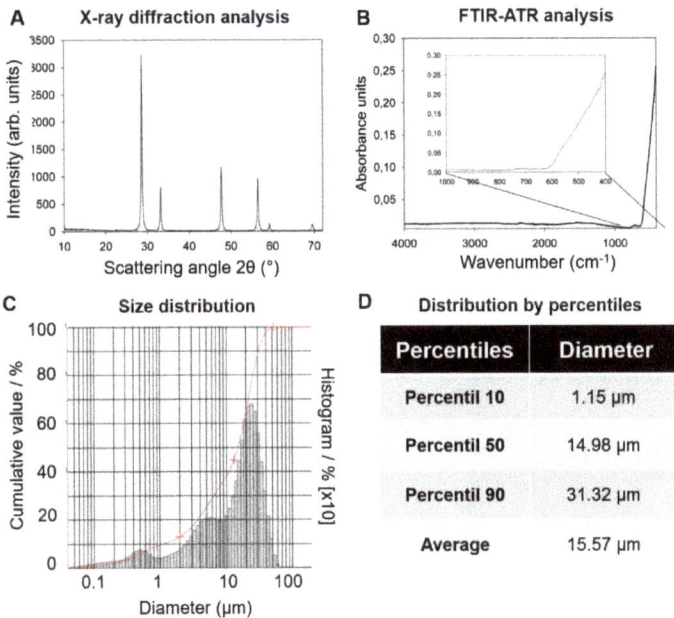

Figure 1. CeO$_2$ particles' characterization: (**A**) X-ray diffraction analysis shows the pattern of the monoclinic phase of crystallinity for CeO$_2$; (**B**) FTIR analysis displays the peaks related to Ce-O bonds at 400 cm^{-1}; (**C**) and (**D**) show the non-normal distribution size of CeO$_2$ and the values of size distribution.

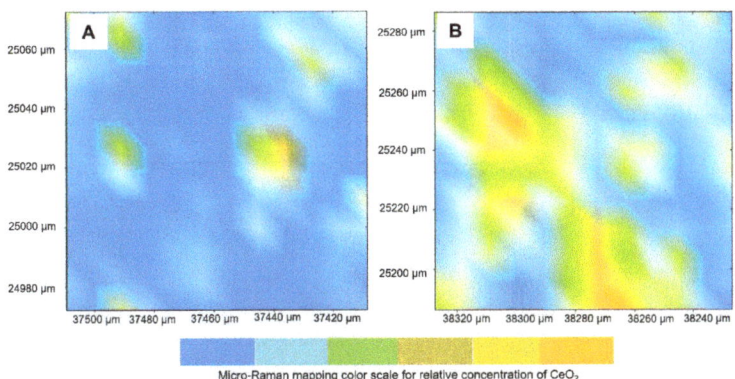

Figure 2. Micro-Raman analysis of the dental adhesive surfaces containing 0.36 (**A**) and 0.76 (**B**) vol.% of CeO$_2$. Both images display the intensity of the CeO$_2$ peak at 464 cm^{-1}. The higher the load of CeO$_2$ in the dental adhesive, the more areas in yellow are observed.

Figure 3 indicates the results of the radiopacity of each experimental adhesive resin containing a different percentage of volume fraction of CeO$_2$. In Figure 3A, an illustrative radiograph displays the location of the dental adhesive layer underneath the composite restorative material. The arrows indicate the radiolucent areas corresponding to the adhesive layer. The difference of radiopacity between the composite resin and the adhesive layer can be clearly observed. In Figure 2B, the mean and standard deviation of radiopacity is expressed in mm of aluminum. The control group, which is an unfilled adhesive, showed the lower mean value of radiopacity, without statistical difference for the groups with 0.36 vol.% and 0.72 vol.% of CeO_2. From the addition of 1.44 vol.% of CeO_2, there was increased radiopacity of the adhesive in comparison to the control group ($p < 0.05$). The group with 5.76 vol.% of CeO_2 showed the highest value of radiopacity among all groups ($p < 0.05$). The incorporation of 4.32 vol.% and 5.76 vol.% presented values of more than 1 mm of aluminum.

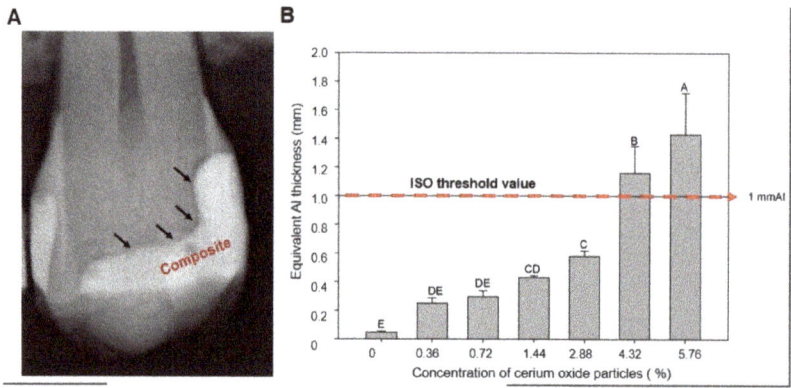

Figure 3. The dental radiograph displays the layer of dental adhesive applied under the composite resin (**A**) The arrows guide the visualization of the radiolucent adhesive layer. (**B**) The radiopacity of the experimental dental adhesives according the to the increasing concentration of cerium oxide. Different letters indicate statistical differences among groups ($p < 0.05$).

Figure 4 shows the results of the degree of conversion analysis of the experimental dental adhesives. The uncured adhesive samples were directly dispensed on the attenuated total reflection (ATR) device of FTIR to analyze the conversion of carbon-carbon double bonds in the aliphatic chain. The values ranged from 61.52 (±0.33) % for the control group to 47.90 (±1.64) % for 5.76 vol.% of CeO_2. From the incorporation of 2.88 vol.% of CeO_2, the degree of conversion reduced in comparison to the control group ($p < 0.05$). The lowest value of the degree of conversion was found for the highest load of CeO_2 incorporated in the adhesive ($p < 0.05$).

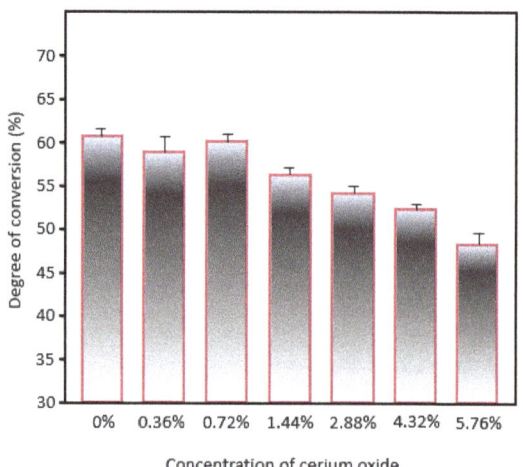

Figure 4. Degree of conversion of the experimental dental adhesives containing CeO_2. Different letters indicate statistical differences among groups ($p < 0.05$).

3. Discussion

In this in vitro study, CeO_2 particles were explored as potential radiopacifier for dental adhesives since high atomic weight and density can provide a suitable level of radiopacity. CeO_2 was first chemically characterized by XRD, Raman, and FTIR to be incorporated for the first time in a dental adhesive resin. In the current investigation, CeO_2 successfully increased the radiopacity of the adhesive, maintaining a suitable degree of monomer conversion.

CeO_2 particles used in this study presented a monoclinic crystalline phase with particular chemical groups and the average particle size of around 16 µm. This oxide was incorporated in an experimental adhesive resin formulated as previously reported, with conventional dental methacrylate monomers [19,20]. The presence of a radiopacifying agent is essential to enable the identification of restorative materials and dissimilarity from pathological processes in the adjacent areas [21]. Recurrent caries and marginal gaps are very often the reasons for the replacement of composite restorations [22]. The misdiagnosis and treatment decision for unnecessary replacement of restorations lead to additional loss of sound tooth tissue, with an increased cost and discomfort for the patient [23]. Therefore, the restorative materials should have optimal radiopacity for accurate radiographic differentiation for existing restorations and recurrent dental caries, supporting clinical follow-ups. Substantial changes in radiopacity are compulsory by the International Organization for Standardization (ISO) for dental restorative materials [24]. The compliance with the ISO 4049 requires a minimum radiopacity of restorative materials higher than that of dentin and greater or equivalent thickness of aluminum (with ≥98% purity) [24]. In the present study, the increased load of CeO_2, by the percentage of volume, was added to an experimental dental adhesive, providing a material with proper radiopacity. Our results have shown that CeO_2 at 4.32 vol.% had radiopacity higher than 1 mm of Al.

Besides the increase of radiopacity, the addition of inorganic fillers in monomeric blends may improve polymers properties, such as the elastic modulus, tensile strength, fracture toughness, Knoop hardness, and stability against solvents [25]. However, the higher filler addition can also decrease the degree of monomer's conversion [19]. The degree of conversion is a valuable chemical property for polymers, and it may be associated with the stability of the restoration over time [20]. Therefore, a reliable polymerization is desired for adhesive resins to reduce hydrolytic degradation in the clinical setting [26,27]. Here, we observed that the incorporation of CeO_2 at a load that reaches the high radiopacity level reduced the degree of conversion of the dental polymer. Previous studies have reported similar outcomes in the investigations of radiopacifying agents containing different kinds of heavy metals [28]. Marins et al. [29] evaluated the addition of niobium pentoxide (Nb_2O_5) nanoparticles to the dental adhesive as radiopacifiers and observed that the degree of conversion decreased with the addition of particles at percentage mass fraction equal or greater than 10%. In other studies, Nb_2O_5 and Ta_2O_5 showed a decrease in the degree of conversion up to 5 wt.% [19,20]. Our results are also in agreement with those of Amirouche-Korichi et al. [28], where progressive decreases of the degree of conversion were linearly related to the filler contents.

The outcome observed for the degree of conversion may be attributed to the high refractive index of CeO_2 (approx. η = 2.2 to 2.8) [30], which may have decreased the accessibility of light energy inside the polymer. The limited mobility of monomer chains by the incorporation of the opaque fillers has also been considered as a contributing factor for the decrease in the monomer's conversion [31]. Another consideration for observed decay in the degree of conversion with regard to the dental adhesive with loadings higher than 2.88 vol.% relies on particle size. The fillers used in our study are micro fillers with an average size of around 16 µm. Previous reports support the effect of particle size in microns and highlight that the volume occupied by the particles may compromise the polymerization rates of the material [32–34]. Further evaluation of the polymerization behavior of nanosized CeO_2 could be interesting to address this subject. Moreover, it would be noteworthy to evaluate the effect of micro-sized CeO_2 in the adhesive with thinner samples or in situ in dentin.

Moreover, we suggest further biological studies on dental adhesives containing CeO_2 in micro and nanoscale, since this oxide has emerged as a noteworthy agent for bioactive (such as scaffolds and bioglasses) and antimicrobial materials [35–37]. In this context, a dental adhesive with such properties could be a promising strategy to assist in decreasing the incidence of recurrent caries and improving the remineralization process after selective removal of carious dentin. In this study, despite the decreased degree of conversion observed by increasing CeO_2 addition, all groups showed values around 50%, which is in accordance with commercial dental adhesives [38]. Therefore, CeO_2 may be a promising alternative filler for biopolymers.

4. Materials and Methods

4.1. X-ray Diffraction Analysis of CeO_2

CeO_2 particles purchased were analyzed via X-ray diffraction to detect the crystalline phases of the powder. diffractometer (PW 1730/1 model, Philips, Santa Clara, CA, USA) was operated at 40 kV and 40 mA with CuKa radiation. The scanning rate used was 0.058/min, with 2 s of time-steps at 0.02° each, from 5° to 60° [18].

4.2. FTIR Analysis of CeO_2

Fourier Transform Infrared Spectroscopy (FTIR) was used to chemically characterize the powder of CeO_2. The analysis was performed using the spectrophotometer Vertex 70 (Bruker Optics, Ettlingen, Germany) with an attenuated total reflectance device (ATR). CeO_2 powder was placed on the ATR, and the analysis performed using 20 scans, 4 cm^{-1} of resolution in 5000 to 400 cm^{-1}, and Opus 6.5 software (Bruker Optics, Ettlingen, Germany).

4.3. Particle Size Distribution of CeO_2

CeO_2 particles were dispersed in water with sonication for 60 s. Then, particle size was analyzed via laser diffraction particle size analyzer (CILAS 1180, Cilas, Orleans, France) according to a previous study [18].

4.4. Preparation of Dental Adhesives

The adhesive resins were formulated by mixing 50 wt.% bisphenol A glycol dimethacrylate (BisGMA), 25 wt.% triethylene glycol dimethacrylate (TEGDMA), and 25 wt.% 2-hydroxyethyl methacrylate (HEMA). As a photoinitiator system, camphorquinone and ethyl 4-dimethylaminobenzoate were added at 1 mol%, each one in the base resin. As polymerization inhibition, 0.01 wt.% of butylated hydroxytoluene was incorporated. CeO_2 was added at 0.36, 0.72, 1.44, 2.88, 4.32, and 5.76 vol.%. The resins adhesives were hand-mixed for 5 min, sonicated during 180 s, and hand-mixed again for 5 min. One group without CeO_2 was used as control.

4.5. Qualitative Analysis of CeO_2 into Dental Adhesives

To identify the presence of CeO_2 in dental adhesives, two groups containing this filler were evaluated via micro-Raman Spectroscopy (Senterra, Bruker Optics, Ettlingen, Germany). One sample from the group with 0.36 vol.% and another from the group with 5.76 vol.% were prepared using a polyvinylsiloxane mold. The samples were photoactivated for 20 s on each side using a light-emitting diode dental curing unit (Radii Cal, SDI, Melbourne, Australia) with 1200 mW/cm^2. An area of 110 μm × 110 μm of the surfaces was analyzed via Opus 7.5 software (Opus 7.5, Bruker Optics, Ettlingen, Germany). The analyses were performed with a wavelength of 785 nm, with 5 s and two co-additions. The peak correspondent to CeO_2 (464 cm^{-1}) was used for integration.

4.6. Radiopacity Evaluation

The adhesive resins were evaluated for their radiopacity, according to the International Organization of Standardization (ISO) 4049/2009 guidelines [24]. Five samples per group (n = 5) with disc-shaped were prepared with 10.0 mm (±0.5 mm) in diameter and 1.0 mm (±0.1 mm) in thickness with photoactivation for 20 s on each side. The images were made using a phosphor plate for the digital system (VistaScan; Dürr Dental GmbH & Co. KG, Bietigheim-Bissingen, Germany) at 70 kV, 8 mA, and 0.4 s of exposure time. A focus-film distance of 400 mm was used in the assays. One specimen per group was positioned on the film for each X-ray exposition. An aluminum step-wedge (99.12 wt.% of aluminum, thickness from 0.5 mm to 5.0 mm, in increments of 0.5 mm) was exposed with the samples for each image specimens in all images. The images were saved in TIFF, analyzed using Photoshop software (Adobe Systems Incorporated, San Jose, CA, USA), and the mean and standard deviation of the grey levels (pixel density) were measured.

4.7. Degree of Conversion

To analyze the degree of conversion of the adhesives, FTIR-ATR was used according to a previous study [39]. Three drops per group (n = 3, 3 µL each one) were directly dispensed on ATR diamond crystal. An adjustable holder stand was used to fix the light-curing unit and to standardize the distance between its tip and the top of the samples. Two spectra were acquired per group from 4000 to 400 cm^{-1} with a resolution of 4 cm^{-1}. The first spectrum was obtained before the photoactivation. Then, the samples were photoactivated for 20 s and analyzed again. To calculate the monomer conversion, the peak related to the carbon-carbon (C=C) double bond in the aromatic ring of BisGMA (1610 cm^{-1}) was used as an internal standard. The peak related to C=C in the aliphatic chains (1640 cm^{-1}) was used along with the C=C at 1610 cm^{-1} according to the following equation:

$$\textbf{Degree of conversion } (\%) = 100 \times \left(\frac{I\_1640-cured / I\_1610-cured}{I\_1640-uncured / I\_1610-uncured} \right)$$

4.8. Statistical Analysis

CeO$_2$ characterization was descriptively analyzed, as well as the identification of CeO$_2$ in the polymerized adhesives. Shapiro-Wilk test was used to evaluate the data normality. The data of radiopacity was evaluated via ANOVA on ranks and Dunn's test. The data of the degree of conversion was analyzed via one-way ANOVA and Tukey's post-hoc test. Both analyses were performed at 0.05 level of significance ($p < 0.05$).

5. Conclusions

CeO$_2$ was investigated for the first time as a promising filler to improve the radiopacity of dental adhesives. The particles were chemically analyzed and then used in experimental adhesives formulation. Through of incorporation of CeO$_2$ particles at 1.44 vol.%, the dental adhesives were able to show increased radiopacity and a proper degree of conversion. The restorative materials here proposed could be a reliable strategy to assist clinicians diagnose recurrent caries.

Author Contributions: I.M.G., contributed to data analysis, data interpretation, and drafted the manuscript; V.C.B.L., contributed to design, data acquisition, data analysis, and data interpretation; A.S.T., contributed to data analysis and data interpretation; C.P.B., contributed to data analysis and data interpretation; S.M.W.S., contributed to design and data interpretation; M.A.M., contributed to data interpretation, drafted the manuscript, and critically revised the manuscript; F.M.C., contributed to conception, design, data acquisition, analysis, and interpretation, and drafted and critically revised the manuscript. All authors gave final approval and agreed to be accountable for all aspects of the work.

Funding: This research received no external funding.

Acknowledgments: This study was financed in part by the Coordenação de Aperfeiçoamento de Pessoal de Nível Superior—Brasil (CAPES)—Finance Code 001 (scholarship).

Conflicts of Interest: The authors declare no conflict of interest.

References

1. Alzraikat, H.; Burrow, M.F.; Maghaireh, G.A.; Taha, N.A. Nanofilled Resin Composite Properties and Clinical Performance: A Review. *Oper. Dent.* **2018**, *43*, E173–E190. [CrossRef] [PubMed]
2. Rodrigues, S.A.; Scherrer, S.S.; Ferracane, J.L.; Della Bona, A. Microstructural Characterization and Fracture Behavior of a Microhybrid and a Nanofill Composite. *Dent. Mater.* **2008**, *24*, 1281–1288. [CrossRef] [PubMed]
3. Oztas, B.; Kursun, S.; Dinc, G.; Kamburoglu, K. Radiopacity evaluation of composite restorative resins and bonding agents using digital and film x-ray systems. *Eur. J. Dent.* **2012**, *6*, 115–122. [CrossRef] [PubMed]
4. Keenan, J.R.; Keenan, A.V. Accuracy of dental radiographs for caries detection. *Evid. Based Dent.* **2016**, *17*, 43–49. [CrossRef] [PubMed]
5. Pekkan, G.; Ozcan, M. Radiopacity of different shades of resin-based restorative materials compared to human and bovine teeth. *Gen. Dent.* **2012**, *60*, e237–e243. [PubMed]
6. Raitz, R.; Moruzzi, P.D.; Vieira, G.; Fenyo-Pereira, M. Radiopacity of 28 Composite Resins for Teeth Restorations. *J. Contemp. Dent. Pract.* **2016**, *17*, 136–142. [CrossRef]
7. Brouwer, F.; Askar, H.; Paris, S.; Schwendicke, F. Detecting Secondary Caries Lesions: A Systematic Review and Meta-analysis. *J. Dent. Res.* **2016**, *95*, 143–151. [CrossRef]
8. Aoyagi, Y.; Takahashi, H.; Iwasaki, N.; Honda, E.; Kurabayashi, T. Radiopacity of experimental composite resins containing radiopaque materials. *Dent. Mater. J.* **2005**, *24*, 315–320. [CrossRef]
9. Schwendicke, F.; Tzschoppe, M.; Paris, S. Radiographic caries detection: A systematic review and meta-analysis. *J. Dent.* **2015**, *43*, 924–933. [CrossRef]
10. Melo, M.A.S.; Guedes, S.F.F.; Xu, H.H.K.; Rodrigues, L.K.A. Nanotechnology-based restorative materials for dental caries management. *Trends Biotechnol.* **2013**, *31*, 459–467. [CrossRef]
11. Espelid, I.; Tveit, A.B.; Erickson, R.L.; Keck, S.C.; Glasspoole, E.A. Radiopacity of restorations and detection of secondary caries. *Dent. Mater. Off. Publ. Acad. Dent. Mater.* **1991**, *7*, 114–117. [CrossRef]
12. Melo, M.A.S.; Weir, M.D.; Rodrigues, L.K.A.; Xu, H.H.K. Novel calcium phosphate nanocomposite with caries-inhibition in a human in situ model. *Dent. Mater.* **2013**, *29*, 231–240. [CrossRef]
13. Younis, A.; Chu, D.; Li, S. Cerium Oxide Nanostructures and their Applications. *Funct. Nanomater.* **2016**, *32*, 521–529.
14. Scire, S.; Palmisano, L. Cerium Oxide (CeO_2): Synthesis, Properties and Applications—1st Edition. Available online: https://www.elsevier.com/books/cerium-oxide-ceo2-synthesis-properties-and-applications/scire/978-0-12-815661-2 (accessed on 13 December 2019).
15. Hosseini, A.; Sharifi, A.M.; Abdollahi, M.; Najafi, R.; Baeeri, M.; Rayegan, S.; Cheshmehnour, J.; Hassani, S.; Bayrami, Z.; Safa, M. Cerium and yttrium oxide nanoparticles are neuroprotective. *Biol. Trace Elem. Res.* **2015**, *164*, 80–89. [CrossRef]
16. Rajeshkumar, S.; Naik, P. Synthesis and biomedical applications of Cerium oxide nanoparticles—A Review. *Biotechnol. Rep.* **2017**, *17*, 1–5. [CrossRef]
17. Akça, B.; Erzeneoğlu, S.Z. The Mass Attenuation Coefficients, Electronic, Atomic, and Molecular Cross Sections, Effective Atomic Numbers, and Electron Densities for Compounds of Some Biomedically Important Elements at 59.5 keV. Available online: https://www.hindawi.com/journals/stni/2014/901465/ (accessed on 14 January 2020).
18. Mullan, B.F.; Madsen, M.T.; Messerle, L.; Kolesnichenko, V.; Kruger, J. X-ray attenuation coefficients of high-atomic-number, hexanuclear transition metal cluster compounds: A new paradigm for radiographic contrast agents. *Acad. Radiol.* **2000**, *7*, 254–259. [CrossRef]
19. Leitune, V.C.B.; Collares, F.M.; Takimi, A.; de Lima, G.B.; Petzhold, C.L.; Bergmann, C.P.; Samuel, S.M.W. Niobium pentoxide as a novel filler for dental adhesive resin. *J. Dent.* **2013**, *41*, 106–113. [CrossRef]
20. Garcia, I.M.; Leitune, V.C.B.; Ferreira, C.J.; Collares, F.M. Tantalum oxide as filler for dental adhesive resin. *Dent. Mater. J.* **2018**, *37*, 897–903. [CrossRef]
21. Gu, S.; Rasimick, B.J.; Deutsch, A.S.; Musikant, B.L. Radiopacity of dental materials using a digital X-ray system. *Dent. Mater. Off. Publ. Acad. Dent. Mater.* **2006**, *22*, 765–770. [CrossRef]
22. Demarco, F.F.; Collares, K.; Correa, M.B.; Cenci, M.S.; de Moraes, R.R.; Opdam, N.J. Should my composite restorations last forever? Why are they failing? *Braz. Oral Res.* **2017**, *31*, e56. [CrossRef]
23. Kanzow, P.; Wiegand, A.; Schwendicke, F. Cost-effectiveness of repairing versus replacing composite or amalgam restorations. *J. Dent.* **2016**, *54*, 41–47. [CrossRef]

24. 14:00–17:00 ISO 4049:2009. Available online: http://www.iso.org/cms/render/live/en/sites/isoorg/contents/data/standard/04/28/42898.html (accessed on 16 January 2020).
25. Belli, R.; Kreppel, S.; Petschelt, A.; Hornberger, H.; Boccaccini, A.R.; Lohbauer, U. Strengthening of dental adhesives via particle reinforcement. *J. Mech. Behav. Biomed. Mater.* **2014**, *37*, 100–108. [CrossRef]
26. Abdalla, A.I.; Feilzer, A.J. Four-year water degradation of a total-etch and two self-etching adhesives bonded to dentin. *J. Dent.* **2008**, *36*, 611–617. [CrossRef]
27. Malacarne, J.; Carvalho, R.M.; de Goes, M.F.; Svizero, N.; Pashley, D.H.; Tay, F.R.; Yiu, C.K.; de Oliveira Carrilho, M.R. Water sorption/solubility of dental adhesive resins. *Dent. Mater. Off. Publ. Acad. Dent. Mater.* **2006**, *22*, 973–980. [CrossRef]
28. Amirouche-Korichi, A.; Mouzali, M.; Watts, D.C. Effects of monomer ratios and highly radiopaque fillers on degree of conversion and shrinkage-strain of dental resin composites. *Dent. Mater. Off. Publ. Acad. Dent. Mater.* **2009**, *25*, 1411–1418. [CrossRef]
29. Marins, N.H.; Meereis, C.T.W.; Silva, R.M.; Ruas, C.P.; Takimi, A.S.; Carreño, N.L.V.; Ogliari, F.A. Radiopaque dental adhesive with addition of niobium pentoxide nanoparticles. *Polym. Bull.* **2018**, *75*, 2301–2314. [CrossRef]
30. Debnath, S.; Islam, M.R.; Khan, M.S.R. Optical properties of CeO2 thin films. *Bull. Mater. Sci.* **2007**, *30*, 315–319. [CrossRef]
31. Garoushi, S.; Vallittu, P.K.; Watts, D.C.; Lassila, L.V.J. Effect of nanofiller fractions and temperature on polymerization shrinkage on glass fiber reinforced filling material. *Dent. Mater. Off. Publ. Acad. Dent. Mater.* **2008**, *24*, 606–610. [CrossRef]
32. Ordinola-Zapata, R.; Bramante, C.M.; García-Godoy, F.; Moldauer, B.I.; Gagliardi Minotti, P.; Tercília Grizzo, L.; Duarte, M.A.H. The effect of radiopacifiers agents on pH, calcium release, radiopacity, and antimicrobial properties of different calcium hydroxide dressings. *Microsc. Res. Tech.* **2015**, *78*, 620–625. [CrossRef]
33. Martins, G.C.; Reis, A.; Loguercio, A.D.; Zander-Grande, C.; Meier, M.; Mazur, R.F.; Gomes, O.M.M. Does Making An Adhesive System Radiopaque by Filler Addition Affect Its Bonding Properties? *J. Adhes. Dent.* **2015**, *17*, 513–519.
34. Collares, F.M.; Ogliari, F.A.; Lima, G.S.; Fontanella, V.R.C.; Piva, E.; Samuel, S.M.W. Ytterbium trifluoride as a radiopaque agent for dental cements. *Int. Endod. J.* **2010**, *43*, 792–797. [CrossRef]
35. Farias, I.A.P.; Dos Santos, C.C.L.; Sampaio, F.C. Antimicrobial Activity of Cerium Oxide Nanoparticles on Opportunistic Microorganisms: A Systematic Review. *BioMed Res. Int.* **2018**, *2018*, 1923606. [CrossRef]
36. Varini, E.; Sánchez-Salcedo, S.; Malavasi, G.; Lusvardi, G.; Vallet-Regí, M.; Salinas, A.J. Cerium (III) and (IV) containing mesoporous glasses/alginate beads for bone regeneration: Bioactivity, biocompatibility and reactive oxygen species activity. *Mater. Sci. Eng. C Mater. Biol. Appl.* **2019**, *105*, 109971. [CrossRef]
37. Walkey, C.; Das, S.; Seal, S.; Erlichman, J.; Heckman, K.; Ghibelli, L.; Traversa, E.; McGinnis, J.F.; Self, W.T. Catalytic Properties and Biomedical Applications of Cerium Oxide Nanoparticles. *Environ. Sci. Nano* **2015**, *2*, 33–53. [CrossRef]
38. Gaglianone, L.A.; Lima, A.F.; Gonçalves, L.S.; Cavalcanti, A.N.; Aguiar, F.H.B.; Marchi, G.M. Mechanical properties and degree of conversion of etch-and-rinse and self-etch adhesive systems cured by a quartz tungsten halogen lamp and a light-emitting diode. *J. Mech. Behav. Biomed. Mater.* **2012**, *12*, 139–143. [CrossRef]
39. Maktabi, H.; Ibrahim, M.; Alkhubaizi, Q.; Weir, M.; Xu, H.; Strassler, H.; Fugolin, A.P.P.; Pfeifer, C.S.; Melo, M.A.S. Underperforming light curing procedures trigger detrimental irradiance-dependent biofilm response on incrementally placed dental composites. *J. Dent.* **2019**, *88*, 103110. [CrossRef]

© 2020 by the authors. Licensee MDPI, Basel, Switzerland. This article is an open access article distributed under the terms and conditions of the Creative Commons Attribution (CC BY) license (http://creativecommons.org/licenses/by/4.0/).

Article

Physicochemical, Mechanical, and Antimicrobial Properties of Novel Dental Polymers Containing Quaternary Ammonium and Trimethoxysilyl Functionalities

Diane R. Bienek [1,*], Anthony A. Giuseppetti [1], Stanislav A. Frukhtbeyn [1], Rochelle D. Hiers [2], Fernando L. Esteban Florez [2], Sharukh S. Khajotia [2] and Drago Skrtic [1]

1. ADA Foundation, Research Division, Frederick, MD 21704, USA; giuseppettia@ada.org (A.A.G.); frukhtbeyns@ada.org (S.A.F.); dskrtic@verizon.net (D.S.)
2. College of Dentistry, University of Oklahoma Health Sciences Center, Oklahoma City, OK 73117, USA; Shelley-Hiers@ouhsc.edu (R.D.H.); fernando-esteban-florez@ouhsc.edu (F.L.E.F.); Sharukh-Khajotia@ouhsc.edu (S.S.K.)
* Correspondence: bienekd@ada.org; Tel.: 1-301-694-2999

Received: 4 November 2019; Accepted: 11 December 2019; Published: 18 December 2019

Abstract: The aims of this study were to evaluate the physicochemical and mechanical properties, antimicrobial (AM) functionality, and cytotoxic potential of novel dental polymers containing quaternary ammonium and trimethoxysilyl functionalities (e.g., N-(2-(methacryloyloxy)ethyl)-N,N-dimethyl-3-(trimethoxysilyl)propan-1-aminium iodide (AM_{sil1}) and N-(2-(methacryloyloxy)ethyl)-N,N-dimethyl-11-(trimethoxysilyl)undecan-1-aminium bromide (AM_{sil2})). AM_{sil1} or AM_{sil2} were incorporated into light-cured (camphorquinone + ethyl-4-N,N-dimethylamino benzoate) urethane dimethacrylate (UDMA)/polyethylene glycol-extended UDMA/ethyl 2-(hydroxymethyl)acrylate (EHMA) resins (hereafter, UPE resin) at 10 or 20 mass %. Cytotoxic potential was assessed by measuring viability and metabolic activity of immortalized mouse connective tissue and human gingival fibroblasts in direct contact with monomers. AM_{sil}–UPE resins were evaluated for wettability by contact angle measurements and degree of vinyl conversion (DVC) by near infra-red spectroscopy analyses. Mechanical property evaluations entailed flexural strength (FS) and elastic modulus (E) testing of copolymer specimens. The AM properties were assessed using *Streptococcus mutans* (planktonic and biofilm forms) and *Porphyromonas gingivalis* biofilm. Neither AM_{sil} exhibited significant toxicity in direct contact with cells at biologically relevant concentrations. Addition of AM_{sil}s made the UPE resin more hydrophilic. DVC values for the AM_{sil}–UPE copolymers were 2–31% lower than that attained in the UPE resin control. The mechanical properties (FS and E) of AM_{sil}–UPE specimens were reduced (11–57%) compared to the control. Compared to UPE resin, AM_{sil1}–UPE and AM_{sil2}–UPE (10% mass) copolymers reduced *S. mutans* biofilm 4.7- and 1.7-fold, respectively ($p \leq 0.005$). Although not statistically different, *P. gingivalis* biofilm biomass on AM_{sil1}–UPE and AM AM_{sil2}–UPE copolymer disks were lower (71% and 85%, respectively) than that observed with a commercial AM dental material. In conclusion, the AM function of new monomers is not inundated by their toxicity towards cells. Despite the reduction in mechanical properties of the AM_{sil}–UPE copolymers, AM_{sil2} is a good candidate for incorporation into multifunctional composites due to the favorable overall hydrophilicity of the resins and the satisfactory DVC values attained upon light polymerization of AM_{sil}-containing UDMA/PEG-U/EHMA copolymers.

Keywords: antimicrobial effect; biofilms; cytotoxicity; dental resins; physicochemical properties; mechanical properties; quaternary ammonium methacrylates

1. Introduction

As people live longer and retain more of their own teeth, the incidence of dental caries, especially root caries, increases. Currently, the prevalence of root caries in older adults ranges from 29% to 89% [1]. The expected aging of the population will further increase root caries occurrences. Therefore, implementing more effective prevention strategies and/or developing new treatments for root caries is prudent. Often, compromised integrity of the conventional restorative/tooth interface [2,3] ultimately results in bacterial microleakage and secondary caries. Class V restoratives may release fluoride ions which, at adequate concentrations, protect teeth from demineralization and possibly contribute to regeneration of mineral lost to caries. Fluoride release, however, does not provide an effective antimicrobial (AM) protection, although fluoride can have some AM properties [4,5]. The majority of the contemporary dental restoratives do not possess substantial AM properties [6] verifiable in clinical trials [7]. To improve the longevity of repair, the restorative material should be AM. Adding AM function to dental materials typically focuses on release/slow release of various low molecular weight AM agents [5,8–15]. However, mechanism(s) of their actions are elusive, and there are concerns about their toxicity to human cells, the development of tolerances (in the case of antibiotics), and long-term efficacies. Moreover, the release of these agents can compromise mechanical performance of the restoratives and, if the dose or release kinetics are not properly controlled, can induce toxicity to the surrounding tissues [16].

Antimicrobial polymeric materials with quaternary ammonium (QA) salts have been widely applied to a variety of antimicrobial-relevant areas (reviewed by [17,18]). QA methacrylates are known for their AM action against both Gram-positive and Gram-negative bacteria. Studies have indicated that QA compounds destroy bacterial cell membrane integrity and eventually lead to cell death [19–21]. The proposed mechanism of action is the electrostatic interaction between positively charged molecules and negatively charged microbial cell membranes. So far, QA methacrylates have not been successfully incorporated into dental restorative(s) to yield a sustained AM function [22]. Historically, the most attention has been given to methacryloyloxydodecyl pyrimidinium bromide (MDPB) and its acrylamide copolymer [6,23]. MDBP has been commercialized and suggested to be potentially applicable to various restoratives. However, due to their poor color stability, MDBP-based materials can only be used for aesthetically inferior restorations. To widen the utility of QAs in restorative dentistry, various QAs have been formulated into bonding agents and dental resin composites [20,24–26]. Successful incorporation of these new AM QAs into polymeric phases of composite materials would be a major step in creating new Class V restoratives that are clinically effective against secondary caries. We are advancing the development of Class V restorative materials by introducing bioactive amorphous calcium phosphate (ACP) filler into polymer-based restorative in parallel to the AM monomer, thus creating a multifunctional AM and remineralizing materials (hereafter AMRE). ACP has been indicated as a precursor to hydroxyapatite formation both in vitro and in vivo [27–33]. ACP also exhibited favorable in vivo osteoconductivity compared to hydroxyapatite [34]. Based on our group's knowledge of ACP chemistry and our understanding of structure/composition/property relationships existing in ACP polymeric systems [35–39], we have undertaken a task to formulate AMRE polymeric composites that maintain a desired state of supersaturation with respect to hydroxyapatite and efficiently restore mineral-depleted tooth structures while providing sustained AM protection. Prior to formulating AMRE composites, it is essential to evaluate the AM-containing resins (no ACP filler) to establish the effect of AM monomers on basic biological, physicochemical, and mechanical properties of copolymers.

AM_{sil} syntheses and subsequent validation protocols of novel dental monomers containing QA and trimethoxysilyl functionalities are described in detail by Okeke et al. (2019) [40]. The motivation for synthesizing AM_{sil1} and AM_{sil2} was to develop coupling agents capable of conferring the AM properties and coupling with both ACP phase and polymer phase of Class V resin-based composites. This study reports on the incorporation of these two polymerizable QA monomers with different alkyl chain lengths (e.g., N-(2-(methacryloyloxy)ethyl)-N,N-dimethyl-3-(trimethoxysilyl)propan-1-aminium iodide (AM_{sil1}) and N-(2-(methacryloyloxy)ethyl)-N,N-dimethyl-11-(trimethoxysilyl)undecan-1-aminium bromide (AM$_{sil2}$)) (Figure 1) into UDMA/poly(ethylene glycol)-extended UDMA (PEG-U)/ethyl 2-(hydroxymethyl) acrylate (EHMA) resins (hereafter UPE resin) and the biological, physicochemical, and mechanical screening of AM–UPE copolymers. Working hypotheses were that AM$_{sil}$ monomers will show minimal or no toxicity towards immortalized mouse subcutaneous connective tissue fibroblasts (CCL1) or human gingival fibroblasts (HGF) and that AM$_{sil}$–UPE copolymers will have similar physicochemical and mechanical properties compared to the parent UPE copolymers. AM$_{sil1}$– and AM$_{sil2}$–UPE copolymers were assessed for their AM activity against *Streptococcus mutans* and *Porphyromonas gingivalis*, which are model microorganisms for dental caries [41,42] and periodontal disease [43,44], respectively.

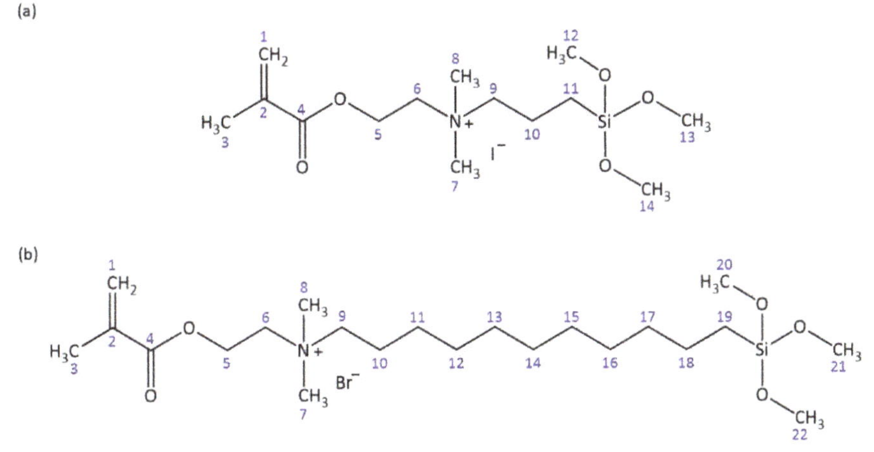

Figure 1. Skeletal structural formulas of AM$_{sil1}$ (a) and AM$_{sil2}$ (b) monomers.

2. Results

2.1. Structural Verification of AM$_{sil}$s

The AM$_{sil}$ structures were verified by ^{1}H and ^{13}C NMR (Tables 1 and 2). In brief, both ^{1}H and ^{13}C NMR spectra of AM$_{sil}$s confirmed the successful quaternization of 2-(dimethylamino)ethyl methacrylate (DMAEMA) precursor. By performing the syntheses in chloroform, methoxy groups of (3-iodopropyl)trimethoxy silane (IPTMS) and (11-bromoundecyl)trimethoxy silane (BrUDTMS) were protected from hydrolysis to which they are prone in aqueous environment.

Table 1. Assignments of ^{13}C and ^{1}H NMR chemical shifts of AM$_{sil1}$.

Atom #	^{13}C Chemical Shift, ppm	^{1}H Chemical Shift, ppm	# of H's	Signal Splitting
1	126.6 (CH$_2$)	5.77, 6.09	1, 1	singlets
2	135.3 (C)		0	
3	17.8 (CH$_3$)	1.92	3	singlet
4	165.8 (C)		0	
5	58.0 (CH$_2$)	4.52	2	multiplet
6	61.7 (CH$_2$)	3.70	2	multiplet
7, 8	50.6 (CH$_3$)	3.09	6	singlet
9	65.9 (CH$_2$)	3.34	2	multiplet
10	15.6 (CH$_2$)	1.72	2	multiplet
11	5.2 (CH$_2$)	0.54	2	multiplet
12, 13, 14	50.1 (CH$_3$)	3.51	9	singlet

Atom numbering is illustrated in Figure 1.

Table 2. Assignments of ^{13}C and ^{1}H NMR chemical shifts of AM$_{sil2}$.

Atom #	^{13}C Chemical Shift, ppm	^{1}H Chemical Shift, ppm	# of H's	Signal Splitting
1	126.5 (CH$_2$)	5.76, 6.08	1, 1	singlets
2	135.3 (C)		0	
3	17.8 (CH$_3$)	1.91	3	singlet
4	165.8 (C)		0	
5	58.1 (CH$_2$)	4.52	2	multiplet
6	61.6 (CH$_2$)	3.70	2	multiplet
7, 8	50.4 (CH$_3$)	3.09	6	singlet
9	63.8 (CH$_2$)	3.36	2	multiplet
10	21.7 (CH$_2$)	1.67	2	multiplet
11–18	22.1, 25.7, 28.4, 28.6, 28.7, 28.8, 28.9, 32.3 (CH$_2$)	1.25	16	multiplet
19	8.6 (CH$_2$)	0.57	2	multiplet
20, 21, 22	49.9 (CH$_3$)	3.46	9	singlet

Atom numbering is illustrated in Figure 1.

2.2. Biocompatibility Testing

2.2.1. AM$_{sil1}$

AM$_{sil1}$ concentration did not exert a statistically significant effect on the viability of CCL1 cells (Figure 2). However, the exposure time reduced the number (1.2-fold difference of mean, $p \leq 0.002$) of live cells. Similarly, AM$_{sil1}$ concentration did not affect the metabolic activity of CCL1 cells. Time of exposure had a modest effect ($p \leq 0.02$) on CCL1 metabolic activity, although no significant paired comparisons (within each concentration) were observed.

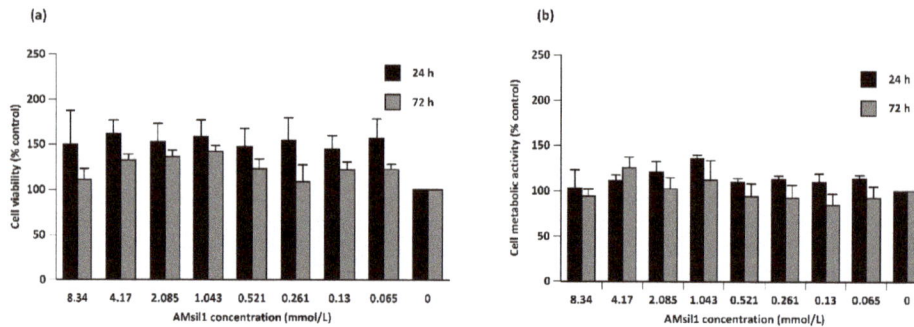

Figure 2. Percent control value of viability (a) and metabolic activity (b) of CCL1 cells exposed to 2-fold serial dilutions of AM$_{sil1}$ (≤8.34 mmol/L) for 24 or 72 h. Data represent mean ± standard error for five independent replicates tested in triplicate.

Independent of time, AM$_{sil1}$ concentration exerted a main effect ($p \leq 0.001$) on the number of viable HGFs (Figure 3). Paired comparisons indicated that at lower AM$_{sil1}$ concentrations (≤0.13 mmol/L), the number of live HGFs decreased. Independent of AM$_{sil1}$ concentration, exposure time reduced (19.73 difference of means, $p \leq 0.001$) HGF viability. Monomer concentration or time of exposure did not statistically affect the metabolic activity of HGF cells (data not shown).

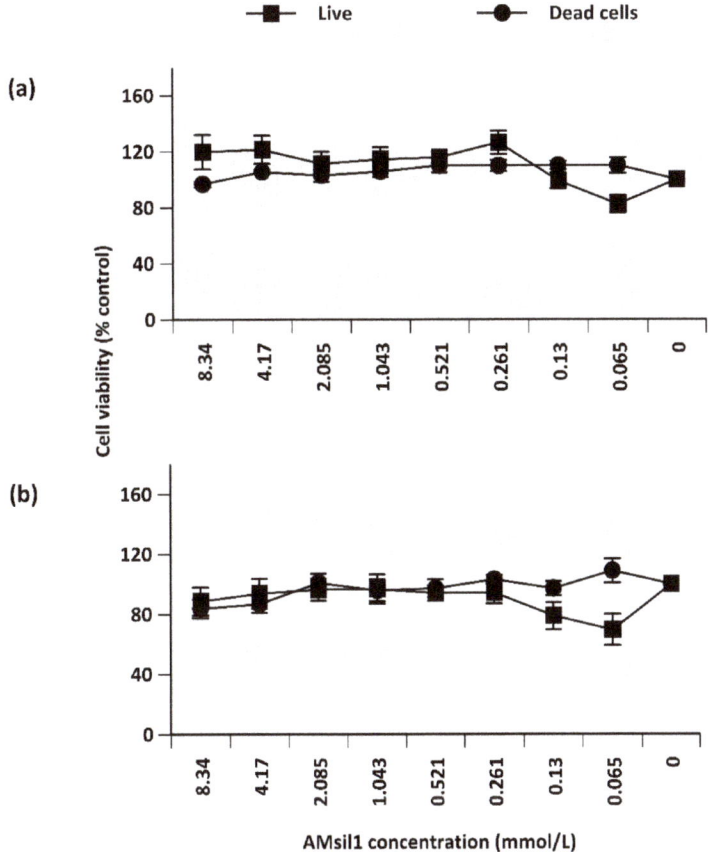

Figure 3. Percent control value of viability of human gingival fibroblast (HGF) cells exposed to 2-fold serial dilutions of AM_{sil1} (≤8.34 mmol/L) for (a) 24 h or (b) 72 h. Data represent mean ± standard error for five independent replicates tested in triplicate.

2.2.2. AM_{sil2}

Regardless of time, AM_{sil2} concentration exerted a main effect ($p \leq 0.001$) on the number of viable CCL1s (Figure 4). Paired comparisons indicated that at 24 h exposure, CCL1 cells exposed to ≥3.64 mmol/L were lower ($p \leq 0.05$) than the number of live cells exposed to all lower concentrations. At 72 h exposure, the percentage of live CCL1 cells exposed to ≥1.82 mmol/L AM_{sil2} was at least 3- to 4-fold lower ($p \leq 0.001$) compared to all concentrations ≤ 0.91 mmol/L.

Like cell viability, AM_{sil2} concentration exerted an effect ($p \leq 0.001$) on CCL1 metabolic activity. Viability and metabolic activity of CCL1 cells exposed to AM_{sil2} showed a strong positive linear correlation at 24 h ($R^2 = 0.91$, $p \leq 0.0005$) and 72 h ($R^2 = 0.93$, $p \leq 0.0005$) (data not shown).

AM_{sil2} exhibited a concentration effect ($p \leq 0.001$) on HGF cell viability (Figure 5). Paired comparisons indicated that exposure to ≥1.82 mmol/L AM_{sil2} reduced the number of live HGFs by more than 3-fold ($p \leq 0.05$) compared to the control group. When considering the effect of time regardless of AM_{sil2} concentration, the number of live cells was consistently lower (~23% difference of means). Like viability, AM_{sil2} concentration exhibited an effect ($p \leq 0.001$) on HGF metabolic activity. At 24 and 72 h, viability and metabolic activity of HGF cells exposed to AM_{sil2} showed a positive linear correlation ($R^2 = 0.94$; ($p \leq 0.0005$) and $R^2 = 0.77$; ($p \leq 0.001$), respectively) (data not shown).

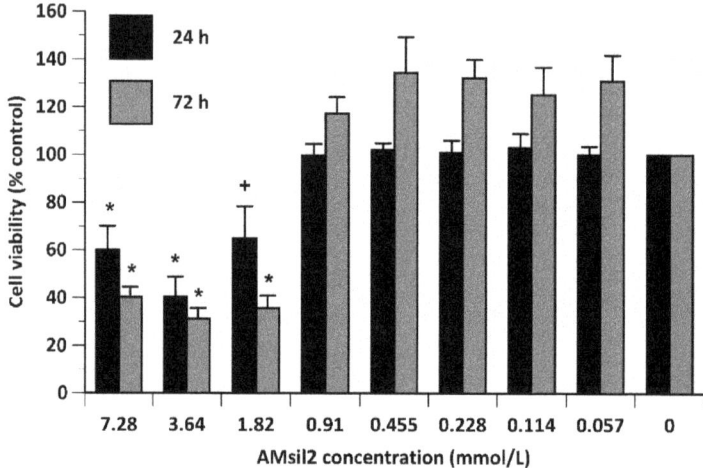

Figure 4. Percent control value of viability of CCL1 cells exposed to 2-fold serial dilutions of AM_{sil2} (≤7.28 mmol/L) for 24 or 72 h. Data represent mean ± SEM for five independent replicates tested in triplicate. * indicates $p ≤ 0.05$ when compared to concentrations ≤ 0.91 mmol/L within the same time period. + indicates $p ≤ 0.05$ when compared to 0.455, 0.228, or 0.114 mmol/L concentrations within same time period.

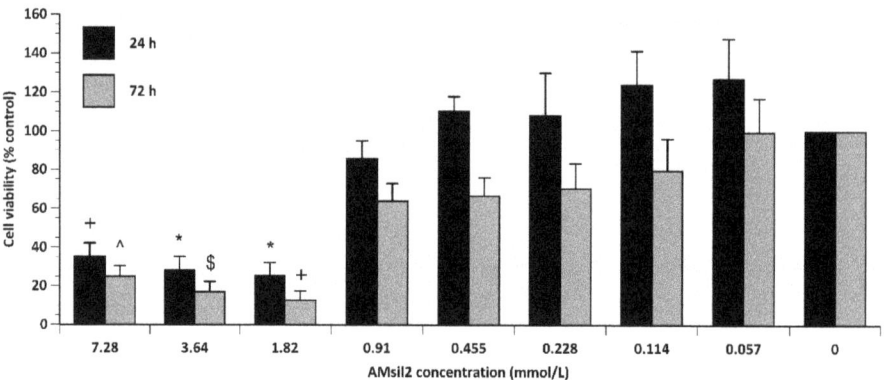

Figure 5. Percent control value of viability of HGF cells exposed to 2-fold serial dilutions of AM_{sil2} (≤7.28 mmol/L) for 24 h or 72 h. Data represent mean ± SEM for five independent replicates tested in triplicate. + indicates $p ≤ 0.05$ when compared to concentrations ≤ 0.455 mmol/L within same time period. * indicates $p ≤ 0.05$ when compared to concentrations ≤ 0.91 mmol/L within same time period. ^ indicates $p ≤ 0.05$ when compared to concentrations ≤ 0.114 mmol/L within same time period. $ indicates $p ≤ 0.05$ when compared to concentrations ≤ 0.228 mmol/L within same time period.

For both AM_{sil}s, control wells (with or without cells) in which the tetrazolium salt reagent was omitted resulted in negligible optical density values. Positive control wells containing unexposed cells (i.e., no AM_{sil}s) that were given an equal volume of culture medium were not significantly different from cells that were previously treated with the viability stain (data not shown).

2.3. Hydrophobicity/Hydrophilicity of the Resins

Copolymers comprised of UPE resin with added AM_{sil}s generally exhibited lower contact angles (CAs) (Figure 6), suggesting change in their hydrophilic/hydrophobic balance toward more hydrophilic surfaces. At 10 mass % monomer in the resin, CAs of both AM_{sil1}–UPE and AM_{sil2}–UPE copolymers (46.9 ± 5.9° and 37.4 ± 9.2°, respectively) were significantly lower (23% and 38% reduction, respectively; $p \leq 0.01$) than the CA of the UPE control 60.8 ± 5.1°. The apparent increase in the CA in going from 10% to 20% AM_{sil} in the resin was significant only for AM_{sil2} ($p \leq 0.035$). The overall order of the decreasing relative hydrophilicity (evidenced by the increasing CA values) of the examined UPE-based copolymers was as follows: (10% AM_{sil2}–UPE $\geq$ 10% AM_{sil1}–UPE) > (20% AM_{sil2}–UPE = 20% AM_{sil1}–UPE) > UPE control.

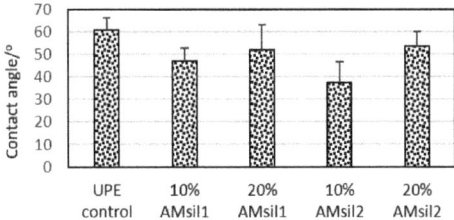

Figure 6. The contact angle (CA) values of AM_{sil}–UPE and UPE control indicative of the changes in resin's overall hydrophilicity/hydrophobicity upon introduction of AM_{sil} monomers at 10 and 20 mass % relative to UPE. Shown are mean values + standard deviation of four repetitive measurements in each experimental group.

2.4. Effect of AM_{sil}s on Degree of Vinyl Conversion (DVC)

Introduction of 10% and 20% AM_{sil1} into UPE reduced ($p \leq 0.05$) the mean vinyl moiety conversion upon photopolymerization by 31% and 20%, respectively (Figure 7). No significant effect was observed with the increasing levels of AM_{sil1} in the resin. Although reduced, the DVC observed amongst the AM_{sil2} groups was not statistically different from one another or the UPE control group.

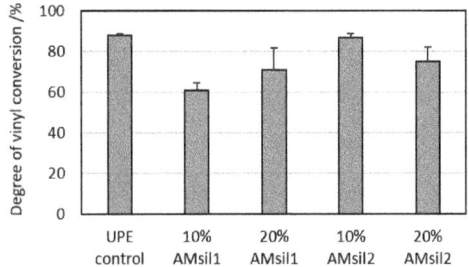

Figure 7. The values for degree of vinyl conversion (DVC) attained 24 h post-cure in AM_{sil}–UPE copolymers compared to no-AM UPE control. Shown are mean values + standard deviation of three repetitive measurements.

2.5. Mechanical Properties of AM_{sil}–UPE Copolymers

The FS and E of AM_{sil}–UPE copolymers were, generally, diminished compared to the UPE resin control (Figure 8). The extent of reduction in FS and E varied with the type and the concentration of AM_{sil}. In all AM_{sil}–UPE formulations, the FS values were significantly ($p \leq 0.05$) lower than the UPE control counterparts. In both 10 mass % AM_{sil} formulations, the E was reduced, although not statistically significant. At 20 mass %, the E of both AM_{sil} formulations were notably lower ($p \leq 0.0008$) than the UPE resin control. Both FS and E reductions ranged from moderate (11–13%) for 10 mass % AM_{sil} formulations to substantial (25–57%) for 20 mass % AM_{sil} formulations.

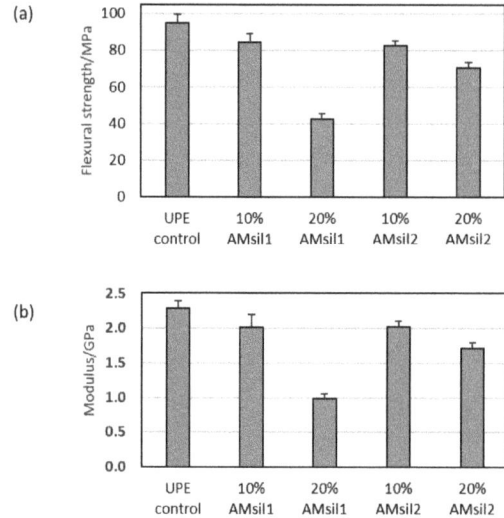

Figure 8. (a) Flexural strength and (b) tensile elasticity of AM$_{sil}$–UPE copolymers in comparison with the UPE control. Indicated are mean values + standard deviation of three specimens.

2.6. Bacterial Testing

For planktonic bacterial testing, the number of *S. mutans* colony-forming units/mL observed amongst the AM$_{sil}$ groups were not statistically different from one another or the control groups (UPE only and commercial control). However, compared to UPE resin, AM$_{sil1}$–UPE and AM$_{sil2}$–UPE (10% mass) copolymers reduced the colonization of *S. mutans* biofilm 4.7- and 1.7-fold, respectively ($p \leq 0.002$) (Figure 9). *S. mutans* biofilms exposed to AM$_{sil1}$–UPE were at least 2.8-fold lower ($p \leq 0.005$) than that observed with AM$_{sil2}$–UPE.

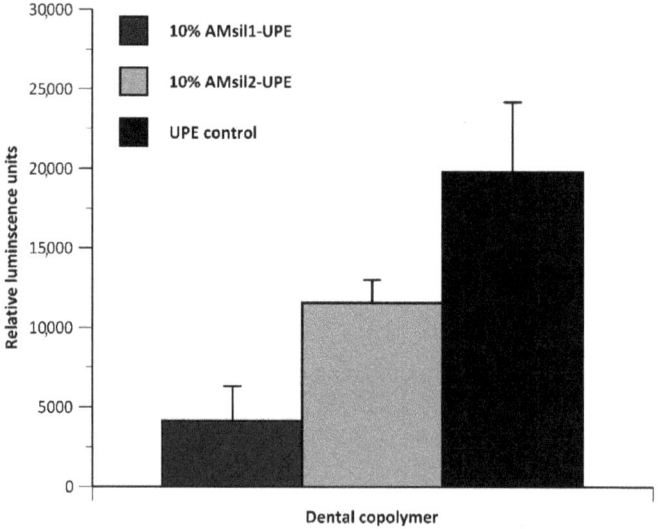

Figure 9. *Streptococcus mutans* biofilm growth inhibition by the experimental AM$_{sil}$s–UPE (10 mass %) copolymers compared to UPE control resin. Bar height indicates mean + standard deviation of 5 specimens/group.

P. gingivalis biofilm biomass on copolymer disks exposed to AM$_{sil1}$–UPE and AM$_{sil2}$–UPE were lower (71% and 85%, respectively) than that observed with the commercial control, albeit not statistically different ($p \leq 0.07$) (Figure 10).

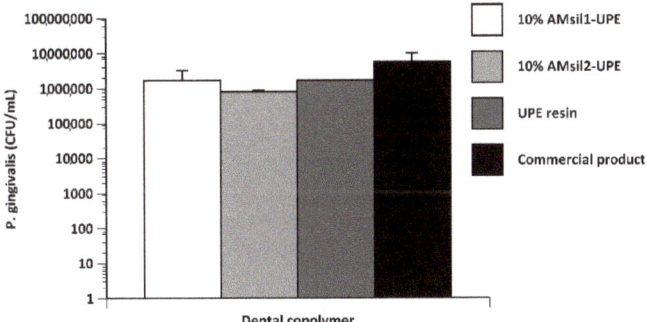

Figure 10. *Porphyromonas gingivalis* biofilm growth inhibition by the experimental AM$_{sil}$s–UPE (10 mass %) copolymers compared to UPE control resin. Bar height indicates mean + standard deviation of 5 specimens/group.

3. Discussion

We considered cytotoxicity of any AM monomer to be a major determinant of whether the materials that incorporate the monomer in their resin phase are worthy of further study. We believe that direct contact cellular testing of the new agent must be done at biologically relevant eluent concentrations. In this study, we employed conditions that reflect the accelerated leachability of UPH resins and included AM$_{sil}$ concentrations that significantly exceed the upper thresholds established experimentally in the previous work [45]. The direct toxicity of AM$_{sil}$s towards CCL1 cells and/or HGFs was demonstrated to be marginal or undetectable, except at the higher concentrations of monomers tested. These high AM$_{sil}$ levels correspond to unrealistically high levels of the unreacted monomers and are highly unlikely to ever be registered clinically. Our cytotoxicity results support the basic hypothesis of the study and suggest that, from a biotoxicity viewpoint, AM$_{sil}$s can be safely utilized in design of AM new materials. Leachability studies of AM–UPE formulations employing high-performance liquid chromatography are currently underway in our laboratory, and are expected to confirm the conclusions derived from the tests based on the accelerated UPH leachability study.

Upon introduction of AM$_{sil}$ into the UPE resin, a shift towards lower CA values, consistent with the moderate increase in the overall hydrophilicity, was seen in all AM$_{sil}$–UPE formulations. The detected range of CAs in AM$_{sil}$–UPE resins (37.4–53.3°) correlates very well with the range of CAs typical for the commercial resin composites (37.4–53.3°) [46]. The lowest CAs detected in 10% AM$_{sil2}$–UPE formulation make this resin a good candidate for the incorporation of the ACP filler in future design of AMRE composites. ACP-filled composite requires sufficient water absorption to initiate water-catalyzed transformation of ACP during which the remineralizing calcium and phosphate ions are released, by diffusion, into surrounding mineral-deficient tooth structures. There, they regenerate these mineral-depleted structures via redeposition of hydroxyapatite [47]. Taken together, the enhanced wettability should ease a diffusion of water into AMRE composite and result in the subsequent release of calcium and phosphate ions from the composite needed for demineralization prevention and/or active remineralization at the restoration site.

The range of DVC values attained in AM$_{sil}$–UPE copolymers (60.7–86.7%), dependent on both the monomer type and its quantity in the resin, were higher or equal to the DVC reported for 2,2-bis[p-(2-hydroxy-3-methacryloxypropoxy)phenyl]propane (bis-GMA)-based resins/composites with incorporated QA ionic dimethacrylate (67.9–70.7%) [25]. AM$_{sil2}$ copolymers reached significantly higher DVC values (75.2–86.7%) compared to their AM$_{sil1}$ counterparts (60.7–70.9%). In AM$_{sil2}$–UPE formulations, the inclusion of AM monomer apparently did not affect the high levels of DVC typically seen in UDMA-based resins [35,36]. This phenomenon has been attributed to the chain transfer reactions caused by UDMA's –NH– groups, resulting in increased mobility of the resin network's radical sites [48]. The DVCs attained in AM_{sil2}–UPE copolymers suggest limited mobility of cross-linked polymer matrix, thus reducing the likelihood of unreacted monomers leaching out to a minimum. The observed DVC decrease in AM_{sil1}–UPE formulations compared to those of AM_{sil2}–UPE is yet to be explained.

The results of FS and E tests indicated a reduction of the copolymers' mechanical properties in going from the UPE control to 10% AM_{sil}–UPE to 20% AM_{sil}–UPE. The reduction was far more pronounced in AM_{sil1} series (50%) compared to the AM_{sil2} series (15%). This overall reduction in mechanical properties, particularly in AM_{sil2}–UPE copolymers, should not disqualify this monomer from further exploration, as AM agent in multifunctional AMRE composites. To compensate for the reduction in the mechanical properties, incorporation into composites of the reinforcing fillers in addition to ACP should be considered.

AM_{sil}s integrated into UPE resin were effective in reducing ($p \leq 0.002$) S. mutans biofilms. Compared to the commercial control, P. gingivalis biofilm biomass was notably lower (71–85%) on AM_{sil}s–UPE copolymer disks, however, not statistically different. As Gram-positive bacteria have peptidoglycan with long anionic polymers, called teichoic acids [49] (i.e., yielding a higher cell surface net negative charge than Gram-negative organisms), one could anticipate S. mutans to be more susceptible to AM_{sil}s. Notwithstanding, quaternary ammonium compound AM functionality can also be affected by the type of counter-ion [50], pendant active groups [51], molecular weight, and length of the alkyl chains [52].

AM_{sil}s show promise, as S. mutans planktonic and biofilm forms were reduced. Nonetheless, this reduction was only significant ($p \leq 0.002$) for biofilms. When comparing planktonic and biofilm responses, similar trends were observed with an endodontic sealer. For example, bacteria tested with a resin-based root canal sealer did not statistically reduce the planktonic forms, while notably decreasing bacteria in monospecies biofilms [53]. These authors attributed this to the release of substances during the setting process. This attribute is unlikely applicable to our material, as DVC was high and aqueous extraction was conducted for 72 h. In another report, Staphylococcus aureus biofilms were demonstrated to be more susceptible to killing than the planktonic form of the same strain [54].

Although AM functionality is observed, we believe that the full potential of the AM_{sil} monomers has not yet been realized. As reported for other quaternary ammonium compounds [47], the current AM_{sil}–UPE copolymer formulations are likely to have N^+ charges randomly distributed throughout the material. As the mechanism of AM action is contingent on contact, it would be advantageous to develop fabrication methods that would favor charge density at the materials surface. Further, others have demonstrated that proteins can diminish the AM capability of quaternary ammonium methacrylates (reviewed by [6]). Currently, there is insufficient information concerning the interaction of proteins with quaternary ammonium methacrylates. Elucidation of the protein–material interactions would yield valuable information to develop strategies to maximize AM efficacy of materials with a charge-based AM mechanism of action.

In conclusion, our novel AM dental monomers (AM_{sil1} and AM_{sil2}) exhibited minimal or no toxicity upon direct contact with biologically relevant concentrations, while reducing S. mutans and P. gingivalis biofilm forms. AM_{sil}s made the UDMA/PEG-U/EHMA resin more hydrophilic. This would be an advantageous feature in AMRE composites that require water to induce their remineralizing effects. At 10 mass % level of AM_{sil} monomer, DVC of the ensuing AM_{sil}–UPE copolymers was only marginally lower than in UPE control and still exceeded DVCs typically seen in the commercial composites based on bis-GMA/triethyleneglycol dimethacrylate (TEGDMA) resins. The mechanical properties of AM_{sil}–UPE copolymers were reduced (11–57%) compared with the UPE control. The extent of reduction depended on both the type and the concentration of AM_{sil} monomer in the resin. This finding should not disqualify the AM_{sil}–UPE resins from use in AMRE composites intended for Class V restorations where the mechanical stability is not a critical factor.

4. Materials and Methods

4.1. Monomer Synthesis

The synthesis and validation protocols for the AM_{sil}s are described in detail by Okeke et al. (2019) [40]. In brief, AM_{sil1} and AM_{sil2} were synthesized at 50–55 °C by reacting equimolar amounts of tertiary amine, DMAEMA, with IPTMS and BrUDTMS, respectively, in the presence of chloroform and butylated hydroxytoluene. DMAEMA, IPTMS, and butylated hydroxytoluene were purchased from Sigma, St. Louis, MO, USA. BrUDTMS was purchased from Gelest Inc., Morrisville, PA, USA. Reactants and solvents (chloroform, diethyl ether, hexane; Sigma, St. Louis, MO, USA) used during synthesis and the subsequent purification were used as received, without further purification. The reaction yields were 94.8% and 36.0% for AM_{sil1} and AM_{sil2}, respectively. Due to the generally hygroscopic nature of QA monomers, the AM_{sil}s were stored under vacuum (25 mm Hg) before being used for resin formulation and/or copolymer disk specimen preparation.

4.2. Structural Verification

Purified monomers were characterized by ^1H and ^{13}C NMR spectroscopy as described [40]. Briefly, spectra were obtained using a Bruker Advance II (600 MHz) spectrometer equipped with a Broadband Observe room temperature probe (Bruker, Corp., Billerica, MA, USA). Monomers were dissolved in deuterated dimethyl sulfoxide containing tetramethylsilane.

4.3. Experimental Resin Formulation

UPE resin was formulated from the commercially available monomers UDMA, PEG-U, and EHMA at 2.8:1.0:1.7 mass ratio (corresponds to the average mass ratio of the ternary UDMA-based formulations explored by our group so far). A conventional visible light initiator system comprised of camphorquinone and ethyl-4-N,N-dimethylamino benzoate (4EDMAB) was introduced to the resin at concentrations of 0.2 mass % camphorquinone and 0.8 mass % 4EDMAB. AM_{sil}s were blended into UPE resin to yield (AM_{sil1} or AM_{sil2})–UPE resin with 10 or 20 mass % of AM component. Addition of AM_{sil1} or AM_{sil2} to the light-activated UPE resin took place in the absence of blue light. The rationale for the chosen levels of AM monomers is based on the previously reported AM activities of similar QA methacrylates in camphorquinone /4EDMAB-activated bis-GMA/TEGDMA resins [25]. Once all components were introduced, the mixture was stirred magnetically (38 rad/s) at 22 °C until a uniform consistency was achieved. CLEARFIL SE Protect BOND (Kurary America, Inc., New York, NY, USA) was used as a comparative commercial AM material. This resin was prepared using equal quantities of primer (containing MDPB) and bonding (containing bis-GMA-HEMA) agent.

4.4. Biocompatibility Tests

Direct contact cytotoxicity of AM_{sil}s was determined following described protocols [55,56]. Briefly, immortalized mouse subcutaneous connective tissue fibroblasts (NCTC clone 929 [L-cell, L-929, Strain L derivative]; American Type Culture Collection (ATCC), Manassas, VA, USA) (CCL1) or HGF (Applied Biological Materials, Inc., Richmond, BC, Canada) were exposed to 2-fold serial dilutions (AM_{sil1}: ≤ 8.34 mmol/L; AM_{sil2}: ≤ 7.28 mmol/L). Chosen concentrations corresponded to approx. 7% mass fraction of AM_{sil1} or AM_{sil2} in the copolymer resin and a maximum of 2% leaching. To allow for the possibility of restoration multiplicity and variable size, a 2-fold greater dilution was also included in the testing. These calculations are based on the accelerated leachability study of UDMA/PEG-U/2-hydroxyethyl methacrylate (HEMA) resin (abbreviated UPH; a close analog to UPE resin used in this study) and ACP-UPH composites [57]. After 24 and 72 h incubation, cells were assessed for cell viability (LIVE/DEAD® Viability/Cytotoxicity kit, Life Technologies, Corp., Grand

Island, NY, USA) and metabolic activity (CellTiter® AQueous One Solution Reagent; Promega, Corp., Madison, WI, USA). Controls were without the AM_{sil}s and/or cells. The CCL1 cells and HGFs were maintained, at 37 °C and 5% CO_2, in 10% serum-supplemented Eagle's minimum essential medium (ATCC) and PriGrow III medium (Applied Biological Materials, Inc.), respectively. For experiments, cells were obtained from a subconfluent stock culture. Means were obtained from 5 independent replicates tested in duplicate.

4.5. Contact Angle (CA)

Changes in hydrophilicity/hydrophobicity of UPE resins due to the introduction of AM_{sil}s were assessed by CA measurements (drop shape analyzer DSA100, Krüss GmbH, Hamburg, Germany). Following the deposition of the sessile droplets of the resin on the substrate, they were imaged after 1 min resting time with a charge-coupled device camera at the points of intersection (three-phase contact points) between the drop contour and the projection of the surface (baseline). The CA water values were calculated employing the Krüss Advance software. Four repetitive measurements were performed in each group.

4.6. Copolymer Specimen Preparation

For biotesting, UPE and AM_{sil1}–UPE and $AMsi_{l2}$–UPE copolymer specimens were fabricated by filling circular openings of a flat stainless-steel molds (6 mm diameter, 0.5 mm thickness) with the resins. Each side of the mold was covered with Mylar film and a glass slide, firmly clamped, and then cured (2 min/side: Triad 2000; Dentsply International, York, PA, USA).

Specimens were subjected to extraction in Dulbecco's phosphate-buffered saline lacking both calcium and magnesium (Life Technologies, Grand Island, NY, USA) for 72 h, at 37 °C and 6.3 rad/s using an orbital shaker. After extraction, the disks were dried under vacuum (desiccator; ~22 °C) for 7 days. Specimens were sterilized for 12 h using an Anprolene gas sterilization chamber (Andersen Products, Inc., Haw River, NC, USA). Prior to bacterial testing, specimens were degassed for ≥5 days under vacuum (desiccator; ~22 °C).

4.7. Degree of Vinyl Conversion (DVC)

DVC of UPE and (AM_{sil1} or AM_{sil2})–UPE resins was determined by collecting the near-IR (NIR) spectra (Nexus; ThermoFisher, Madison, WI, USA) before and 24 h after the light cure and calculating the reduction in =C–H absorption band at 6165 cm^{-1} in the overtone region in going from monomers to polymers. By maintaining a constant specimen thickness, a need for an invariant internal standard was eliminated. The DVC was calculated as

$$DVC\ (\%) = [(area_{monomer} - area_{polymer})/area_{monomer}] \times 100 \qquad (1)$$

where $area_{polymer}$ and $area_{monomer}$ correspond to the areas under 6165^{-1} absorption peak after and before the polymerization, respectively.

4.8. Mechanical Properties of Copolymers

Test specimens (2 mm × 2 mm × 25 mm) for flexural strength (FS) and elastic modulus (E) determinations were photopolymerized in the same manner as the copolymer disks for biological testing. Polymerized specimens did not undergo any additional treatment. The FS and E of UPE and (AM_{sil1} or AM_{sil2}) UPE copolymer specimens was tested employing the Universal Testing Machine (Instron 5500R, Instron Corp., Canton, MA, USA). The load was applied (crosshead speed of 1 mm/min) to the center of a specimen positioned on a test device with supports 20 mm apart. The FS and E of the specimens (three replicates/experimental group) were calculated as instructed in the ISO4049:2009 document.

4.9. Bacterial Testing

4.9.1. Planktonic

Testing of *Streptococcus mutans* UA-159 (ATCC® 700610) planktonic forms was according to described methods [47,58]. Briefly, bacterial cultures with an optical density of 1.2–1.3 at 600 nm (Unico® 1200 Spectrophotometer, United Products & Instruments, Inc., Dayton, NJ, USA) were diluted and seeded onto copolymer disks at a density of ~3 × 10^7 CFU/disk. Another disk was placed atop to maximize contact. After a 2 h incubation (37 °C, 5% CO_2) the samples were placed in 1 mL of Todd Hewitt broth, rigorously mixed, and utilized to prepare a 10-fold dilution series. A 100 µL aliquot of the resulting suspensions were spread onto the surface of THB agar plates. After incubation (~20 h at 37 °C, 5% CO_2), colony-forming units were enumerated using an IncuCount Colony Counter (Revolutionary Science, Shafer, MN, USA). UPE resin disks and HemCon® Dental Dressing (HemCon Medical Technologies, Inc., Portland, OR, USA) were used as negative and positive controls, respectively. Around 30 to 300 CFU per spread plate was the range considered countable. Notwithstanding, agar plates streaked with neat solutions of some groups yielded less than the lower limit of detection. Such data were reported as less than the limit of quantification. The number of CFU/mL was calculated as CFU number/(volume plated × dilution factor). Mean counts were obtained from five independently tested copolymer disk sandwiches.

4.9.2. Biofilm

A bioluminescent *S. mutans* strain JM 10 (derivative of wild type UA159 [59] was used to assess the AM properties of $AM_{sil}s$. Methods of the real-time bioluminescence assay were as described [41,58].

Porphyromonas gingivalis, strain FDC 381 (ATCC® BAA-1703) was propagated in Becton Dickinson BBL chopped meat carbohydrate, pre-reduced II broth, using a shaking incubator (37 °C, anaerobic conditions). Three-day cultures were diluted in broth to approximate 5 × 10^6 CFU/mL. Copolymer disks, vertically supported in a 24-well plate, were immersed in 1.6 mL of the bacterial suspension. In anaerobic conditions, the plate was incubated at 37 °C for 4 days. The copolymer disks were washed thrice in sterile 0.89% NaCl solution. Thereafter, the biofilm was displaced from the copolymer disks by transferring them to a sterile glass tube containing 1 mL of saline, vortexed (1 min), sonicated (10 min), and vortexed (1 min). Each disk was visually examined to ensure that the biomass was removed. The resulting suspensions were used to make 10-fold serial dilutions and subsequently spread onto the surface of Brucella agar with hemin and vitamin K1 (Sigma-Aldrich, St. Louis, MO, USA) plates. After incubation (3 days at 37 °C, anaerobic conditions), colony-forming units were enumerated.

4.10. Statistical Analyses

Analysis of variance and multiple paired comparisons (two-sided, 95% confidence interval) were used to analyze the experimental data as a function of material makeup and/or exposure/incubation time and establish a statistical significance of differences between the experimental groups. Correlation coefficient (r) was calculated to determine the functional dependence between cellular metabolic activity and viability. (SigmaPlot™, Systat Software, San Jose, CA, USA and/or Microsoft Office Excel 2016; Microsoft, Redmond, WA, USA). Graphics were created using Microsoft Office Excel 2016 and/or DeltaGraph6 for Windows® (Red Rock Software, Inc., Salt Lake City, UT, USA).

Author Contributions: D.R.B. conceptualized and conducted biotesting, fabricated copolymer specimens, and conducted statistical analyses. A.A.G. conducted biotesting, physicochemical/mechanical testing of copolymers. S.A.F. conducted monomer synthesis/validation, fabrication of copolymer resins, and measured their physicochemical/mechanical properties. S.S.K., F.L.E.F., and R.D.H. conducted S. mutans biofilm analyses. D.S. conceptualized the AM monomers, performed literature compilation, and statistical data interpretation. All authors contributed to the preparation/review of this manuscript. All authors have read and agreed to the published version of the manuscript.

Funding: This study was supported by the National Institute for Dental and Craniofacial Research (grant R01 DE26122), American Dental Association (ADA), and ADA Foundation.

Acknowledgments: Assistance of U.C. Okeke (formerly ADA Foundation) in monomer synthesis/characterization are gratefully acknowledged. Technical expertise of H. Kim (Integrated Pharma Services) to assess AM properties of *P. gingivalis* is appreciated. Authors gratefully acknowledge donation of UDMA and PEG-U from Esstech, Essington, PA, USA.

Conflicts of Interest: D.R.B., A.A.G., S.A.F., and D.S. are employees of the non-profit ADA Foundation, which has applied for a patent describing synthesis and uses of polymerizable multifunctional antimicrobial quaternary ammonium monomers.

Disclaimer: The sole purpose of identifying certain commercial materials and equipment in this article was to adequately define the experimental protocols. Such identification, in no instance, implies recommendation or endorsement by the ADA or ADA Foundation, or means that the material/equipment specified is the best available for the purpose.

Abbreviations

ACP	amorphous calcium phosphate
AM	antimicrobial
AMRE	antimicrobial and remineralizing
AM_{sil1}	N-(2-(methacryloyloxy)ethyl-N,N-dimethyl-3-(trimethoxysilyl)propan-1-aminium iodide
AM_{sil2}	N-(2-(methacryloyloxy)ethyl-N,N-dimethyl-3-(trimethoxysilyl)undecan-1-aminium bromide
Bis-GMA	2,2-bis[p-(2-hydroxy-3-methacryloxypropoxy)phenyl]propane
BrUDTMS	(11-bromoundecyl)trimethoxy silane
CA	contact angle
CCL1	immortalized mouse subcutaneous connective tissue fibroblasts
CQ	camphorquinone
DMAEMA	2-(dimethylamino)ethyl methacrylate
DVC	degree of vinyl conversion
4EDMAB	ethyl-4-N,N-dimethylamino benzoate
EHMA	ethyl 2-(hydroxymethyl) acrylate
HGF	human gingival fibroblasts
IPTMS	(3-iodopropyl)trimethoxy silane
MDBP	methacryloyloxydodecyl pyrimidinium bromide
NMR	nuclear magnetic resonance
PEG-U	poly(ethylene glycol)-extended UDMA
QA	quaternary ammonium
TEGDMA	triethyleneglycol dimethacrylate
UDMA	urethane dimethacrylate
UPE	UDMA/PEG-U/EHMA resin

References

1. Gluzman, R.; Katz, R.V.; Frey, B.J.; McGowan, R. Prevention of root caries: A literature review of primary and secondary preventive agents. *Spec. Care Dentist.* **2013**, *33*, 133–140. [CrossRef] [PubMed]
2. Cramer, N.B.; Stansbury, J.W.; Bowman, C.N. Recent advances and developments in composite dental restorative materials. *J. Dent. Res.* **2011**, *90*, 402–416. [CrossRef] [PubMed]
3. Ferracane, J.L. Resin composite—state of the art. *Dent. Mater.* **2011**, *27*, 29–38. [CrossRef] [PubMed]
4. Moreau, J.L.; Xu, H.H. Fluoride releasing restorative materials: Effects of pH on mechanical properties and ion release. *Dent. Mater.* **2010**, *26*, e227–e235. [CrossRef] [PubMed]
5. Wiegand, A.; Buchalla, W.; Attin, T. Review on fluoride-releasing restorative materials–fluoride release and uptake characteristics, antibacterial activity and influence on caries formation. *Dent. Mater.* **2007**, *23*, 343–362. [CrossRef] [PubMed]
6. Imazato, S. Bio-active restorative materials with antibacterial effects: New dimension of innovation in restorative dentistry. *Dent. Mater. J.* **2009**, *28*, 11–19. [CrossRef] [PubMed]
7. Pereira-Cenci, T.; Cenci, M.S.; Fedorowicz, Z.; Marchesan, M.A. Antibacterial agents in composite restorations for the prevention of dental caries. *Cochrane Database Syst. Rev.* **2009**. [CrossRef]

8. Dallas, P.; Sharma, V.K.; Zboril, R. Silver polymeric nanocomposites as advanced antimicrobial agents: Classification, synthetic paths, applications, and perspectives. *Adv. Colloid Interface Sci.* **2011**, *166*, 119–135. [CrossRef]
9. Jedrychowski, J.R.; Caputo, A.A.; Kerper, S. Antibacterial and mechanical properties of restorative materials combined with chlorhexidines. *J. Oral Rehabil.* **1983**, *10*, 373–381. [CrossRef]
10. Kawahara, K.; Tsuruda, K.; Morishita, M.; Uchida, M. Antibacterial effect of silver-zeolite on oral bacteria under anaerobic conditions. *Dent. Mater.* **2000**, *16*, 452–455. [CrossRef]
11. Knetsch, M.L.; Koole, L.H. New strategies in the development of antimicrobial coatings: The example of increasing usage of silver and silver nanoparticles. *Polymers* **2011**, *3*, 340–366. [CrossRef]
12. Osinaga, P.W.; Grande, R.H.; Ballester, R.Y.; Simionato, M.R.; Delgado Rodrigues, C.R.; Muench, A. Zinc sulfate addition to glass-ionomer-based cements: Influence on physical and antibacterial properties, zinc and fluoride release. *Dent. Mater.* **2003**, *19*, 212–217. [CrossRef]
13. Syafiuddin, T.; Hisamitsu, H.; Toko, T.; Igarashi, T.; Goto, N.; Fujishima, A.; Miyazaki, T. In vitro inhibition of caries around a resin composite restoration containing antibacterial filler. *Biomaterials* **1997**, *18*, 1051–1057. [CrossRef]
14. Takahashi, Y.; Imazato, S.; Kaneshiro, A.V.; Ebisu, S.; Frencken, J.E.; Tay, F.R. Antibacterial effects and physical properties of glass-ionomer cements containing chlorhexidine for the ART approach. *Dent. Mater.* **2006**, *22*, 647–652. [CrossRef] [PubMed]
15. Yoshida, K.; Tanagawa, M.; Atsuta, M. Characterization and inhibitory effect of antibacterial dental resin composites incorporating silver-supported materials. *J. Biomed. Mater. Res.* **1999**, *47*, 516–522. [CrossRef]
16. Weng, Y.; Guo, X.; Chong, V.J.; Howard, L.; Gregory, R.L.; Xie, D. Synthesis and evaluation of a novel antibacterial dental resin composite with quaternary ammonium salts. *J. Biomed. Sci. Eng.* **2011**, *4*, 147. [CrossRef]
17. Makvandi, P.; Jamaledin, R.; Jabbari, M.; Nikfarjam, N.; Borzacchiello, A. Antibacterial quaternary ammonium compounds in dental materials: A systematic review. *Dent. Mater.* **2018**, *34*, 851–867. [CrossRef] [PubMed]
18. Xue, Y.; Xiao, H.; Zhang, Y. Antimicrobial polymeric materials with quaternary ammonium and phosphonium salts. *Int. J. Mol. Sci.* **2015**, *16*, 3626–3655. [CrossRef]
19. Gottenbos, B.; van der Mei, H.C.; Klatter, F.; Nieuwenhuis, P.; Busscher, H.J. In vitro and in vivo antimicrobial activity of covalently coupled quaternary ammonium silane coatings on silicone rubber. *Biomaterials* **2002**, *23*, 1417–1423. [CrossRef]
20. Lee, S.B.; Koepsel, R.R.; Morley, S.W.; Matyjaszewski, K.; Sun, Y.; Russell, A.J. Permanent, nonleaching antibacterial surfaces. 1. Synthesis by atom transfer radical polymerization. *Biomacromolecules* **2004**, *5*, 877–882. [CrossRef]
21. Lu, G.; Wu, D.; Fu, R. Studies on the synthesis and antibacterial activities of polymeric quaternary ammonium salts from dimethylaminoethyl methacrylate. *React. Funct. Polym.* **2007**, *67*, 355–366. [CrossRef]
22. Li, F.; Chen, J.; Chai, Z.; Zhang, L.; Xiao, Y.; Fang, M.; Ma, S. Effects of a dental adhesive incorporating antibacterial monomer on the growth, adherence and membrane integrity of Streptococcus mutans. *J. Dent.* **2009**, *37*, 289–296. [CrossRef] [PubMed]
23. Thome, T.; Mayer, M.P.; Imazato, S.; Geraldo-Martins, V.R.; Marques, M.M. In vitro analysis of inhibitory effects of the antibacterial monomer MDPB-containing restorations on the progression of secondary root caries. *J. Dent.* **2009**, *37*, 705–711. [CrossRef] [PubMed]
24. Antonucci, J.M. Polymerizable biomedical composition. U.S. Patent 8,217,081 B2, 10 July 2012.
25. Antonucci, J.M.; Zeiger, D.N.; Tang, K.; Lin-Gibson, S.; Fowler, B.O.; Lin, N.J. Synthesis and characterization of dimethacrylates containing quaternary ammonium functionalities for dental applications. *Dent. Mater.* **2012**, *28*, 219–228. [CrossRef]
26. Li, F.; Chai, Z.G.; Sun, M.N.; Wang, F.; Ma, S.; Zhang, L.; Fang, M.; Chen, J.H. Anti-biofilm effect of dental adhesive with cationic monomer. *J. Dent. Res.* **2009**, *88*, 372–376. [CrossRef]
27. Boskey, A.L. Amorphous calcium phosphate: The contention of bone. *J. Dent. Res.* **1997**, *76*, 1433–1436. [CrossRef]
28. Dorozhkin, S.V. Biocomposites and hybrid biomaterials based on calcium orthophosphates. *Biomatter* **2011**, *1*, 3–56. [CrossRef]
29. Dorozhkin, S.V. Calcium orthophosphates: Occurrence, properties, biomineralization, pathological calcification and biomimetic applications. *Biomatter* **2011**, *1*, 121–164. [CrossRef]

30. He, G.; Dahl, T.; Veis, A.; George, A. Nucleation of apatite crystals in vitro by self-assembled dentin matrix protein 1. *Nat. Mater.* **2003**, *2*, 552–558. [CrossRef]
31. Tsuji, T.; Onuma, K.; Yamamoto, A.; Iijima, M.; Shiba, K. Direct transformation from amorphous to crystalline calcium phosphate facilitated by motif-programmed artificial proteins. *Proc. Natl. Acad. Sci. USA* **2008**, *105*, 16866–16870. [CrossRef]
32. Weiner, S. Transient precursor strategy in mineral formation of bone. *Bone* **2006**, *39*, 431–433. [CrossRef] [PubMed]
33. Weiner, S.; Sagi, I.; Addadi, L. Structural biology. Choosing the crystallization path less traveled. *Science* **2005**, *309*, 1027–1028. [CrossRef] [PubMed]
34. Tadic, D.; Peters, F.; Epple, M. Continuous synthesis of amorphous carbonated apatites. *Biomaterials* **2002**, *23*, 2553–2559. [CrossRef]
35. Antonucci, J.M.; Skrtic, D. Fine-tuning of polymeric resins and their interfaces with amorphous calcium phosphate. A strategy for designing effective remineralizing dental composites. *Polymers* **2010**, *2*, 378–392. [CrossRef]
36. Skrtic, D.; Antonucci, J.M. Dental composites based on amorphous calcium phosphate - resin composition/physicochemical properties study. *J. Biomater. Appl.* **2007**, *21*, 375–393. [CrossRef]
37. Skrtic, D.; Antonucci, J.M.; Eanes, E.D. Amorphous calcium phosphate-based bioactive polymeric composites for mineralized tissue regeneration. *J. Res. Natl. Inst. Stan.* **2003**, *108*, 167–182. [CrossRef]
38. Skrtic, D.; Antonucci, J.M.; Eanes, E.D.; Eidelman, N. Dental composites based on hybrid and surface-modified amorphous calcium phosphates. *Biomaterials* **2004**, *25*, 1141–1150. [CrossRef]
39. Zhang, F.; Allen, A.J.; Levine, L.E.; Vaudin, M.D.; Skrtic, D.; Antonucci, J.M.; Hoffman, K.M.; Giuseppetti, A.A.; Ilavsky, J. Structural and dynamical studies of acid-mediated conversion in amorphous-calcium-phosphate based dental composites. *Dent. Mater.* **2014**, *30*, 1113–1125. [CrossRef]
40. Okeke, U.C.; Synder, C.R.; Frukhtbeyn, S.A. Synthesis, purification and characterization of polymerizable multifunctional quaternary ammonium compounds. *Molecules* **2019**, *24*, 1464. [CrossRef]
41. Esteban Florez, F.L.; Hiers, R.D.; Smart, K.; Kreth, J.; Qi, F.; Merritt, J.; Khajotia, S.S. Real-time assessment of Streptococcus mutans biofilm metabolism on resin composite. *Dent. Mater.* **2016**, *32*, 1263–1269. [CrossRef]
42. Forssten, S.D.; Bjorklund, M.; Ouwehand, A.C. Streptococcus mutans, caries and simulation models. *Nutrients* **2010**, *2*, 290–298. [CrossRef] [PubMed]
43. Fenesy, K.E. Periodontal disease: An overview for physicians. *Mt. Sinai J. Med.* **1998**, *65*, 362–369. [PubMed]
44. Rafiei, M.; Kiani, F.; Sayehmiri, K.; Sayehmiri, F.; Tavirani, M.; Dousti, M.; Sheikhi, A. Prevalence of Anaerobic Bacteria (P.gingivalis) as Major Microbial Agent in the Incidence Periodontal Diseases by Meta-analysis. *J. Dent.* **2018**, *19*, 232–242.
45. Antonucci, J.M.; Davis, C.H.; Sun, J.; O'Donnell, J.N.; Skrtic, D. Leachability and Cytotoxicity of an Experimental Polymeric ACP Composite. *PMSE Prepr. Am. Chem. Soc. Div. Polym. Mater. Sci. Eng. Meet.* **2011**, *104*, 250–252.
46. Da Silva, E.M.; Almeida, G.S.; Poskus, L.T.; Guimaraes, J.G. Relationship between the degree of conversion, solubility and salivary sorption of a hybrid and a nanofilled resin composite. *J. Appl. Oral Sci.* **2008**, *16*, 161–166. [CrossRef] [PubMed]
47. Bienek, D.R.; Giuseppetti, A.A.; Skrtic, D. Amorphous calcium phosphates as bioactive filler in polymeric dental composites. In *Calcium Phosphates-From Fundamentals to Applications*; Dutour-Sikiric, M., Furedi-Milhofer, H., Eds.; IntechOpen Limited: London, UK, 2019. [CrossRef]
48. Sideridou, I.; Tserki, V.; Papanastasiou, G. Effect of chemical structure on degree of conversion in light-cured dimethacrylate-based dental resins. *Biomaterials* **2002**, *23*, 1819–1829. [CrossRef]
49. Silhavy, T.J.; Kahne, D.; Walker, S. The bacterial cell envelope. *Cold Spring Harb. Perspect. Biol.* **2010**, *2*, a000414. [CrossRef]
50. Chen, C.Z.; Beck-Tan, N.C.; Dhurjati, P.; van Dyk, T.K.; LaRossa, R.A.; Cooper, S.L. Quaternary ammonium functionalized poly (propylene imine) dendrimers as effective antimicrobials: Structure–activity studies. *Biomacromolecules* **2000**, *1*, 473–480. [CrossRef]
51. Ikeda, T.; Hirayama, H.; Yamaguchi, H.; Tazuke, S.; Watanabe, M. Polycationic biocides with pendant active groups: Molecular weight dependence of antibacterial activity. *Antimicrob. Agents Chemother.* **1986**, *30*, 132–136. [CrossRef]

52. Ikeda, T.; Yamaguchi, H.; Tazuke, S. Molecular weight dependence of antibacterial activity in cationic disinfectants. *J. Bioact. Compat. Polym.* **1990**, *5*, 31–41. [CrossRef]
53. Kapralos, V.; Koutroulis, A.; Ørstavik, D.; Sunde, P.T.; Rukke, H.V. Antibacterial Activity of Endodontic Sealers against Planktonic Bacteria and Bacteria in Biofilms. *J. Endod.* **2018**, *44*, 149–154. [CrossRef] [PubMed]
54. Harrison, J.J.; Ceri, H.; Stremick, C.; Turner, R.J. Differences in biofilm and planktonic cell mediated reduction of metalloid oxyanions. *FEMS Microbiol. Lett.* **2004**, *235*, 357–362. [CrossRef] [PubMed]
55. Bienek, D.R.; Frukhtbeyn, S.A.; Giuseppetti, A.A.; Okeke, U.C.; Pires, R.M.; Antonucci, J.M.; Skrtic, D. Ionic dimethacrylates for antimicrobial and remineralizing dental composites. *Ann. Dent. Oral Disord.* **2018**, *1*, 108.
56. Bienek, D.R.; Frukhtbeyn, S.A.; Giuseppetti, A.A.; Okeke, U.C.; Skrtic, D. Antimicrobial monomers for polymeric dental restoratives: Cytotoxicity and physicochemical properties. *J. Funct. Biomat.* **2018**, *9*, 20. [CrossRef] [PubMed]
57. Skrtic, D.; Antonucci, J.M. Bioactive polymeric composites for tooth mineral regeneration: Physicochemical and cellular aspects. *J. Funct. Biomat.* **2011**, *2*, 271–307. [CrossRef]
58. Bienek, D.R.; Giuseppetti, A.A.; Okeke, U.C.; Frukhtbeyn, S.A.; Dupree, P.J.; Khajotia, S.S.; Esteban Florez, F.L.; Hiers, R.D.; Skrtic, D. Antimicrobial, biocompatibility, and physicochemical properties of novel adhesive methacrylate dental monomers. *J. Bioact. Compat. Polym.* **2019**, in press.
59. Merritt, J.; Kreth, J.; Qi, F.; Sullivan, R.; Shi, W. Non-disruptive, real-time analyses of the metabolic status and viability of Streptococcus mutans cells in response to antimicrobial treatments. *J. Microbiol. Methods* **2005**, *61*, 161–170. [CrossRef]

© 2019 by the authors. Licensee MDPI, Basel, Switzerland. This article is an open access article distributed under the terms and conditions of the Creative Commons Attribution (CC BY) license (http://creativecommons.org/licenses/by/4.0/).

Article

Understanding the Role of Shape and Composition of Star-Shaped Polymers and their Ability to Both Bind and Prevent Bacteria Attachment on Oral Relevant Surfaces

Hamid Mortazavian [1], Guillaume A. Picquet [2], Jānis Lejnieks [1], Lynette A. Zaidel [2], Carl P. Myers [2,*] and Kenichi Kuroda [1,*]

1. Department of Biologic and Materials Sciences & Prosthodontics, School of Dentistry, University of Michigan, Ann Arbor, MI 48109, USA; mortazavian.hamid@gmail.com (H.M.); jaanislejnieks@gmail.com (J.L.)
2. Oral Care Early Research, Colgate-Palmolive Company, Piscataway, NJ 08855, USA; guillaume_picquet@colpal.com (G.A.P.); Lynette_Zaidel@colpal.com (L.A.Z.)
* Correspondence: carl_myers@colpal.com (C.P.M.); kkuroda@umich.edu (K.K.)

Received: 25 November 2019; Accepted: 13 December 2019; Published: 17 December 2019

Abstract: In this study, we have prepared a series of 4- and 6-arm star-shaped polymers with varying molecular weight and hydrophobicity in order to provide insight into the role and relationship that shape and composition have on the binding and protecting of oral relevant surfaces (hydroxyapatite, HAP) from bacteria colonization. Star-shaped acrylic acid polymers were prepared by free-radical polymerization in the presence of chain transfer agents with thiol groups, and their binding to the HAP surfaces and subsequent bacteria repulsion was measured. We observed that binding was dependent on both polymer shape and hydrophobicity (star vs. linear), but their relative efficacy to reduce oral bacteria attachment from surfaces was dependent on their hydrophobicity only. We further measured the macroscopic effects of these materials to modify the mucin-coated HAP surfaces through contact angle experiments; the degree of angle change was dependent on the relative hydrophobicity of the materials suggesting future in vivo efficacy. The results from this study highlight that star-shaped polymers represent a new material platform for the development of dental applications to control bacterial adhesion which can lead to tooth decay, with various compositional and structural aspects of materials being vital to effectively design oral care products.

Keywords: polymer; dental; antibacterial; antifouling; hydroxyapatite; star-shaped; hydrophobicity; acrylic acid; oral; composition

---

1. Introduction

The control and reduction of oral biofilm formation [1], initiated by bacterial species living in polymicrobial, pathogenic colonies at or below the gingival margin [2], are critical steps toward the prevention of dental caries and periodontal diseases [3–5]. While many methods have been proposed to prevent or treat these biofilms [6–10], one promising strategy is the use of synthetic polymer additives that bind to the tooth surface to act as a barrier or deterrent to the deposition of planktonic bacteria through either lethal [11–13] or non-lethal [14,15] mechanisms. Dental materials, especially those delivered from common over the counter products such as toothpaste or mouthwash, must effectively be multifunctional materials in that they must (1) deposit and stick to a tooth surface, (2) act as a barrier against bacteria attachment, (3) be robust against external challenges such as food and drink in order to not require constant reapplication, and (4) perform steps 1–3 in the presence of the salivary pellicle. Conventional polymers used for this strategy include poly (methyl vinyl ether/maleic acid) (Gantrez) and cross-linked micron-sized polyacrylic acid particles (carbopol). Additional polymers such as polyaspartate adhered to hydroxyapatite (HAP) and reduced the attachment of *Streptococcus sanguinis* [16]. While these materials do show the ability to reduce bacterial attachment to the tooth surface, the relationship between a polymer's structure and composition to maximize efficacy, i.e., to bind to an enamel surface and provide anti-attachment properties, remains disconnected.

We have previously shown that star-shaped polymers with pre-assembled poly(hydroxyethyl methacrylate) (HEMA) chains formed stable polymer coatings on polyethylene terephthalate surfaces [17]. Consistent with similar reports [18–32], these star-shaped architectures provided brush-like structures of highly packed polymer chains that could physically repel bacteria to result in a significant reduction of attached bacteria. While this approach using water-insoluble polymers was suitable for hydrophobic resin surfaces, those potentially delivered from common oral care products are required to have significant water solubility. Materials, such as poly(acrylic acid) (PAA) are indeed water soluble, and have been known to bind to the tooth or HAP surfaces through interactions between the anionic carboxylate (COO-) groups in the polymer side chains and cationic calcium ions at the enamel surface [33–35].

As illustrated in Figure 1, we have prepared a small library of linear and star-shaped water-soluble polymers comprising of acrylic acid (AA) and methyl acrylate (MA) in order to draw a relationship between polymer structure (linear vs. 4-arm vs. 6-arm) and polymer composition (variable hydrophobicity) with their ability to bind to oral-relevant surfaces and provide bacterial anti-attachment in model systems. The specific goal of this report is to increase fundamental understanding of polymer-tooth surface interactions toward the development of new polymer platforms and products for anti-bacterial attachment activity.

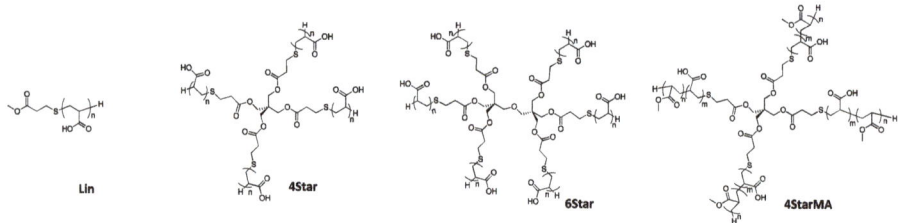

Figure 1. Star-shaped and linear poly(acrylic acid)s used in this study. The chemical structures of linear poly(acrylic acid) polymers (Lin), 4-arm (4Star) and 6-arm (6Star) star-shaped poly(acrylic acid), and 4-arm star-shaped copolymers with methyl acrylate (4StarMA).

2. Results and Discussion

2.1. Synthesis of 4- and 6-Arm Star-Shaped Polymers and Characterization

We wanted to test the free-radical polymerization in the presence of chain transfer agents (CTAs) as a facile synthetic strategy to prepare star-shaped polymers (Figure 2A) (See Tables S1–S6 in Supplementary Materials for polymerization conditions). In this polymerization, the thiol of a CTA reacts (R-SH) with the radical at the polymer chain and terminates the chain propagation by transferring the hydrogen atom (Figure S1). At the same time, the thiyl radical (R-S•) is generated and initiates new polymerization with the remaining monomers. This chain transfer cycle continues to consume all the remaining monomers, and the molecular weight of polymers can be determined by the relative reactivity of the radicals to CTAs compared to the monomers (chain transfer coefficient, C_{tr} in Equation (2)) (See Section 3.5. Analysis of Polymerization Process in Materials and Methods for Equation (2)) and the molar ratio of CTA to monomers ([SH]/[monomer]). We have previously prepared star-shape polymers with 10–12 polymer chain arms by crosslinking the end groups of pre-existing polyHEMA polymer chains [17]. However, this preparation method required multiple synthesis and purification steps. To that end, we synthesized a series of polymers with a range of molecular weights by altering the ratio of CTA to monomers to determine if this synthetic method can provide star-shaped polymers with sufficient size control. In general, star-shaped polymers have been synthesized by living radical polymerization methods, which provided well-defined polymers [36]. However, we chose free-radical polymerization with thiol CTAs because we are interested in a facile approach, potentially capable of large-scale production.

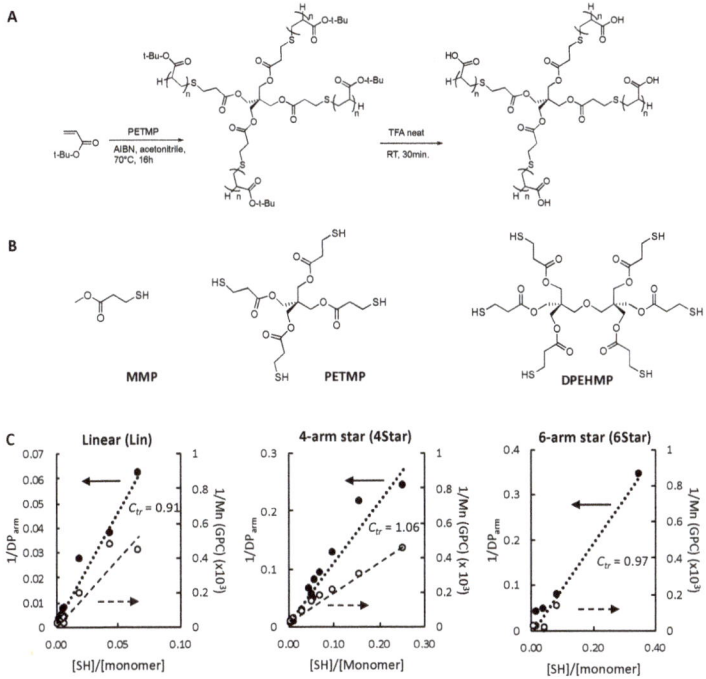

Figure 2. Synthesis of star-shaped polymers. (A) Synthesis of 4-arm star-shaped polymers, (B) Chemical structures of chain transfer agents. MMP: methyl 3-mercaptopropionate, PETMP: pentaerythritol tetrakis(3-mercaptopropionate), DPEHMP: dipentaerythritol hexakis(3-mercaptopropionate). (C) Mayo plots. The broken lines present the results of line fitting. C_{tr} was determined from the slope of the line. [SH]/[monomer] = (The number of thiol groups in a CTA) × [CTA]/[monomer].

The CTAs are small compounds with 4 or 6 thiol groups (PETMP and DPEHMP) (Figure 2B), which serve as core molecules to initiate propagation of polymer chains and yield star-shaped polymer structures. The mono-thiol chain transfer agent MMP provided linear polymers. We used *tert*-butyl acrylate (tBuA) as a protecting group of acrylic acid to facilitate polymer synthesis, and characterization through GPC and NMR spectroscopy. The average degree of polymerization of each arm (DP_{arm}) was determined by comparing the integrated peak area from the polymer backbone to that of CTAs in the ^{1}H NMR spectrum (See Experimental for details). As the ratio of CTA to monomers was increased, the molecular weight of polymers decreased, giving a series of star-shaped and linear polymers with M_n of ~2000 to 200,000 g/mole (See Table S7 for the molecular weight of polymers). It should be noted that the molecular weights of polymers were further measured by size exclusion chromatography (SEC), which separates materials of various masses by the hydrodynamic volume of polymer chains. In general, the hydrodynamic volume of star-shaped polymer is smaller than that of linear polymer with same absolute molecular weight [37]. Therefore, the comparison of molecular weights between the linear and star-shaped polymers by SEC alone would not be sufficient.

To probe the chain transfer polymerization, we examined the relationship between the ratio of the thiol groups to monomers and the polymer chain length. The Mayo plots (1/DP or 1/Mn (GPC) vs. [SH]/[monomer] based on Equation (2) for linear, 4-arm, and 6-arm star-shaped polymers showed linear correlations (Figure 2C), and the C_{tr} value of each CTA was determined as the slope of fitted lines in the Mayo plots (C_{tr} = 0.91 (Lin), 1.06 (4Star), 0.97, (6Star)). In addition, the plots of $1/M_n$ (determined by GPC) against [SH]/[Monomer] also showed linear correlations (Figure 2C). These results suggested that the polymerization was driven by independent chain transfer processes initiated by each thiol group of the CTAs. This further suggested that a polymer chain grew from each CTA arm (formation of star-shaped polymers), and the average polymer chain arm length could be controlled by varying the ratio of CTA to monomers.

The protected t-BuA polymers were treated with TFA to yield acrylic acid polymers (Figures 1 and 2). Because the resultant acrylic acid polymers were no longer soluble in GPC solvent (THF), the molecular weights and distribution of polymers were not determined. The DP of resultant acrylic acid polymers could also not be relatively determined by ^{1}H NMR analysis because the signals from the CTA agents were very small or not detected, which is likely due to low solubility of polymers in solvent. The linear, 4-arm, and 6 arm star-shaped polymers are denoted as Lin-X, 4Star-X, and 6Star-X, respectively, where X indicates the DP of each arm determined for the protected t-BuA polymers.

We also extended the synthetic approach to the preparation of hydrophobic random copolymers. tert-Bu acrylate (tBuA) was co-polymerized with methyl acrylate (MA) to give random copolymers with acidic carboxylic and methyl (ester) groups in the side chains (4StarMA) (Table S8). The mole percentages of MA in the polymers were close to the initial feed ratios, indicating that the MA monomers were quantitatively incorporated to the polymer chains. The copolymers are denoted as 4StarMAY-X, where X and Y indicates the DP of each arm and mole percentage of MA in a polymer, respectively.

2.2. Binding of Star-Shaped Polymers to HAP

We first investigated the binding behaviors of star-shaped polymers onto HAP as a tooth surface model as it has a similar chemical composition to enamel [38,39]. The intrinsic binding properties of polymers can provide useful insights into the relationship between polymer structure and surface activity. Specifically, the binding constant of the polymers and the maximum amount of binding sites on hydroxyapatite surfaces would represent the polymers binding properties. Such information would be helpful to predict polymer activity and design new polymers for subsequent improvements.

We synthesized rhodamine-labeled polymers (Figure 3A and Table S9) for the fluorescence-based binding assay described below. The assay used HAP powder dispersed in an aqueous solution as a model for HAP surfaces for polymer binding [40,41]. This assay provided a facile high-throughput method to determine the amounts of polymers that remained free in supernatant at equilibrium (C_{eq}) and adsorbed onto the HAP surface (q) (Figure 3B). The amount of polymers adsorbed onto the HAP surface was increased as the polymer concentration was increased and appears to level off at high concentrations (Figure 3B). The adsorption isotherms may be represented by the following equation for the Langmuir adoption model:

$$\frac{C_{eq}}{q} = \frac{C_{eq}}{q_{max}} + \frac{K_d}{q_{max}} \quad (1)$$

where q_{max} and K_d are the maximum amount of adsorbed polymers and dissociation constant, respectively [42–44]. The data were well fitted by Equation (1) (Figure 3C). The q_{max} and K_d values were calculated from the slope and intercept of each plot (Table S10). To compare the molecular behaviors of the star-shaped and linear polymers with different molecular sizes, we use the q_{max} and K_d values given in molar concentrations (µmol/g HAP and µM) for discussion, which present the binding behaviors of each polymer molecule. It should be noted that we used the DP of protected tBu polymers to calculate the M_n values of de-protected polymers because of the difficulty to determine the DP of de-protected polymers by ^{1}H NMR as described above. The M_n values were used to convert the q_{max} and K_d values given in weight-based concentrations to molar concentrations.

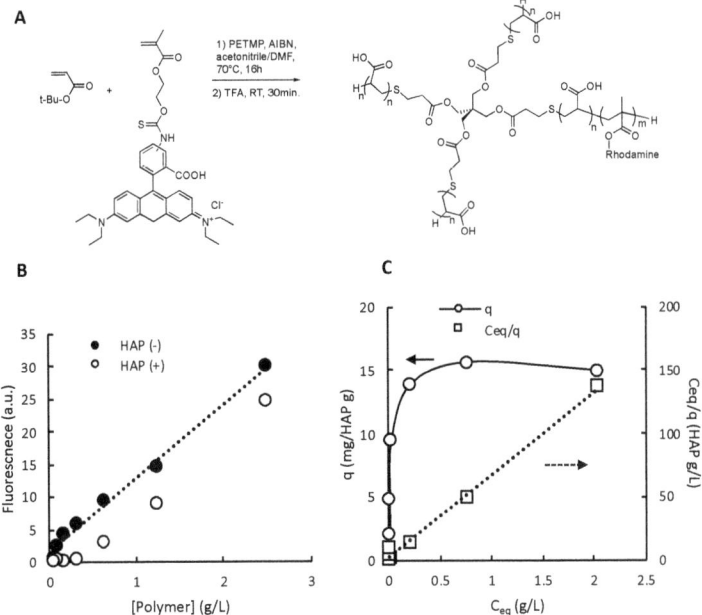

Figure 3. Synthesis of fluorescent dye-labeled polymers and their adsorption on hydroxyapatite powder (HAP). (**A**) Synthesis of rhodamine-labeled 4-arm star-shaped polymers. The rhodamine monomer (0.1 mol.% to the total number of monomers) was polymerized with t-Bu methacrylate. (**B**) Adsorption of 4-arm star-shaped polymer F-4Star-192 on HAP surfaces. Fluorescence intensities from supernatants of polymer assay solutions with and without HAP. (**C**) Adsorption isotherm and linear Langmuir plot. q: the amounts of polymers adsorbed onto the HAP surface. C_{eq}: the polymer concentration of supernatant at equilibrium.

First, we examined the binding properties of star-shaped and linear polymers which have a range of DPs of each polymer arm in order to evaluate the effects of assembly of polymer chains on their HAP binding as well as the effect of polymer arm length (DP) on their binding behavior. In general, the q_{max} values for 4- and 6-arm star-shaped polymers were smaller than that of the linear polymer (Figure 4A). This is likely because of the larger molecular sizes of star-shaped polymers, which occupy larger areas on the hydroxyapatite surface than the linear polymer, such that fewer star-shaped polymers could be bound to the hydroxyapatite surface. On the other hand, the dissociation constant K_d values of star-shaped polymers were smaller than that of the linear polymer (Figure 4B), indicating that the star-shaped polymers adsorbed on the hydroxyapatite surface more strongly than the linear polymer. This can be explained by the large polymer sizes of star-shaped polymers which have more contact points on the hydroxyapatite surface for binding. The 4- and 6-arm star-shaped polymers with DP~120 showed the similar q_{max} and K_d, indicating that these polymers occupy similar areas on the HAP surfaces and have similar binding.

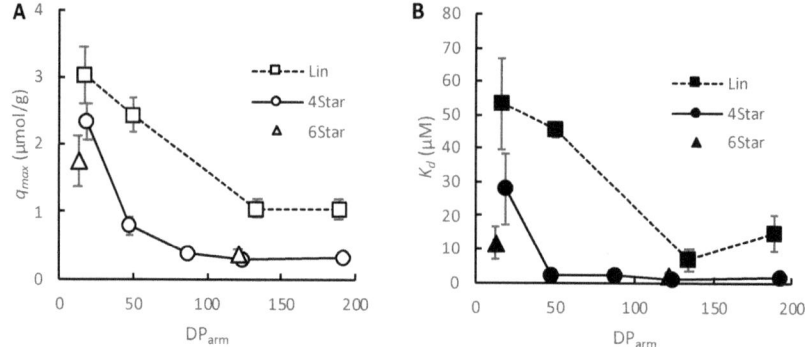

Figure 4. HAP adsorption of linear and 4-ram star-shaped polymers. (**A**) q_{max} and (**B**) K_d were determined by the Langmuir plot.

Regarding the effect of polymer arm length (DP) on their binding behavior, the q_{max} and K_d value of 4-arm star-shaped polymers decreased as the DP of arms increased and leveled off at large DPs (Figure 4A). On the other hand, the K_d value also leveled off for the polymers with large DPs (Figure 4B). The q_{max} and K_d values of linear polymers also decreased and appeared to level off at large DPs. These results suggest that the maximum number of adhered polymers and their binding affinity did not increase once the size of polymers became sufficiently large. This leveling-off of HAP binding behavior of the polymers may be explained by the following model. The anionic carboxylic groups of the polymer side chains are the binding ligand to HAP surfaces through electrostatic interactions. Therefore, as the polymer chains become longer, having more carboxylic side chains, the binding affinity of polymers for HAP would increase. However, the binding of carboxylic side chains to HAP surfaces requires the polymer chains to be flattened and/or stretched on the HAP surface, which is not favorable because of the large entropic penalty. Therefore, the binding of polymers would be determined by the balance between the two driving forces to maximize the number of binding sites by carboxylic groups on the HAP surface (enthalpy gain) and minimize the strain on polymer chains (entropic penalty). As the DP of the polymers increase, the number of carboxylic side chain groups increase, thus increasing their binding. However, once the polymers are long enough, the polymer chains would be difficult to be constrained on the HAP surface because of the entropic penalty, resulting in the leveling of q_{max} and K_d.

The effect of hydrophobic side chains on polymers binding to HAP surfaces was also examined. The random copolymers with hydrophobic monomer MA showed maximum points in the q_{max} and K_d values as the composition of MA was increased (Figure 5). This binding behavior with maximum points may be explained by the interplay between the electrostatic binding of carboxylate groups to HAP and the intramolecular and intermolecular associations of MA groups. Increasing the MA composition reduces the number of carboxylic side chains, which may in turn reduce the binding affinity of polymers (higher K_d). On the other hand, the hydrophobic groups may associate intramolecularly (within the same star-shaped polymer), which may prevent the extension of polymer chains for binding, resulting in low binding affinity (higher K_d). Based on this model, the increase in the K_d values for the low percentage of MA may indicate that the intramolecular association and/or reduced number of acidic groups are dominant, but low K_d value for the polymer with 55% MA indicates the intermolecular hydrophobic association between star-shaped polymers may play an important role to stabilize the polymer layer. On the other hand, the q_{max} also slightly increased, indicating the conformation of bound polymer chains are more compact (smaller occupied surface area). The polymers with 55% MA showed lower q_{max}, indicating the polymer chains are more expanded likely because of increased intermolecular associations of MA groups between the star-shaped polymers, which is in good agreement with the low K_d value. These results suggest that the binding behaviors of polymers to HAP surfaces can be controlled by their hydrophobicity, but is a less contributing factor than the overall shape (linear vs. star) of the polymer.

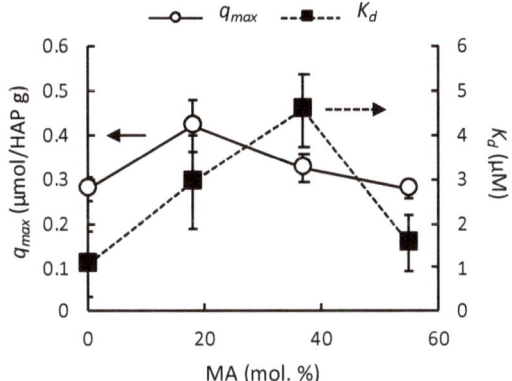

Figure 5. Effect of hydrophobic monomer composition on HAP adsorption of 4-arm star-shaped polymers.

2.3. Anti-Bacterial Attachment Activity of Linear and Star-Shaped Polymers

We have shown that shape significantly impacts the binding constant of polymers to HAP surfaces and that higher DPs are required for linear polymers than star-shaped in order to reach minimum K_d values. As such, we selected a subset of linear, star-shaped, and hydrophobic polymers based on their respective DP values that were similar to their tightly-binding fluorescent counterparts (DP = 100–200) in order to choose those materials with the strongest affinity to the HAP surface (Figure 6 and Table 1). In this way, we have effectively normalized to the polymers' K_d such that any differences in anti-attachment or contact angle measurements could be ascribed to the shape and composition of the polymers rather than simply their lack of presence through dissociation from the HAP surface. Table 1 lists the non-fluorescent polymers chosen for further evaluation. Additionally, we have intentionally chosen materials that do not kill bacteria but instead repel. Surface modifications that reduce bacteria deposition and colonization through contact kill or simple cell repulsion will ultimately appear the same, i.e., both surfaces will have a sufficiently reduced amount of living bacteria. We acknowledge that a combination of mechanisms, that is, kill + repel, would likely have the greatest efficacy, however by using materials known to not kill, but rather repel, we can isolate the mechanism to a single mode, subsequently making for more easily understood results. Consumer product constraints surrounding materials that reduce bacteria population through bactericidal mechanisms will inherently certain elicit regulatory restrictions. It is therefore important to have an understanding on materials that act as a non-lethal, almost mechanical, barrier only.

HAP-coated substrates were used as an enamel surface model to test the attachment of a mixture of oral bacteria *Actinomyces viscosus* and *Streptococcus oralis*. These bacteria are known as early colonizers of the oral biofilm formation [45,46], so significant reductions of these species are suggestive of efficacy on full healthy oral biofilms. In general, all of the polymers reduced the attachment of the bacteria onto the HAP surfaces by 17–54% relative to untreated control. However, what is immediately clear is that the six samples separated into two distinct groupings based on their relative hydrophobicity (Figure 6). Among the acrylic acid homopolymers, the linear (Lin-211) and 6-arm (6Star-129) star-shaped polymers showed a 30–32% reduction, while the 4-arm star (4Star-165) had only a 17% reduction. In contrast, all hydrophobic random copolymers gave higher percent reductions regardless of their shape. The linear copolymer (LinMA48-194), and 4-arm star-shaped copolymers (4StarMA56-215 and 4StarMA34-171) showed percent reductions of 43–54%, and were statistically superior to their homopolymer counterparts.

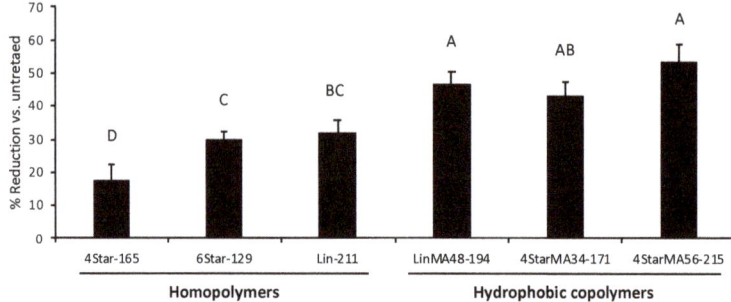

Figure 6. Anti-bacterial attachment activity of linear and star-shaped polymers. The activity of polymers was assessed by the percent reduction in bacterial attachment relative to control (untreated HAP surface). The data and error bars represent the average of 12 replicates with 95% confidence limits. The alphabetical letters on the bars present statistical grouping.

Table 1. Polymers selected for anti-bacterial adhesion assay and contact angle experiments and analogue polymers for HAP-binding assay.

Polymers Selected for Anti-Bacterial Adhesion and Contact Angle Experiments				Polymer Analogues for HAP-Binding Assay				
Polymer	DP_{arm}	M_n [a]	% Reduction	Polymer	DP_{arm}	M_n [a]	q_{max} (μmol/HAP g)	K_d (μM)
Lin-211	211	15,300	31.7 ± 4.2	F-Lin-189	189	13,700	1.05 ± 0.14	14.8 ± 5.5
4Star-165	165	48,000	17.6 ± 5.0	F-4Star-192	192	55,800	0.33 ± 0.06	1.4 ± 0.4
6Star-129	129	56,600	29.9 ± 2.5	F-6Star-121	121	53,100	0.38 ± 0.07	2.1 ± 0.6
LinMA48-194	194 [b]	15,400	46.3 ± 4.3	F-LinMA51-189	189 [b]	15,100	2.83 ± 0.11	14.0 ± 2.0
4StarMA34-171	171	53,000	43.2 ± 4.2	F-4StarMA37-185	185	57,700	0.33 ± 0.03	4.6 ± 0.8
4StarMA56-215	215	69,200	53.5 ± 5.5	F-4StarMA55-149	149	48,000	0.28 ± 0.02	1.6 ± 0.6

[a] The M_n (the number average molecular weight) of the polymers was calculated based on the DP of protected tBu polymers and molecular weights of chain transfer agent and acrylic acrylate; [b] The theoretical DP calculated based on the Mayo equation using the C_{tr} value and [SH]/[monomer].

This data is suggestive of several points in regard to the material characteristics required for anti-attachment properties. First, while there may be small differences between the efficacies of various shapes within the homo vs. copolymer families, their effect is diminished by the presence of hydrophobic monomers within the random polymer chain. This is to say that bacteria are less able to attach themselves to the HAP surface when that surface is coated with a hydrophobic polymer, regardless of whether that polymer is linear or star-shaped. Second, the statistical groupings of the star-shaped hydrophobic polymers were near equivalent, with 34 and 56 mol.% of methyl acrylate providing similar effects. This implies that a "hydrophobic" polymer provides better anti-attachment, but no significant increases in this effect were observed over the ranges evaluated. In the future, we will explore this facet more closely as it may be possible to draw a true correlation here. In addition, the maximum amount of adsorbed polymers q_{max} and dissociation constant K_d of the homopolymer and hydrophobic copolymer with the same shape (linear, 4-, 6-arm star-shape) are very similar (Table 1). This suggests that the enhanced effect of anti-bacterial adhesion by the hydrophobicity of copolymers is not due to the difference in their inherent binding properties (the amount of polymers adhered) to HAP surfaces, but it could be related to the physicochemical properties of polymers or polymer conformations on the surface. It has been previously reported that random, block, cross-linked amphiphilic copolymers effectively prevent protein adhesion and bacterial adhesion [47–54]. The proposed mechanism is that these polymers form phase separated nano-scale domains, which reduce protein adsorption and subsequent bacterial attachment [48]. These domains are smaller than the hydrophobic/hydrophilic domains of proteins so that it is difficult for proteins to adopt their conformations to match with the surface domains. While it is not clear at this point, we speculate that the copolymers in our study may also form such segregated microdomains by association of hydrophobic side chains, which may be a more dominant factor for bacterial attachment than polymer shapes.

The ability for a polymer to provide anti-attachment effects to an oral surface can only happen if the material sufficiently first binds to the surface. The data above has demonstrated that the chemistries needed to bind and repel are not the same. A multi-arm star-shaped polymer had significantly better binding to HAP, but in contrast composition played no such dominant role. Hydrophobicity did not dramatically decrease K_d, however its presence significantly decreased bacteria attachment. Taken together, a hydrophobic star-shaped material would be the ideal polymer system to both bind to enamel and repel bacteria in our model systems.

2.4. Water Contact Angle

The polymer binding isotherms combined with bacteria anti-attachment clearly indicate the tunability and functionality of these polymer systems. However, these materials would need to perform in the presence of the salivary pellicle that coats all oral surfaces in order to provide sufficient effects in-vivo. The pellicle is a complex mixture of proteins, deposited to the surface of enamel by salivary flow [52], and fundamentally examining and predicting the interactions of star-shaped polymers with a pellicle surface is a sufficient and ongoing challenge for our group. Polymers will interact with a pellicle layer in different ways, depending on the dominate chemistry. For example, a recent publication [55] showed that polyanions and polycations interacted differently with the pellicle, which included their penetration depth relative to the HAP-pellicle interface. Another described changes in pellicle thickness as well as antimicrobial functionality as a function of polymer deposition and interaction with pellicle-coated HAP surfaces [12]. This effect can be rationally extended to variations in hydrophobicity and shape. While the presence of a pellicle would add significant complexity, and is outside of the scope of our current study, we did choose to examine how these polymers affected the *macroscopic* properties of HAP through contact angle measurements after pre-treatment with artificial saliva in order to demonstrate a small facet of in-vivo activity. Mucin-based artificial saliva has been a substitute for human saliva in dental research [56], and we found that HAP discs first treated with artificial saliva produced sufficient surfaces allowing for consistent measurements. Significant changes in surface energies, exhibited by major differences in water droplet contact angle, would indicate positive interactions between polymers and mucin-coated surfaces. This experiment represents our first bridging data between fundamental studies and practical applications.

We generally observed that following treatment with polymer solutions, an increase in CA was observed for most samples by >7°, indicating that the polymer-treated surfaces were more hydrophobic than the untreated control (Figure 7). The magnitude of this difference also reflected the compositional changes within the polymers themselves. 4StarMA56-171, for example, had the highest contact angle of 87.6°, an effect attributed to the 56% MA concentration within the star shaped material. Example images of the droplets can be seen in Figure 7, illustrating that these materials are effective at altering the surface characteristics of HAP.

Within this series, however, the linear hydrophobic polymer LinMA48-194, exhibited a lower CA than the untreated control, even though the polymer contains 48% methyl acrylate. The hydrophobic side chains might stabilize the polymer coatings on the HAP surface by the hydrophobic interactions with HAP and/or between the polymer chains. Such polymer network anchored on the HAP surface might retain more water and therefore exhibit higher hydrophilicity, as compared to homopolymer Lin-211 which increased CA. On the other hand, the hydrophobic star-shaped polymers exhibited larger contact angles than LinMA48-194. It may be possible that linear polymer chains can adopt a conformation on the surface such that the hydrophobic side chains face down toward the HAP, and the hydrophilic (carboxylate) face up. The formation of such amphiphilic polymer conformation would be more efficient than the star-shaped polymers which have denser polymer chains, giving more constraints to conformational change. Because of the difference in the HAP pre-treatment (mucin-coated or non-coated), we cannot directly compare these results to those of binding and anti-bacterial attachment. However, the results suggest that the polymers are capable of altering the surface properties of HAP even in the presence of a protein layer.

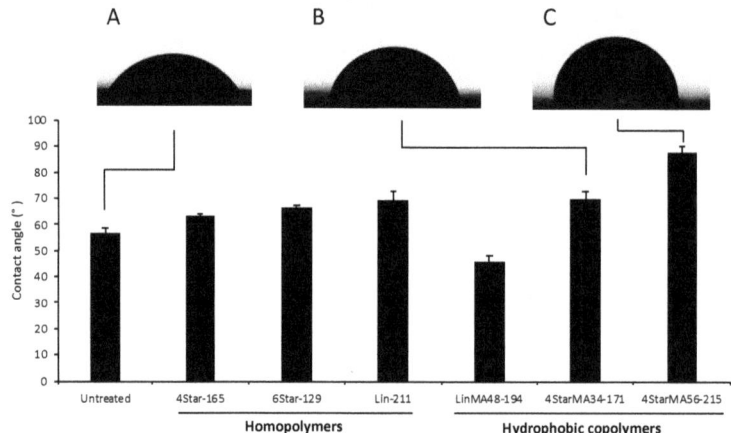

Figure 7. Water contact angle of polymer treated HAP surfaces pre-treated with artificial saliva. Example contact angle images of treated HAP surfaces: (**A**) untreated; (**B**) 4StarMA34-171; (**C**) 4StarMA56-215.

3. Materials and Methods

3.1. Materials

2,2′-azobisisobutyronitrile (AIBN) and pentaerythritol tetrakis(3-mercaptopropionate) (PETMP) was purchased from Sigma-Aldrich Co. LLC. (St. Louis, MO, USA). Dipentaerythritol hexakis(3-mercaptopropionate) (DPEHMP) was purchased from TCI America (Montgomeryville, PA, USA). Methacryloxyethyl thiocarbamoyl rhodamine B was purchased from Polysciences (Warrington, PA, USA). Trifluoroacetic acid (TFA) and solvents were purchased from Thermo Fisher Scientific, Inc. (Waltham, MA, USA). tert-Butyl acrylate, methyl acrylate, and methyl mercaptopropionate (MMP) were purchased from Acros Organics (Morris County, NJ, USA). The inhibitors of these monomers were removed by passing through alumina before use. Other chemicals and solvents were used without further purification. 1H NMR was performed using a Varian MR400 (400 MHz, Agilent Scientific Instruments, Santa Clara, CA, USA) and analyzed using VNMRJ 3.2 (Agilent Scientific Instruments, Santa Clara, CA, USA) and MestReNova. Gel permeation chromatography (GPC) analysis was performed using a Waters 1515 HPLC instrument (Milford, MA, USA) using THF as an eluent, equipped with Waters Styragel (7.8 × 300 mm) HR 0.5, HR 1, and HR 4 columns in sequence and detected by a differential refractometer (RI). Sintered HAP discs (0.5 cm in diameter) were purchased from Himed, Inc. (Old Bethpage, NY, USA).

3.2. Synthesis of tBu PAA Homopolymers

tert-Butyl acrylate (t-BuA), AIBN, and chain transfer agent (CTA) (MMP, PETMP, or DPEHMP) in acetonitrile were mixed in a flask (See Table S1 for the polymerization conditions). The oxygen of the reaction mixture was removed by bubbling nitrogen gas for 10 min, and the reaction solution was stirred at 70 °C for 16 h. The reaction was cooled to room temperature. The solvent was removed by evaporation under reduced pressure. The resultant residue was dissolved in diethyl ether, and the polymer was isolated by precipitation in a methanol:water [50:50 (v/v)] mixture. The yield of purification was >90% for most cases. The polymer arm length (DP) was calculated by comparing the integrated peaks of -OCH2- group of chain transfer agent to the -CH- polymer backbone. The number average molecular weight (Mn) was calculated using the DP and molecular weights of monomers and

CTAs. Gel permeation chromatography molecular mass results were determined using a calibration curve based on the standard samples of polystyrene. 1H NMR (CDCl$_3$, 400 MHz) δ: 4.21–4.06 (s, 2H, –OCH$_2$– of PETMP), 2.85–2.51 (brs, 4H, –SCH$_2$CH$_2$–), 2.37-2.07 (brs, 1H, –CH–,), 1.97–1.14 (brs, 11H, –CH$_3$ and –CH$_2$–).

The tBu groups of polymers were then removed by the addition of trifluoroacetic acid (TFA) (5 mL to 1 g of polymer). After stirring for 30 min, TFA was removed by blowing with nitrogen gas in a closed container, and the gas was passed through a base (NaOH) aqueous solution to trap TFA. The residue was dissolved in methanol, and deprotected polymers were isolated by precipitating in excess diethyl ether. Subsequently, the precipitate was dissolved in distilled water and lyophilized to yield a powdery product. 1H NMR (DMSO, 400 MHz) 2.4-2.0 (brs, 1H, –CH–,), 1.8-1.2 (brs, 2H, –CH$_2$–).

3.3. Synthesis of Random Copolymers with MA

The PAA random copolymers with methacrylate (MA) were synthesized by the same method with the tBu PAA homopolymers as described above. See Table S2 in Supporting Information for the monomer feed compositions and reaction conditions.

3.4. Synthesis of Rhodamine B-Labeled Polymers

The rhodamine B-labeled copolymers were synthesized using methacryloxyethyl thiocarbamoyl rhodamine B (0.1 mol.% to the total amount of monomers) by the same method as described above with the tBu PAA homopolymers. See Supporting Information for the detailed procedure, polymerization conditions, and monomer feed compositions (Tables S3–S6). The Mayo plots showed linear correlations, and the Ctr values of each thiol group of linear and 4-arm polymers are 0.91 and 0.97 (Figure S2 and Table S11).

3.5. Analysis of Polymerization Process

In general, DP_{arm} of polymer prepared in the presence of thiol groups as a CTA may be presented by the Mayo equation [38]:

$$\frac{1}{DP_{arm}} = \frac{1}{DP_0} + C_{tr}\frac{[SH]}{[Monomer]} \qquad (2)$$

where DP_0, C_{tr}, [CTA] and [Monomer] represent the DP of each polymer arm in the absence of CTA, chain transfer coefficient, initial mole concentration of thiol groups, and mole concentration of monomers, respectively. According to the Mayo equation, the plot of 1/DP would be proportional to [SH]/[Monomer], and the slop presents C_{tr}.

3.6. HAP Binding Assay

Fluorescence spectroscopy was used to evaluate the binding capacity of rhodamine-labeled polymers onto HAP powder. The polymer solutions in 10 mM phosphate buffer with 150 mM NaCl with different concentrations (pH = 7, adjusted by NaOH aq., 0.5 mL, 0.04, 0.08, 0.16, 0.31, 0.63, and 1.25 g/L) were mixed with HAP (30 mg/mL) in a 1.5 mL tube. The solution was gently shaken using a mechanical shaker for 2 h at room temperature and then centrifuged at 10,000 rpm for 10 min. The fluorescence emission intensities of the supernatant were measured (excitation wavelength = 553 nm, emission wavelength = 627 nm) and compared with those for samples with same concentration of polymers without HAP.

3.7. Anti-Bacterial Adhesion Assay

HAP coated MBEC™ lids were treated by polymer solutions in MilliQ water (1 wt.%, pH 6.5 adjusted with NaOH or HCl) and allowed to shake in the incubator at 37 °C for 1 h. Following treatment, excess polymer solution was removed from the MBEC™ lids by submerging in Trypticase soy broth (TSB) for 10–15 s for three cycles, replacing the TSB broth for each new cycle. The MBEC™ lids were then incubated with freshly prepared overnight cultures of mixed *Actinomyces viscosus* (ATCC#43146, American Type Culture Collection, Manassas, VA, USA) and *Streptococcus oralis* (ATCC#35037, American Type Culture Collection, Manassas, VA, USA) for 3 h at 37 °C. After incubation the MBECTM lids were submerged in TSB and sonicated two times for 2 min each time in order to detach the HAP-bound bacteria into the TSB. The BacTiter-Glo Microbial Cell Viability Assay was utilized on the re-suspended TSB to determine the percent reduction in the cell viability. The percent reduction was calculated by the following equation based on the luminescent output of bacteria removed from untreated surfaces and polymer-treated surfaces: % reduction = 100 × (bacteria attached on untreated surface—bacteria attached on polymer-treated surface)/bacteria on untreated surface.

Bartlett's test ($p = 0.265$) suggested that any variations are not significant, and the samples have equal variances. Therefore one-way analysis of variance (ANOVA) was used to assess the treatment effect and determine the statistical differences between the various sets. A Tukey multiple comparison test was used to assess pairwise treatment differences. A $p < 0.05$ was used to indicate significant statistical differences.

3.8. Contact Angle Measurements

Contact angle was performed on an Attension Theta instrument from Biolin Scientific (Stockholm, Sweden). Data was analyzed using One Attension software v 2.9. Briefly, 1.0 wt.% polymer solutions in MilliQ water were prepared, and their pH adjusted to 6.5 with concentrated NaOH or HCl. Because of the immediate absorption of solution droplets into hydroxyapatite, surface modification was required prior to treatment with polymer solutions in order to obtain stable droplets for comparison. Sintered HAP was first treated with modified artificial saliva [39] for 1 h (see Supporting Information). After this time, the discs were soaked in 2 mL of polymer solution for three hours on an orbital shaker. The discs were removed and rinsed slightly to remove excess or loosely bound material, and then dried overnight. Contact angle measurements of a 3 µL droplet on four separate HAP discs were collected and averaged to provide statistical significance.

4. Conclusions

In summary of the present study, we synthesized linear, 4- and 6-arm star-shaped polymers based on acrylic acid using chain transfer agents with corresponding thiol groups in order to provide insight into the types of polymers that could both bind to HAP and repel bacteria from the surface. We have found that polymer shape was more important to HAP surface binding than polymer composition (hydrophobicity). However, polymer composition played a larger role than polymer shape (linear vs. star-shape) when providing anti-bacterial protection. This information will be important for targeted properties (binding, anti-bacterial attachment, wettability, etc.,) and further design polymers for dental applications. In this study, our focus was the synthesis of star-shaped polymers and initial evaluation of their physical and biological properties in order to test new polymer platforms for dental applications. The oral environment is quite dynamic and subject to continuously changing environments due to salivary flow, food and drink intake, and the resulting fluctuating pH. Therefore, the efficiency of dental materials to provide benefits, as delivered through common oral care products, must be investigated through delivery, substantivity, and efficacy. While limited to very simple systems here, this approach is critical to build an understanding of dental materials as it can more effectively isolate and identify specific modes of action in addition to chemical or physical barriers to the effectiveness of these materials. Our future research will focus on further developing an understanding of star-shaped polymers in reference to artificial-saliva and human-saliva-coated HAP surfaces, including binding activity and bacterial anti-attachment properties.

Supplementary Materials: The following are available online at http://www.mdpi.com/2079-4983/10/4/56/s1. Figure S1: Chain transfer process in free-radical polymerization. Figure S2: The relationships of 1/DP (A) and 1/Mn (B) with SH/monomer ratio for RhB-labeled PAA (tBu)-protected polymers. Table S1: Polymerization conditions of tBu PAA polymers. Table S2: Polymerization conditions for hydrophobic random copolymers. Table

S3: Polymerization conditions for rhodamine-labeled tBu linear polymers. Table S4: Polymerization conditions for rhodamine-labeled tBu 4-arm star-shaped polymers. Table S5: Polymerization conditions for rhodamine-labeled tBu 6-arm star-shaped polymers. Table S6: Polymerization conditions of rhodamine-labeled tBu linear and 4-arm PAA/MA random copolymer. Table S7: Polymer characterization of tBu-protected polymers. Table S8: Polymer characterization of tBuA-MA random copolymers. Table S9: Polymer characterization of F-labeled polymers. Table S10: Polymer characterization and Langmuir constants. Table S11: Chain transfer constants for tBu polymers.

Author Contributions: Conceptualization, L.A.Z., C.P.M., and K.K.; investigation, H.M., G.A.P., J.L., and C.P.M.; writing—original draft preparation, C.P.M., and K.K.; writing—review and editing, C.P.M., and K.K.; supervision, C.P.M. and K.K.; project administration, C.P.M. and K.K.; funding acquisition, K.K.

Funding: Kenichi Kuroda, Hamid Mortazavian, and Janis Lejnieks acknowledge the funding from Colgate-Palmolive (Star-Shaped Polymer Architecture for Anti-Attachment, Anti-Stain, and Actives Delivery).

Acknowledgments: We would like to thank Rehana Begum-Gafur and Mark Vandeven for their help with statistical analysis of anti-attachment microbial data, and Donghui Wu for helpful discussions.

Conflicts of Interest: The authors declare no conflict of interest.

References

1. Wu, H.; Moser, C.; Wang, H.-Z.; Høiby, N.; Song, Z.-J. Strategies for combating bacterial biofilm infections. *Int. J. Oral Sci.* **2014**, *7*, 1–7. [CrossRef] [PubMed]
2. Teles, R.; Teles, F.; Frias-Lopez, J.; Paster, B.; Haffajee, A. Lessons learned and unlearned in periodontal microbiology. *Periodontol 2000* **2013**, *62*, 162. [CrossRef]
3. Koo, H.; Falsetta, M.L.; Klein, M.I. The Exopolysaccharide Matrix: A Virulence Determinant of Cariogenic Biofilm. *J. Dental Res.* **2013**, *92*, 1065–1073. [CrossRef] [PubMed]
4. Sanz, M.; Beighton, D.; Curtis, M.A.; Cury, J.A.; Dige, I.; Dommisch, H.; Ellwood, R.; Giacaman, R.A.; Herrera, D.; Herzberg, M.C.; et al. Role of microbial biofilms in the maintenance of oral health and in the development of dental caries and periodontal diseases. Consensus report of group 1 of the Joint EFP/ORCA workshop on the boundaries between caries and periodontal disease. *J. Periodontol.* **2017**, *44*, S5–S11. [CrossRef] [PubMed]
5. Bowen, W.H.; Burne, R.A.; Wu, H.; Koo, H. Oral Biofilms: Pathogens, Matrix, and Polymicrobial Interactions in Microenvironments. *Trends Microbiol.* **2018**, *26*, 229–242. [CrossRef] [PubMed]
6. Chaves, P.; Oliveira, J.; Haas, A.; Beck, R.C.R. Applications of Polymeric Nanoparticles in Oral Diseases: A Review of Recent Findings. *Curr. Pharm. Des.* **2018**, *24*, 1377–1394. [CrossRef]
7. Fernandes, T.; Bhavsar, C.; Sawarkar, S.; D'Souza, A. Current and novel approaches for control of dental biofilm. *Int. J. Pharm.* **2018**, *536*, 199–210. [CrossRef]
8. Hu, X.Q.; Huang, Y.Y.; Wang, Y.G.; Wang, X.Y.; Hamblin, M.R. Antimicrobial Photodynamic Therapy to Control Clinically Relevant Biofilm Infections. *Front. Microbiol.* **2018**, *9*, 1299. [CrossRef]
9. Koo, H.; Allan, R.N.; Howlin, R.P.; Stoodley, P.; Hall-Stoodley, L. Targeting microbial biofilms: Current and prospective therapeutic strategies. *Nat. Rev. Microbiol.* **2017**, *15*, 740. [CrossRef]
10. Pleszczynska, M.; Wiater, A.; Bachanek, T.; Szczodrak, J. Enzymes in therapy of biofilm-related oral diseases. *Biotechnol. Appl. Biochem.* **2017**, *64*, 337–346. [CrossRef]
11. Gaffar, A.; Solis-Gaffar, M.C.; Tavss, E.; Marcussen, H.W.; Rustogi, K.N. Long-term Antiplaque, Anticalculus, and Antigingivitis Effects of Benzethonium/Polymer Complex in Beagle Dogs. *J. Dent. Res.* **1981**, *60*, 1897–1903. [CrossRef] [PubMed]
12. Lee, H.S.; Myers, C.; Zaide, L.; Nalam, P.C.; Caporizzo, M.A.; Daep, C.A.; Eckmann, D.M.; Masters, J.G.; Composto, R.J. Competitive Adsorption of Polyelectrolytes onto and into Pellicle-Coated Hydroxyapatite Investigated by QCM-D and Force Spectroscopy. *Acs Appl. Mater. Interfaces* **2017**, *9*, 13079–13091. [CrossRef] [PubMed]
13. Zhang, Q.M.; Serpe, M.J. Synthesis, Characterization, and Antibacterial Properties of a Hydroxyapatite Adhesive Block Copolymer. *Macromolecules* **2014**, *47*, 8018–8025. [CrossRef]
14. Cui, X.N.; Koujima, Y.; Seto, H.; Murakami, T.; Hoshino, Y.; Miura, Y. Inhibition of Bacterial Adhesion on Hydroxyapatite Model Teeth by Surface Modification with PEGMA-Phosmer Copolymers. *Acs Biomater. Sci. Eng.* **2016**, *2*, 205–212. [CrossRef]

15. Kang, S.; Lee, M.; Kang, M.; Noh, M.; Jeon, J.; Lee, Y.; Seo, J.-H. Development of anti-biofouling interface on hydroxyapatite surface by coating zwitterionic MPC polymer containing calcium-binding moieties to prevent oral bacterial adhesion. *Acta Biomater* **2016**, *40*, 70–77. [CrossRef]
16. Guan, Y.H.; Lath, D.L.; de Graaf, T.; Lilley, T.H.; Brook, A.H. Moderation of oral bacterial adhesion on saliva-coated hydroxyapatite by polyaspartate. *J. Appl. Microbiol.* **2003**, *94*, 456–461. [CrossRef]
17. Totani, M.; Ando, T.; Terada, K.; Terashima, T.; Kim, I.Y.; Ohtsuki, C.; Xi, C.; Kuroda, K.; Tanihara, M. Utilization of star-shaped polymer architecture in the creation of high-density polymer brush coatings for the prevention of platelet and bacteria adhesion. *Biomater. Sci.* **2014**, *2*, 1172–1185. [CrossRef]
18. Muszanska, A.K.; Rochford, E.T.J.; Gruszka, A.; Bastian, A.A.; Busscher, H.J.; Norde, W.; van der Mei, H.C.; Herrmann, A. Antiadhesive Polymer Brush Coating Functionalized with Antimicrobial and RGD Peptides to Reduce Biofilm Formation and Enhance Tissue Integration. *Biomacromolecules* **2014**, *15*, 2019–2026. [CrossRef]
19. Nejadnik, M.R.; van der Mei, H.C.; Norde, W.; Busscher, H.J. Bacterial adhesion and growth on a polymer brush-coating. *Biomaterials* **2008**, *29*, 4117–4121. [CrossRef]
20. Roest, S.; van der Mei, H.C.; Loontjens, T.J.A.; Busscher, H.J. Charge properties and bacterial contact-killing of hyperbranched polyurea-polyethyleneimine coatings with various degrees of alkylation. *Appl. Surf. Sci.* **2015**, *356*, 325–332. [CrossRef]
21. Swartjes, J.; Veeregowda, D.H.; van der Mei, H.C.; Busscher, H.J.; Sharma, P.K. Normally Oriented Adhesion versus Friction Forces in Bacterial Adhesion to Polymer-Brush Functionalized Surfaces Under Fluid Flow. *Adv. Funct. Mater.* **2014**, *24*, 4435–4441. [CrossRef]
22. Mi, L.; Jiang, S.Y. Integrated Antimicrobial and Nonfouling Zwitterionic Polymers. *Angew. Chem. -Int. Ed.* **2014**, *53*, 1746–1754. [CrossRef] [PubMed]
23. Shao, Q.; Jiang, S.Y. Molecular Understanding and Design of Zwitterionic Materials. *Adv. Mater.* **2015**, *27*, 15–26. [CrossRef] [PubMed]
24. Ibanescu, S.A.; Nowakowska, J.; Khanna, N.; Landmann, R.; Klok, H.A. Effects of Grafting Density and Film Thickness on the Adhesion of Staphylococcus epidermidis to Poly(2-hydroxy ethyl methacrylate) and Poly(poly(ethylene glycol)methacrylate) Brushes. *Macromol. Biosci.* **2016**, *16*, 676–685. [CrossRef]
25. Klok, H.A.; Genzer, J. Expanding the Polymer Mechanochemistry Toolbox through Surface-Initiated Polymerization. *Acs Macro Letters* **2015**, *4*, 636–639. [CrossRef]
26. Zoppe, J.O.; Ataman, N.C.; Mocny, P.; Wang, J.; Moraes, J.; Klok, H.A. Surface-Initiated Controlled Radical Polymerization: State-of-the-Art, Opportunities, and Challenges in Surface and Interface Engineering with Polymer Brushes. *Chem. Rev.* **2017**, *117*, 1105–1318. [CrossRef]
27. Gasteier, P.; Reska, A.; Schulte, P.; Salber, J.; Offenhausser, A.; Moeller, M.; Groll, J. Surface grafting of PEO-Based star-shaped molecules for bioanalytical and biomedical applications. *Macromol. Biosci.* **2007**, *7*, 1010–1023. [CrossRef]
28. Heyes, C.D.; Groll, J.; Moller, M.; Nienhaus, G.U. Synthesis, patterning and applications of star-shaped poly(ethylene glycol) biofunctionalized surfaces. *Mol. Biosyst.* **2007**, *3*, 419–430. [CrossRef]
29. Morgese, G.; Trachsel, L.; Romio, M.; Divandari, M.; Ramakrishna, S.N.; Benetti, E.M. Topological Polymer Chemistry Enters Surface Science: Linear versus Cyclic Polymer Brushes. *Angew. Chem. -Int. Ed.* **2016**, *55*, 15583–15588. [CrossRef]
30. Kim, D.-G.; Kang, H.; Choi, Y.-S.; Han, S.; Lee, J.-C. Photo-cross-linkable star-shaped polymers with poly(ethylene glycol) and renewable cardanol side groups: Synthesis, characterization, and application to antifouling coatings for filtration membranes. *Polym. Chem.* **2013**, *4*, 5065–5073. [CrossRef]
31. Kim, D.-G.; Kang, H.; Han, S.; Lee, J.-C. The increase of antifouling properties of ultrafiltration membrane coated by star-shaped polymers. *J. Mater. Chem.* **2012**, *22*, 8654–8661. [CrossRef]
32. Kim, D.G.; Kang, H.; Han, S.; Kim, H.J.; Lee, J.C. Bio- and oil-fouling resistance of ultrafiltration membranes controlled by star-shaped block and random copolymer coatings. *Rsc Advances* **2013**, *3*, 18071–18081. [CrossRef]
33. Fukuda, R.; Yoshida, Y.; Nakayama, Y.; Okazaki, M.; Inoue, S.; Sano, H.; Suzuki, K.; Shintani, H.; Meerbeek, B.V. Bonding efficacy of polyalkenoic acids to hydroxyapatite, enamel and dentin. *Biomaterials* **2003**, *24*, 1867. [CrossRef]
34. Yoshida, Y.; Van Meerbeek, B.; Nakayama, Y.; Snauwaert, J.; Hellemans, L.; Lambrechts, P.; Vanherle, G.; Wakasa, K. Evidence of Chemical Bonding at Biomaterial-Hard Tissue Interfaces. *J. Dental Res.* **2000**, *79*, 714. [CrossRef]

35. McConnell, M.D.; Liu, Y.; Nowak, A.P.; Pilch, S.; Masters, J.G.; Composto, R.J. Bacterial plaque retention on oral hard materials: Effect of surface roughness, surface composition, and physisorbed polycarboxylate. *J. Biomed. Mater. Res. Part A* **2010**, *92*, 1518–1527. [CrossRef]
36. Ren, J.M.; McKenzie, T.G.; Fu, Q.; Wong, E.H.H.; Xu, J.; An, Z.; Shanmugam, S.; Davis, T.P.; Boyer, C.; Qiao, G.G. Star Polymers. *Chem. Rev.* **2016**, *116*, 6836. [CrossRef]
37. Voit, B.I.; Lederer, A. Hyperbranched and Highly Branched Polymer Architectures—Synthetic Strategies and Major Characterization Aspects. *Chem. Rev.* **2009**, *109*, 5924–5973. [CrossRef]
38. Clark, W.B.; Bammann, L.L.; Gibbons, R.J. comparative estimates of bacterial affinities and adsorption sites on hydroxyapatite surfaces. *Infect. Immun.* **1978**, *19*, 846–853.
39. Hillman, J.D.; Vanhoute, J.; Gibbons, R.J. sorption of bacteria to human enamel powder. *Arch. Oral Biol.* **1970**, *15*, 899–903. [CrossRef]
40. Lei, Y.; Wang, T.; Mitchell, J.W.; Qiu, J.; Kilpatrick-Liverman, L. Synthesis of Carboxylic Block Copolymers via Reversible Addition Fragmentation Transfer Polymerization for Tooth Erosion Prevention. *J. Dental Res.* **2014**, *93*, 1264–1269. [CrossRef]
41. Lei, Y.D.; Wang, T.X.; Mitchell, J.W.; Zaidel, L.; Qiu, J.H.; Kilpatrick-Liverman, L. Bioinspired amphiphilic phosphate block copolymers as non-fluoride materials to prevent dental erosion. *Rsc Adv.* **2014**, *4*, 49053–49060. [CrossRef] [PubMed]
42. Misra, D.N. adsorption of LOW-molecular-weight poly(acrylic acid) on hydroxyapatite-role of molecular association and apatite dissolution. *Langmuir* **1991**, *7*, 2422–2424. [CrossRef]
43. Misra, D.N. Adsorption OF low-molecular-weight sodium polyacrylate on hydroxyapatite. *J. Dent. Res.* **1993**, *72*, 1418–1422. [CrossRef] [PubMed]
44. Misra, D.N. Adsorption of polyacrylic acids and their sodium salts on hydroxyapatite: Effect of relative molar mass. *J. Colloid Interface Sci.* **1996**, *181*, 289–296. [CrossRef]
45. Huang, R.; Li, M.; Gregory, R.L. Bacterial interactions in dental biofilm. *Virulence* **2011**, *2*, 444. [CrossRef] [PubMed]
46. Kolenbrander, P.E.; Andersen, R.N.; Blehert, D.S.; Egland, P.G.; Foster, J.S.; Palmer, R.J. Communication among Oral Bacteria. *Microbiol. Mol. Biol. Rev.* **2002**, *66*, 486–505. [CrossRef]
47. Gudipati Chakravarthy, S.; Greenlief, C.M.; Johnson Jeremiah, A.; Prayongpan, P.; Wooley Karen, L. Hyperbranched fluoropolymer and linear poly(ethylene glycol) based amphiphilic crosslinked networks as efficient antifouling coatings: An insight into the surface compositions, topographies, and morphologies. *J. Polym. Sci. Part A Polym. Chem.* **2004**, *42*, 6193–6208. [CrossRef]
48. Kerstetter, J.L.; Gramlich, W.M. Nanometer-scale self-assembly of amphiphilic copolymers to control and prevent biofouling. *J. Mater. Chem. B* **2014**, *2*, 8043–8052. [CrossRef]
49. Krishnan, S.; Ayothi, R.; Hexemer, A.; Finlay, J.A.; Sohn, K.E.; Perry, R.; Ober, C.K.; Kramer, E.J.; Callow, M.E.; Callow, J.A.; et al. Anti-Biofouling Properties of Comblike Block Copolymers with Amphiphilic Side Chains. *Langmuir* **2006**, *22*, 5075–5086. [CrossRef]
50. Martinelli, E.; Agostini, S.; Galli, G.; Chiellini, E.; Glisenti, A.; Pettitt, M.E.; Callow, M.E.; Callow, J.A.; Graf, K.; Bartels, F.W. Nanostructured Films of Amphiphilic Fluorinated Block Copolymers for Fouling Release Application. *Langmuir* **2008**, *24*, 13138–13147. [CrossRef]
51. Weinman, C.J.; Gunari, N.; Krishnan, S.; Dong, R.; Paik, M.Y.; Sohn, K.E.; Walker, G.C.; Kramer, E.J.; Fischer, D.A.; Ober, C.K. Protein adsorption resistance of anti-biofouling block copolymers containing amphiphilic side chains. *Soft Matter* **2010**, *6*, 3237–3243. [CrossRef]
52. Zhao, Z.; Ni, H.; Han, Z.; Jiang, T.; Xu, Y.; Lu, X.; Ye, P. Effect of Surface Compositional Heterogeneities and Microphase Segregation of Fluorinated Amphiphilic Copolymers on Antifouling Performance. *ACS Appl. Mater. Interfaces* **2013**, *5*, 7808–7818. [CrossRef] [PubMed]
53. Gudipati, C.S.; Finlay, J.A.; Callow, J.A.; Callow, M.E.; Wooley, K.L. The Antifouling and Fouling-Release Perfomance of Hyperbranched Fluoropolymer (HBFP)–Poly(ethylene glycol) (PEG) Composite Coatings Evaluated by Adsorption of Biomacromolecules and the Green Fouling Alga Ulva. *Langmuir* **2005**, *21*, 3044–3053. [CrossRef] [PubMed]
54. Imbesi, P.M.; Gohad, N.V.; Eller, M.J.; Orihuela, B.; Rittschof, D.; Schweikert, E.A.; Mount, A.S.; Wooley, K.L. Noradrenaline-Functionalized Hyperbranched Fluoropolymer–Poly(ethylene glycol) Cross-Linked Networks As Dual-Mode, Anti-Biofouling Coatings. *ACS Nano* **2012**, *6*, 1503–1512. [CrossRef]

55. Delvar, A.; Lindh, L.; Arnebrant, T.; Sotres, J. Interaction of Polyelectrolytes with Salivary Pellicles on Hydroxyapatite Surfaces under Erosive Acidic Conditions. *ACS Appl. Mater. Interfaces* **2015**, *7*, 21610–21618. [CrossRef]
56. Ionta, F.Q.; Mendonça, F.L.; de Oliveira, G.C.; de Alencar, C.R.B.; Honório, H.M.; Magalhães, A.C.; Rios, D. In vitro assessment of artificial saliva formulations on initial enamel erosion remineralization. *J. Dent.* **2014**, *42*, 175–179. [CrossRef]

© 2019 by the authors. Licensee MDPI, Basel, Switzerland. This article is an open access article distributed under the terms and conditions of the Creative Commons Attribution (CC BY) license (http://creativecommons.org/licenses/by/4.0/).

MDPI
St. Alban-Anlage 66
4052 Basel
Switzerland
Tel. +41 61 683 77 34
Fax +41 61 302 89 18
www.mdpi.com

Journal of Functional Biomaterials Editorial Office
E-mail: jfb@mdpi.com
www.mdpi.com/journal/jfb

www.ingramcontent.com/pod-product-compliance
Lightning Source LLC
LaVergne TN
LVHW070437100526
838202LV00014B/1614